Smart Card Handbook

Smart Card Handbook

W. Rankl
W. Effing
Giesecke & Devrient GmbH
Munich, Germany

Translated by
Chanterelle Translations
London, UK

JOHN WILEY & SONS
Chichester • New York • Weinheim • Brisbane • Singapore • Toronto

Carl Hanser Verlag © Carl Hanser Verlag, Munich/FRG, 1995.

erman language published by Carl Hanser Verlag, Munich/FRG.

Baffins Lane, Chichester,
West Sussex PO19 1UD, England

National 01243 779777
International (+44) 1243 779777
e-mail (for orders and customer service enquiries): cs-books@wiley.co.uk
Visit our Home Page on http://www.wiley.co.uk
or http://www.wiley.com

Reprinted October 1997
Reprinted December 1997

Other Wiley Editorial Offices

John Wiley & Sons, Inc., 605 Third Avenue,
New York, NY 10158-0012, USA

VCH Verlagsgesellschaft mbH, Pappelallee 3,
D-69469 Weinheim, Germany

Jacaranda Wiley Ltd, 33 Park Road, Milton,
Queensland 4064, Australia

John Wiley & Sons (Canada) Ltd, 22 Worcester Road,
Rexdale, Ontario M9W 1L1, Canada

John Wiley & Sons (Asia) Pte Ltd, 2 Clementi Loop #02-01,
Jin Xing Distripark, Singapore 129809

Library of Congress Cataloging-inPublication Data

Rankl, Wolfgang.
[Handbuch der Chipkarten. English]
Smart card handbook / Wolfgang Rankl, Wolfgang Effing : translated
by Chanterelle Translations
p. cm.
Translation of: Handbuch der Chipkarten
Includes index.
1. Smart cards—Handbooks, manuals, etc. I. Effing, Wolfgang.
II. Title.
TK895.S62R3613 1997
006—dc21 97-1037
CIP

British Library Cataloguing in Publication Data

A catalogue record for this book is available from the British Library

ISBN 0 471 96720 3

Printed and bound in Great Britain by Bookcraft (Bath) Ltd
This book is printed on acid-free paper responsibly manufactured from sustainable forestation,
for which at least two trees are planted for each one used

Contents

FOREWARD TO THE SECOND EDITION XI

SYMBOLS AND NOTATION XV

1 INTRODUCTION 1
1.1 The History of Smart Cards 2
1.2 Applications 4
1.2.1 Memory cards 4
1.2.2 Microprocessor Cards 5
1.3 Standardization 6

2 TYPES OF CARDS 9
2.1 Embossed Cards 9
2.2 Magnetic Stripe Cards 10
2.3 Smart Cards 12
2.3.1 Memory Cards 12
2.3.2 Microprocessor Cards 13
2.3.3 Contactless Smart Cards 14
2.4 Optical Memory Cards 15

3 PHYSICAL AND ELECTRICAL PROPERTIES 17
3.1 Physical properties 17
3.1.1 Formats 18
3.1.2 Cards with contacts 21
3.1.3 Cards without contacts 23
3.1.4 Security features 33
3.2 The Body of the Card 35
3.2.1 Smart Card Material 36
3.2.2 Production methods 37
3.3 Electrical Properties 40
3.3.1 Circuitry 41
3.3.2 Supply voltage 42
3.3.3 Supply current 43
3.3.4 Clock supply 44
3.3.5 Data transmission 45

3.3.6 Booting/Shutdown Sequence 45
3.4 Smart Card Microprocessors 46
3.4.1 Processor types 49
3.4.2 Memory types 50
3.4.3 Supplementary hardware 56

4 FUNDAMENTALS OF INFORMATION TECHNOLOGY 61
4.1 Error Detection and Error Correction Codes 61
4.1.1 XOR checksums 62
4.1.2 CRC checksums 63
4.1.3 Error correction by multiple storage 65
4.2 Encryption 66
4.2.1 Symmetric crypto-algorithms 70
4.2.2 Asymmetric crypto-algorithms 75
4.2.3 Padding 81
4.2.4 Message authentication code 82
4.3 Hash Functions 83
4.4 Random Numbers 84
4.4.1 Random number generation 85
4.4.2 Testing random numbers 87
4.5 Data Structuring 89
4.6 State Automata 92
4.6.1 Fundamentals of automata theory 93
4.6.2 Practical application 94
4.7 SDL Symbolism 96

5 OPERATING SYSTEM ARCHITECTURE 99
5.1 History 100
5.2 Fundamental Principles 101
5.3 Design and Implementation Principles 104
5.4 Program Code Sections 107
5.5 Memory Organization 107
5.6 Data Structures in the Smart Card 111
5.6.1 Types of files 112
5.6.2 File hierarchies 114
5.6.3 File name 114
5.6.4 Addressing 116
5.6.5 File structures 118
5.6.6 Access 122
5.6.7 Attribute 123
5.7 File Management 124
5.8 Execution Control 126
5.9 Atomic Routines 127
5.10 Code Programmed in Circuit 128

6 DATA TRANSMISSION TO THE SMART CARD 133
6.1 Physical Transmission Layer 135
6.2 Transmission Protocols 139
6.2.1 Synchronous data transmission 141
6.2.2 Transmission protocol T=0 147

6.2.3 Transmission protocol T=1 ... 152
6.2.4 Comparison of asynchronous transmission protocols ... 163
6.3 Answer to Reset ... 164
6.4 Protocol Type Selection ... 173
6.5 Message Structure ... 177
6.5.1 Instruction APDU structure ... 178
6.5.2 Response APDU structure ... 180
6.6 Secure Messaging ... 182
6.6.1 The Authentic Procedure ... 185
6.6.2 The combined procedure ... 186
6.6.3 Transmission sequence counter ... 188
6.7 Logical Channels ... 189

7 THE INSTRUCTION SET ... 191
7.1 File Selection ... 194
7.2 Read and Write Instructions ... 196
7.3 Search Instructions ... 204
7.4 File Operations ... 205
7.5 Identification Instructions ... 208
7.6 Authentication Instructions ... 210
7.7 Instructions for Cryptographic Algorithms ... 214
7.8 File Management ... 217
7.9 Instructions for Electronic Purses ... 223
7.10 Credit Card Instructions ... 227
7.11 Completing the Operating System ... 228
7.12 Hardware Testing Instructions ... 230
7.13 Application Specific Instructions ... 233
7.14 Transmission Protocol Instructions ... 234

8 SECURITY METHODS ... 237
8.1 User Identification ... 237
8.1.1 Input of secret numbers ... 238
8.1.2 Biometric methods ... 239
8.2 Authentication ... 246
8.2.1 Unidirectional symmetric authentication ... 247
8.2.2 Mutual symmetric authentication ... 249
8.2.3 Static asymmetric authentication ... 250
8.2.4 Dynamic asymmetric authentication ... 253
8.3 Digital Signature ... 254
8.4 Key Management ... 256
8.4.1 Derived keys ... 257
8.4.2 Key diversification ... 257
8.4.3 Key versions ... 257
8.4.4 Dynamic keys ... 258
8.4.5 Key data ... 258
8.4.6 Example: key manager ... 259
8.5 Smart Card Security ... 261
8.5.1 Technical options for chip hardware ... 261
8.5.1.1 Passive protective mechanisms ... 262
8.5.1.2 Active protective mechanisms ... 263

8.5.2 Software protection mechanisms ... 264
8.5.3 The applications's protective mechanisms ... 266
8.6 Typical Attack and Defence Mechanisms ... 267
8.6.1 Attacks at the physical level ... 268
8.6.2 Attacks at the logical level ... 270

9 QUALITY ASSURANCE AND TESTING ... 273
9.1 Testing the Card's body
9.2 Microprocessor Hardware Tests ... 279
9.3 Testing Microprocessor Software ... 280
9.3.1 Security tests ... 281
9.3.2 Software testing methods ... 283
9.3.3 Dynamic testing of operating systems and applications ... 289

10 SMART CARD MANUFACTURING ... 293

11 SMART CARD TERMINALS ... 307
11.1 Mechanical Features ... 309
11.2 Electrical Properties ... 312
11.3 Security Methods ... 314

12 SMART CARDS IN ELECTRONIC PAYMENT SYSTEMS ... 317
12.1 Card-Based Payment Transactions ... 318
12.1.1 Electronic payment transactions with Smart Cards ... 318
12.1.2 Electronic funds ... 322
12.1.3 Fundamental options for system architecture ... 324
12.2 Prepaid Memory Cards ... 326
12.3 Electronic Purses ... 327
12.3.1 CEN standard prEN 1546 ... 328
12.3.2 The Mondex system ... 342
12.4 Chip-Containing Credit Cards ... 347

13 SAMPLE APPLICATIONS ... 355
13.1 Contact less memory cards in the airline industry ... 355
13.2 Electronic Toll Systems ... 358
13.3 GSM Network ... 362

14 APPLICATION DESIGN ... 369
14.1 General Notes and Data ... 369
14.1.1 Microprocessor ... 369
14.1.2 Application ... 372
14.1.3 System ... 374
14.2 Aids to the generation of Applications ... 375
14.3 Examples of Application Design ... 377
14.3.1 Purse for a gaming machine ... 378
14.3.2 Access control ... 381

15 APPENDIX ... 385
15.1 Glossary ... 385
15.2 Literature ... 396

15.3 Standards List with Commentary 397
15.4 Characteristic Values and Tables 410
15.4.1 Time interval for ATR 410
15.4.2 Conversion table for ATR data elements 410
15.4.3 Calculation table for transmission speed 411
15.4.4 Table for sensing points 412
15.4.5 Table of class bytes used 413
15.4.6 Table of the most important Smart Card instructions 413
15.4.7 Summary of instruction bytes in use 416
15.4.8 Important Smart Card return codes 418
15.4.9 Typical instruction execution times 420
15.4.10 Sample codings of Smart Card instructions 422
15.4.11 Selected chips for memory cards 423
15.4.12 Selected Smart Card microprocessors 427

INDEX 435

Foreword to the Second Edition

This foreword is somewhat unusual, as it has not been written for just any book.

This is a book which sets out to draw together the accumulated knowledge and state-of-the-art information currently available for the new technology known as the Smart Card, and in particular, the Smart Card carrying an integrated circuit or chipcard.

It is easy to predict that the functionality and applications of the new integrated circuit cards will reach far beyond the currently envisaged, eventually perhaps influencing us in our behaviour as citizens.

I use the term 'citizens' rather than 'people' advisedly, as it is clear that in the future, the Smart Card will affect us as members of an organization, as inhabitants of a nation, an economic group or otherwise, as members of a community and therefore as citizens.

As citizens we have to ensure that they do not become instruments of coercion, however subtle, but rather that we exploit the manifold possibilities for using the new technology to serve us in a multitude of ways.

The Smart Card, if used correctly, will not become a tool of oppression, or a means of producing a particular mode of behaviour. On the contrary, it ought to stimulate a freedom of the spirit, and it is this which, in my opinion, constitutes its greatest attraction - not its effectiveness as a transaction/authorization/identification medium, no matter how important these functions may be in future in terms of the economy and the state.

There is thus a temptation to ask: 'Smart Card - quo vadis?'

- Does it represent, in this not-so-brave new world, a new level of monitoring of the individual and therefore even greater control?
- Or might harnessing the potential of the Smart Card be a way of winning back an individual's self-determination or privacy - or at least, of maintaining the status quo?

The authors have purposely restricted the chapter on Applications to an exemplar, as neither they nor any of us may at this time hazard a guess as to what the future holds in terms of application possibilities.

As long ago as 1978, the then Director of the Department of Patents at the French computer giant Honeywell Bull SA, told me: "Smart Cards will one day be as important as computers are today." Even now the truth of that statement is apparent.

The development of a new technique or technology is always a gradual process, and is achieved through the participation of many different people, both men and women, from a wide variety of interest groups, including scientists, engineers, business people, and decision-makers.

What motivates these people?

Curiosity and the quest for knowledge play their part, of course. 'Is it possible, can it be done?' Then there are financial considerations, and beyond that, ever further removed from the actual labour, from the drawing board and the computer, there are the twin forces of money and power.

Progress, sometimes called evolution, is a natural part of life. Perhaps it is the trial-and-error of life. To play the game, we have to allow ourselves to be moved by our curiosity to try new things, knowing that our next trial may well result in error.

Considering the complexity, newness and ever-changing inter-dependencies created by progress, it is easy to see how technological considerations can become mired in the purely speculative. Such considerations seek to encompass a very broad spectrum, from the early stages of development, which might be termed the visible end of the spectrum, to the final goal, with its multiple variations.

Is this not perhaps too wide an area for speculation?

I believe that it is impossible for us to exaggerate the future significance of the Smart Card, or pronounce a final verdict on it at this stage. It is a powerful tool, and in terms of being a semi-anonymous piece of equipment it has become or will become, more important than the PC.

If the Smart Card is developing into a new technology in its own right, ought we to be asking ourselves whether we are opening a Pandora's box? What might it not contain - increased monitoring of the individual, greater control by the state?

'The ID-card is the subtlest tool of terror'. (E.v.Salomon)

To what extent might the Smart Card, used in conjunction with existing computers and networked databanks (with their potential for data comparison) become, in the near or distant future, a kind of ID-card - an instrument of 'the system', of bureaucracy for instance?

Were it dependent on the authority of the card-holder, him- or herself, the Smart Card would become an instrument for personal freedom.

I believe we are at an important crossroads, before which previous generations of engineers, business people and politicians have already stood, confronted with emerging technologies. It is thus apposite to remind readers of a wise Latin saying, *Et respice finem* ('Look to the end'.)

Ask yourselves where your path may be leading you. The Smart Card, in conjunction with certain software, may form the basis of many different systems. When customized for each individual card-holder, it opens up many possibilities. Let us choose the right path, the path of independence! Let us meet on the road to self-determination (that is, if it is still possible to breach the barricades.)

Do not read this book cynically or indifferently, but rather bring a sceptical optimism to it. Let us, as citizens, prove ourselves neither incapable nor unworthy of the proffered emancipation.

This is, indeed, an unusual preface to a scientific text, and one which I have enjoyed writing because I believe that this is a definitive book, and one of great importance, a book that is much needed at this moment in time.

- Definitive, because its authors belong to the chosen few, to the core group of pioneers in the field of miniaturized integrated circuit technology. They are familiar with the most intimate secrets of the new technology, and share them in this text. In earlier times they might have been described as high priests.
- Important, because it virtually provides us with the building blocks for the creation of new combinations and configurations.

This should be considered the definitive work on the subject, serving:

- as a textbook for the beginner
- an encyclopaedia for the curious layman
- a reference book for the professional
- a state-of-the-art description for the inventor
- a foundation for those seeking to create new systems and applications.

The book deserves such accolades - a mere glance at the index shows the amount of information it contains.

Even the late Helmut Gröttrup, who accompanied me on the early steps of my journey, could have nothing more to add.

We owe the authors and publishers a debt of gratitude.

Jürgen Dethloff
February, 1996

Symbols and Notation

The lowest Bit value is always described by the number 1 (analog ISO standard) and not with the number 0.

• 'ABC'	ASCII counter
• '00'	Hexadecimal counter
• 0, 1	Binary counter
• 42	Decomal counter
• Bn	Byte with n number
• bn	Bit with n number
• Dn	Digit with n number
• $\wedge$	logic AND switching
• $\vee$	logic OR switching
• $\|\|$	Interlinking (of data elements)

Abbreviations

A-PET	amorphous polyethylene terephthalate
ABA	American Bankers Association
ABS	Acrylonitrile-Butadiene-Styrene
ACK	acknowledge
AFNOR	Association Française de Normalisation
AID	application identifier
AMA	Automatic Message Accounting
Amd.	amendment
AND	logic AND switching
ANSI	American National Standards Institute
APACS	Association for Payment Clearing Services
APDU	application protocol data unit
A-PET	amorphous polyethyleneenterphtalate
ASC	application specific command
ASCII	American standard code for information exchange
ASIC	application specific integrated circuit

ASN.1	abstract syntax notation one
ATM	automatic teller machine
ATR	answer to reset
BASIC	beginners all-purpose symbolic instruction code
BCD	binary coded digit
BER-TLV	basic endoding rules - tag length value
BGT	block guard time
BS	base station
BWT	block waiting time
CAD	chip accepting device
CASE	computer-aided software engineering
CBC	cipher block chaining
CCD	card coupling device
CCD	charge coupled device
CCITT	Comité Consultatif International Télégraphique et Téléphonique
CCR	chip card reader
CCS	cryptographic checksum
CD	commitee draft
CEN	Comité Européen de Normalisation
CEPT	Conference Européenne des Postes et Télécommunications
CFB	cipher feedback
CHV	card-holder verification
CICC	contactless integrated circuit card
CISC	complex instruction set computer
CLA	class
CLK	clock
CMOS	complementary metal oxide semiconductor
COS	chip operating system
CRC	cyclic redundancy check
CWT	character waiting time
D	day
DAD	destination address
DCS	digital cellular system
DEA	data encryption algorithm
DECT	digital European cordless telephone
DES	data encryption standard
DF	dedicated file
DIL	dual inline
DIN	Deutsche Industrienorm (German industry standard)
DIS	Draft International Standard
DOV	data over voice
DRAM	dynamic random access memory
DSA	digital signature alogorithm

ec	Eurocheque
ECB	electronic code book
EC	error correction code
EDC	error detection code
EEPROM	electrical erasable programmable read only memory
EF	elementary file
EMV	Europay, Mastercard, Visa
EPROM	erasable programmable read only memory
ESD	electrostatic discharge
ETS	European Telecommunication Standard
ETSI	European Telecommunications Standards Institute
etu	elementary time unit
FCB	file control block
FET	field effect transistor
FID	file identifier
FIPS	Federal Information Processing Standard
FRAM	ferroelectric random access memory
GND	ground (earth)
GPS	Global Positioning System
GSM	Global System for Mobile Communications or Groupe Spécial Mobile
I/O	input/output
I^2C	inter-integrated circuit
ICC	integrated circuit card
ID	identifier
IDEA	international data encryption algorithm
IEC	International Electrotechnical Commission
IEEE	Institute of Electrical and Electronics Engineers
IEP	inter-sector electronic purse
IFD	interface device
IFS	information field size
IFSC	information field size for the card
IFSD	information field size for the interface device
IMSI	international mobile subscriber identity
INS	instruction
IPES	improved proposed encryption standard
ISDN	integrated services digital network
ISF	internal secret file
ISO	International Standards Organization
ITSEC	Information Technology Security Evaluation Criteria
IV	initialisation vector
IVU	in-vehicle unit
KFPC	key fault presentation counter
KID	key identifier

Lc	command length
Le	expected length
LEN	length
LRC	longitudinal redundancy check
LSAM	load secure application module
LSB	least significant byte
lsb	least significant bit
M	month
MAC	message authentication code
ME	mobile equipment
MF	master file
MFC	multifunctional card
MIPS	million instructions per second
MLI	multiple laser image
MMU	memory management unit
MOSFET	metal oxide semiconductor field effect transistor
MS	mobile station
MSB	most significant byte
msb	most significant bit
MTBF	mean time between failures
NAD	node address
NBS	US National Bureau of Standards
NCSC	National Computer Security Center
NIST	US National Institute of Standards and Technology
NPU	numeric processing unit
NRZ	non-return to zero
NSA	US National Security Agency
OBU	on-board unit
OFD	output feedback
OR	logic OR switching
OSI	open systems Interconnections
P1, P2, P3	parameter 1,2,3
PB	procedure byte
PC	personal computer
PC	polycarbonate
PCB	protocol control byte
PCMCIA	Personal Computer Memory Card International Association
PCN	personal communications network
PCS	personal communications services
PES	personal encryption standard
PET	polyethylene terephtalate
PETP	semi-crystalline polyethylene terephtatalate
PIN	personal identification number
PIX	proprietary application identifier extension

PLL	phase locked loop
PPM	pulse position modulation
prEN	pré-norme européenne
prETS	pre-European Telecommunication Standard
PROM	programmable read-only memory
PSAM	purchase secure application module
PSK	phase shift keying
PTT	Poste, Télécommunications et Télégraphie (public telephone company)
Pub	Publication
PUK	personal unblocking key
PVC	polyvinyl chloride
PWM	pulse width modulation
RAM	random access memory
REJ	reject
RES	resynchronisation
RFU	reserved for future use
RID	record identifier
RID	registered application provider identifier
RISC	reduced instruction set computer
RND	random number
ROM	read-only memory
RSA	Rivest, Shamir, Adeleman Cryptoalgorithm
SAD	source address
SAM	secure application module
SDL	specification and description language
SIM	subscriber identification module
SIMEG	subscriber identification module expert group
SM	secure messaging
SMD	surface-mounted device
SQL	standard query language
SRAM	static random access memory
STARCOS	Smart Card Chip Operating System
SW1, SW2	status word 1,2
TAB	tape automated bonding
TCSEC	trusted computer system evaluation criteria
TLV	tag, length, value
TMSI	temporary mobile subscriber identity
TPDU	transmission protocol data unit
TTCN	the tree and tabular combined notation
TTL	transistor-transistor logic
UART	universal asynchronous receiver transmitter
WORM	write once read multiple
XOR	exclusive logic OR linkage

WORM	write once read multiple
XOR	exclusive logic OR linkage

Introduction

This book has been written for students, engineers and the technically minded who would like to find out more about Smart Cards. It attempts to cover as much of this wide-ranging topic as possible, and thus provide the reader with an overview of the fundamentals of the field and its state-of-the-art technology.

We have put great emphasis on the practical approach. The wealth of pictures, tables and references to real applications is designed to help the reader become familiar with the subject rather faster than would be possible with a strictly scientific presentation. Hence, the book makes no claims to be scientifically comprehensive, but rather to be useful in practice. This is also the reason why, as far as possible, the explanations steer clear of abstractions. At many points we had to choose between scientific accuracy and ease of understanding, and tried to strike a happy medium between the two. Where this proved impossible, however, we always preferred comprehensibility.

The book is designed so that it can be read right through from front to back; we have tried, as far as possible, to avoid references from early sections to later ones. However, it can also be used as a reference work, using the subject index and glossary. If you want to know more about a specific topic, use the cross-references in the text and the annotated index of standards to help you find the relevant passage in the book or the standard.

At this point, we must mention one small proviso. This book deals mainly with microprocessor cards in the credit card format. Insofar as possible, we have also described memory cards and card terminals. Other types of cards, such as PCMCIA, are not covered.

Unfortunately, a large number of abbreviations have become entrenched in the field of Smart Card technology, as they have in so many areas of technology and in everyday life. This makes it particularly hard for the newcomer to gain familiarity with the subject. Here, too, we have tried to minimize the use of cryptic and often illogical abbreviations. Nevertheless, in many cases, we have had to compromise between the internationally accepted “Smart Card-speak” used by those familiar with the technology, and the vernacular more easily understood by the novice. If we have not always been successful, at least the very comprehensive index of abbreviations should help to overcome any initial barriers in comprehension.

Active learning is indisputably much more effective and interesting than passive learning. Thus, for example, one learns a new language more quickly by spending some time in a country in which the language is spoken. Therefore, we decided to produce a simulation program for PCs which can be used to experiment with a Smart Card's basic functions. You can download the program from the John Wiley & Sons Ltd FTP site, details given in the inside front cover of this book. Good luck and have fun with it!

1.1 THE HISTORY OF SMART CARDS

The proliferation of plastic cards started in the USA in the early 1950s. The low price of the synthetic material PVC enabled the production of robust, long-lasting cards, much more suitable for use in everyday life than the previously conventional paper or cardboard equivalents, which were not equipped for coping with mechanical and climatic damage.

The first all-plastic card for trans-regional payment was issued by Diners Club in 1950. It was designed for an exclusive class of individual, and thus also served as a status symbol, allowing the holder to pay with his "good name" rather than in cash. Acceptance of these cards was initially restricted to the more select restaurants and hotels.

The entry of VISA and Mastercard into the arena led to a very rapid proliferation of plastic money, first in the USA and, a few years later, in Europe and the rest of the world.

Today, plastic cards allow the traveller to shop across the world without cash. The holder is never without means of payment, yet avoids exposure to the risk of loss through theft or other hazards which are difficult to guard against, particularly whilst travelling. Besides, having a card does away with tedious currency exchange when travelling abroad. These unique advantages have helped plastic cards to become widely and rapidly established world-wide. Many hundreds of millions of cards are now produced and issued annually.

At first, the cards' functions were quite simple. Initially, they served as data-carriers protected against forgery and tampering. The general data, such as the issuer's name, was surface-printed while individual data elements, like the cardholder's name or the card number, were embossed. Furthermore, many cards carried a signature field, in which the holder could sign his or her name for reference. Protection against forgery, in these first-generation cards, was provided by visual features, such as security printing and the signature field. As a consequence, the system's security depended quite fundamentally on the quality and care of the retail staff accepting the cards. However, this was not a huge problem due to the cards' initial exclusivity. With increasing proliferation, these rather basic features no longer proved sufficient, all the more so since the danger represented by organized crime was growing apace.

On the one hand, the increasing pressure of handling and bank charges made a machine-readable card necessary; and on the other, the card-issuers' losses due to customers' insolvency and fraud grew from year to year. It became apparent that security measures against fraud and tampering, as well as the cards' functions, had to be extended and improved.

The first improvement consisted of a magnetic stripe on the back of the cards. This allowed further digitized data to be stored in machine-readable form, in addition to the visual data and that obtainable by printing out the embossed data. This type of embossed card with a magnetic stripe is still the most commonly used as a method of payment.

Magnetic stripe technology suffers from a crucial weakness, however, in that the data stored on the stripe can be read, deleted and rewritten at will by anyone with access to the appropriate read/write device. Hence, it is unsuitable for the storage of confidential data. Further measures are needed to ensure confidentiality, as well as to protect against tampering. That is why most systems which employ a magnetic stripe card are connected on-line to the system's host computer. However, this generates considerable costs related to the need for data transmission. In order to keep costs down, a solution had to be found enabling card transactions to be executed off-line, but without putting the system's security at risk.

The development of the Smart Card, in parallel with the expansion of electronic data-processing, opened up completely new possibilities for solving this problem. The huge progress in microelectronics in the 1970s made it possible to integrate data storage and arithmetic logic on a single silicon chip measuring a few square millimetres. The idea of incorporating such an integrated circuit into an ID card was announced, and a patent applied for by the German inventors Jürgen Dethloff and Helmut Grötrupp, as far back as 1968. This was followed by a similar application by Kunitaka Arimura in Japan in 1970. However, the first real progress came with Roland Moreno's announcement of his Smart Card patents in France in 1974. It was only then that the semiconductor industry was able to supply the required integrated circuits at a reasonable price. Nevertheless, many technical problems still had to be solved before the first prototypes, with two partly-integrated chips, became a reliable product which could be manufactured with sufficient quality, in large numbers and at sensible prices.

The great breakthrough was achieved in 1984, when the French PTT (Postal and Telecommunications services) successfully carried out a field trial with telephone cards. Significantly, the breakthrough came not in the traditional bank card market, but in a new application. Such an application enjoyed the great advantage that no account needed to be taken of compatibility with existing systems, and thus the options offered by the new technology could be fully exploited.

A comparative pilot project was conducted in Germany in 1984/85, using telephone cards based on a variety of technologies: magnetic stripes, optical storage (so-called holographic cards) and Smart Cards.

The Smart Card proved the hero of this pilot study. In addition to a high degree of reliability and security against tampering, Smart Card technology promised the greatest flexibility in future applications.

These successful trials of telephone cards, first in France and later in Germany, were followed with breathtaking speed by further developments. By 1986, many millions of French telephone Smart Cards were in circulation. Their number reached nearly 60 million in 1990, and 150 million are projected for 1996.

The German experience was similar, with a time-lag of about three years. Telephone cards incorporating a chip are currently used in over 50 countries.

Progress was significantly slower in the field of bank cards, which can be accounted for by their greater complexity compared with telephone cards. These differences will be described in detail in the following chapters. At this point we merely point out that besides semiconductor technology, developments in modern cryptography have been of decisive importance in the proliferation of bank cards.

With the general expansion of electronic data-processing in the 1960s, the field of cryptography experienced a quantum-leap, so to speak. Modern hardware and software permitted the implementation of complex and demanding mathematical algorithms, which made possible a degree of security unparalleled till then. Moreover, this new method was available to anyone, whereas previously cryptography had been a covert science and the preserve of military and secret services.

With these modern cryptographic procedures, the power of the security mechanisms built into electronic data-processing systems could be proved mathematically. It was no longer necessary to rely on the very subjective assessments used in conventional methods.

The Smart Card was the ideal medium for ensuring a high degree of security, based on cryptography and available to everyone, since it could safely store secret keys and carry out cryptographic algorithms. In addition, Smart Cards are so small and so easy to use that they can become a part of everyday life. Thus, the chip-incorporating bank card made possible the highest degree of security in the field of private payment transactions.

The French banks were the first to introduce this fascinating technology in 1984, following a trial of 60 000 cards in 1982/83. It took another 10 years before all French banks were using chip-incorporating cards. In Germany, Smart Cards will be introduced nationwide in 1997.

Yet another application has ensured that nowadays, almost every German holds a Smart Card: the introduction of the new health insurance card. Over 70 million Smart Cards were issued to all those covered by the statutory scheme.

The Smart Card's high functional flexibility, which allows a card already in service to be reprogrammed for new applications, has opened up completely new fields beyond the card's traditional applications.

1.2 APPLICATIONS

The Smart Card's possible applications are extremely diverse, and are constantly being added to with the increasing arithmetic power and storage capacity of available integrated circuits. Within the confines of this book it is impossible to describe in detail all these applications. Instead, a few typical examples will suffice to demonstrate the Smart Card's fundamental properties. This introductory chapter is only meant to provide an initial overview of the card's flexibility. Chapter 13 will describe in detail several typical applications.

To make this summary easier to follow, it is helpful to divide Smart Cards into memory cards and microprocessor cards.

1.2.1 Memory cards

The first Smart Cards used in large quantities were telephone memory cards. These cards are pre-paid, and the value stored electronically in the chip is deducted during use by the appropriate amount. Naturally, it is necessary to prevent the user from incrementing the stored value, something which is easily done with a magnetic card. All the user would have to do is note the values stored at the time of purchase, and rewrite them to the magnetic stripe after use. The card would then have its original value, and could be reused. This type of tampering, also known as "buffering", is prevented in Smart Cards by a security logic in the chip, which makes it impossible to erase a memory cell once written to. Thus, the reduction in the card's value by the units used is irreversible.

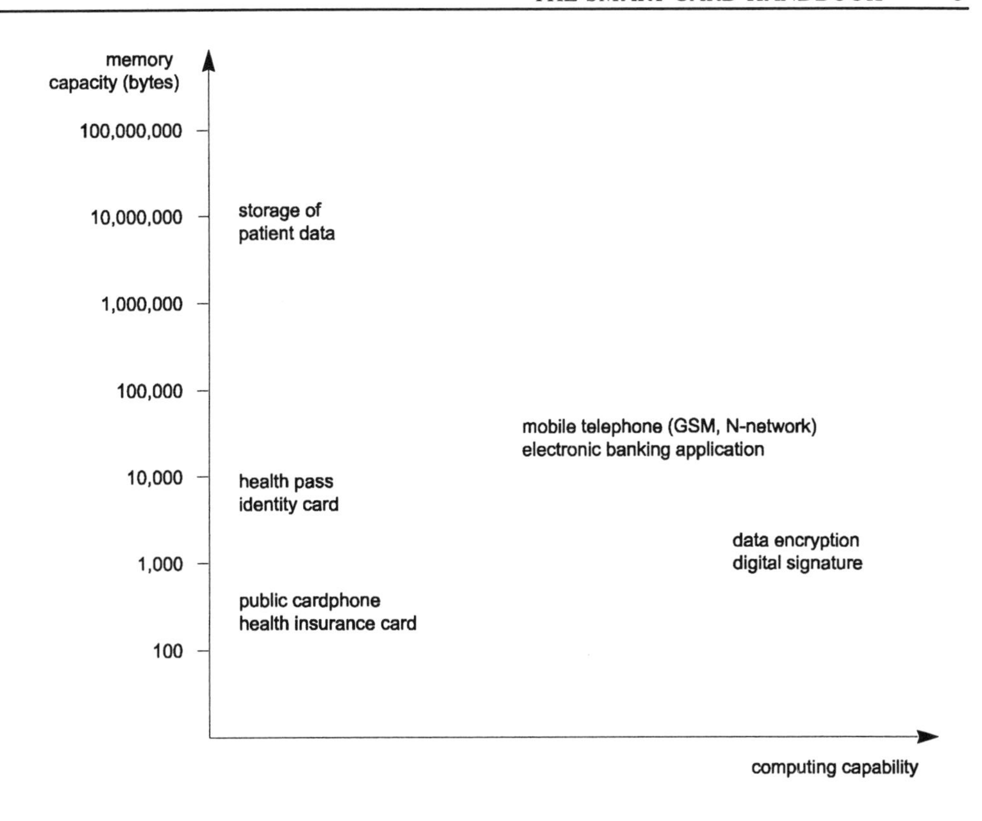

Figure 1.1 Typical Smart Card applications, their memory requirements and arithmetic power

Naturally, this type of Smart Card can be used not just for phone calls but wherever goods or services need to be sold against pre-payment. Examples of possible uses include local public transport, all kinds of vending machines, cafeterias, swimming pools, car parks etc. The advantage of this type of card lies in the simple technology (typically, the chip's surface area is only a few square millimetres), and hence the low cost involved. The disadvantage is that the card cannot be reused but must be disposed of as waste, unless it ends up in a collector's album.

Another typical application of memory cards is the health insurance card, issued in Germany since 1994 to all those insured. The data, previously recorded on the patient's card, is now stored in the chip and also printed on the card, or possibly written on it with a laser. Chip storage makes the cards machine-readable, using simple technology.

To sum up: memory Smart Cards are limited in their functions. The integrated security logic makes it possible to protect stored data against tampering. They can be used as cash-equivalent or ID cards in systems where low cost is a particular consideration.

1.2.2 Microprocessor cards

As already mentioned, the first application of microprocessor cards was in the form of French bank cards. Their ability to store secret keys securely and to execute modern cryptographic algorithms made it possible to implement high-security, off-line payment systems.

Since the microprocessor built into the card can be programmed freely, the functionality of processor cards is only restricted by the available storage space and by the arithmetic unit's power. Thus, the only limits on one's imagination when implementing Smart Card systems are technological ones, and they expand massively with each generation of integrated circuits.

The cost of Smart Cards has fallen drastically since the early 1990s as a result of mass production. Consequently, new applications have been introduced year after year. The one involving mobile phones is of particular significance for the international proliferation of Smart Cards.

Having been successfully tested in the German national C-network for use in mobile sets, the Smart Card was specified as the access medium to the digital European mobile phone system (GSM). On one hand, it proved possible with the Smart Card to achieve a high degree of security when accessing the mobile telephone network. On the other, the Smart Card introduced new options and thus greater advantages in the marketing of mobile phones, since it allowed a simple separation of the sale of sets from that of services by the network operator and service provider. Without the Smart Card, mobile phones would certainly not have spread across Europe so quickly.

Other examples of microprocessor card applications are ID cards, access to restricted areas, access control to computers, protected data storage, electronic signatures and payment cards, as well as multifunctional cards which contain many applications.

The essential advantages of microprocessor cards, then, are their high storage capacity, their secure storage of confidential data and their ability to execute cryptographic algorithms. These advantages make a wide range of new applications possible, in addition to the card's traditional role as a bank card. The Smart Card's potential is not yet exhausted by a long stretch, and in fact is constantly being extended by developments in semiconductor technology.

1.3 STANDARDIZATION

The precondition for further extension of Smart Cards into everyday life, like their current use in Germany in the form of telephone cards and health insurance cards (and as of 1996, in financial transactions), was the creation of national and international standards.

Smart Cards normally form a component in a complex system, as a result of which their proliferation depends critically on the existence of national and international standards and generally recognized specifications. As the Smart Card usually represents the system component which the user holds in his or her hand, it is of enormous importance for the entire system's familiarity and acceptance. However, from a technical and organizational perspective, the Smart Card is usually merely the tip of the iceberg, since behind the card reader there often hide complex and usually networked systems, which make the service possible in the first place.

Let us take as an example the telephone card, which technically is a fairly simple object: in itself it is an almost worthless object, other than as a collectable. It can only perform its service - namely, being able to communicate through public telephones without resorting to coins - once many thousands of sets are installed across the whole relevant region and connected to the network. The high investment needed to set this network up can only be justified if appropriate standards and specifications guarantee the long-term future of such a system. Standards for multifunctional Smart Cards are an indispensable requirement. These are Smart Cards which can be used for various applications: phone calls, electronic payment, electronic driving licences, etc.

Although Smart Card standardization was initiated quite early on - via ISO in 1983 - it could often not keep pace with technical developments. In order to understand why the standardization process is so time-consuming, we need to be clear about what a standard actually is. The question is not as trivial as it may appear at first glance, since the terms "standard", "standardization" and "specification" have been used fairly indiscriminately. To make things clear, let us consider the ISO/IEC definitions:

- **Standard**: A document which is produced by consensus and is adopted by a recognized institution, and which lays down rules, guidelines or features for activities or their outcomes for general and recurring use, whereby an optimum degree of regulation is striven for in a given context. Note: standards should be based on the established results of science, technology and experience, and their objective be the promotion of optimized benefits for society.

And in order to avoid confusion, ISO/IEC have also defined the term "consensus" as follows:

- **Consensus**: General agreement, characterized by the absence of continuing objections to essential components on the part of any significant part of the interested parties, and through a procedure which attempts to take into consideration the views of all relevant parties and to meet all counter-arguments. Note: Consensus does not necessarily mean unanimity.

Although no unanimity is required, the democratic process of consensus naturally needs a little time; all the more so since not only the technical view, but that of all relevant parties, must be heard, as standards have as their objective the promotion and optimization of advantages for the whole of society. Hence, the preparation of an ISO or CEN standard usually takes several years. A frequent consequence of the slowness of this process is that a limited number of interested parties, such as industry, produces its own specifications, the so-called industry standard, in order to make a faster system introduction feasible. Often, these technical specifications later give rise to standards.

However, back to the Smart Card. After more than ten years of standardization, the most important and fundamental ISO standards for Smart Cards are now complete, and form the basis for further, more application-oriented standards, which are currently being prepared by ISO and CEN.

They rely on the earlier ISO standards of the 7810, 7811, 7812 and 7813 series, which define the physical properties of ID cards in the so-called ID-1 format. These standards included embossed cards and cards with magnetic stripes, of the type widely used in credit cards.

As part of Smart Card standardization (the ISO norm terms these "integrated circuit cards" - ICC), attention was paid from the start to compatibility with those existing standards, to promote a smooth transition to Smart Cards from the use of embossed and magnetic stripe cards. This is achieved through the simultaneous integration of all functional elements such as embossing, magnetic stripes, contacts and interface elements for contactless transmission, into a card. Nevertheless, one consequence of this is that sensitive integrated circuits are subjected to heavy loads during embossing or the repeated pressure when embossed data is off-printed. This makes heavy demands on the packaging of the integrated circuits and their incorporation in the cards.

A summary of valid standards with a short description of their contents can be found in section 15.3 of the appendix.

Types of cards

2

As already mentioned in the Introduction, Smart Cards are the youngest member of the ID-1 format family of identification cards, as defined in ISO standard 7810 "Identification Cards - Physical Characteristics". This standard specifies physical properties such as flexibility, temperature resistance and the dimensions of three different card formats (ID-1, ID-2 and ID-3). The Smart Card standard ISO 7816-1 ff. is based on the ID-1 card, widely used nowadays for financial transactions.

This chapter provides an overview of different types of cards in the ID-1 format. In many applications, it is the combination of functions which is of special interest, and that is so whenever an existing system needs to replace the cards formerly employed, like magnetic ones, with Smart Cards. Normally it is impossible to replace an existing infrastructure - in this case, magnetic card terminals - with a new technology overnight.

The solution will generally consist of issuing cards which carry both magnetic stripes and chips during a transitional period, so they can be used both in old and new terminals. Naturally, those new functions which are only possible with a chip cannot be exploited in the magnetic terminals.

2.1 EMBOSSED CARDS

Embossing is the oldest technique for marking ID cards in machine-readable form. The embossed design on a card can be transferred to paper by printing using a simple and cheap device. Visual reading of the embossing is also straightforward. The type and positioning of the embossing are specified in ISO standard 7811 "Identification Cards - Recording Technique". It is divided into five parts, and in addition to embossing also deals with magnetic stripes.

ISO 7811 part 1 specifies the embossed marks, covering their form, size and embossing height.

Part 3 lays down the precise positioning of the marks on the card, and defines two domains. Domain 1 is reserved for the card's ID number, which identifies the card issuer as well as its owner. Domain 2 is provided for further data relating to the owner, e.g. name and address.

At first glance, information transfer by printing off embossed marks may appear quite primitive. However, the simplicity of this technique has made worldwide proliferation possible, including in developing countries. Exploitation of this technique requires neither electric current nor connection to a telephone network.

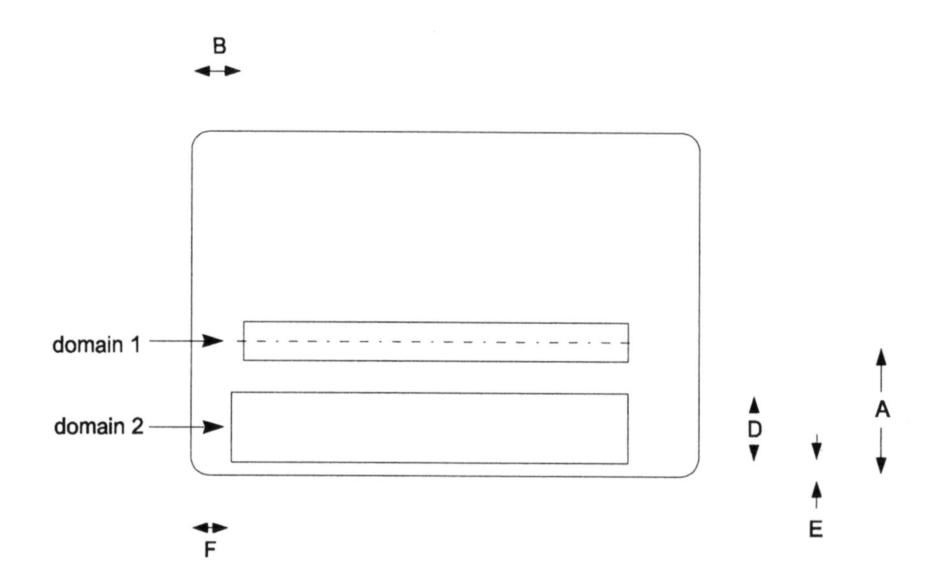

Figure 2.1 Positioning of embossing in accordance with ISO 7811-3.Domain 1 reserved for ID number (19 marks); Domain 2 for name and address (4 · 27 marks). A: 21.42 mm ± 0.12 mm, B: 10.18 mm ± 0.25 mm, D: 14.53 mm, E: 2.41 mm to 3.30 mm, F: 7.65 mm ± 0.25 mm

2.2 MAGNETIC STRIPE CARDS

The fundamental disadvantage of embossed cards is the generation of a flood of paper documents during use, whose handling and evaluation are quite expensive. This problem is obviated by digital encoding of the card's data on a magnetic stripe located on its back.

The magnetic stripe is read by pulling it across a reading head, either manually or automatically, whereby the data is read and stored electronically. Processing this data no longer requires any paper.

Parts 2, 4 and 5 of ISO standard 7811 specify the properties of the magnetic stripe and the coding technique as well as the magnetic track's position. The magnetic stripe may carry a maximum of three tracks.

Tracks 1 and 2 are specified for reading only, while track 3 may also be written to.

Although the stripe's storage capacity is only about 1000 bits and thus not very high, it is nevertheless more than sufficient for storing the information contained in the embossing. Additional data can be read and written on track 3, such as the last transaction data in the case of credit cards.

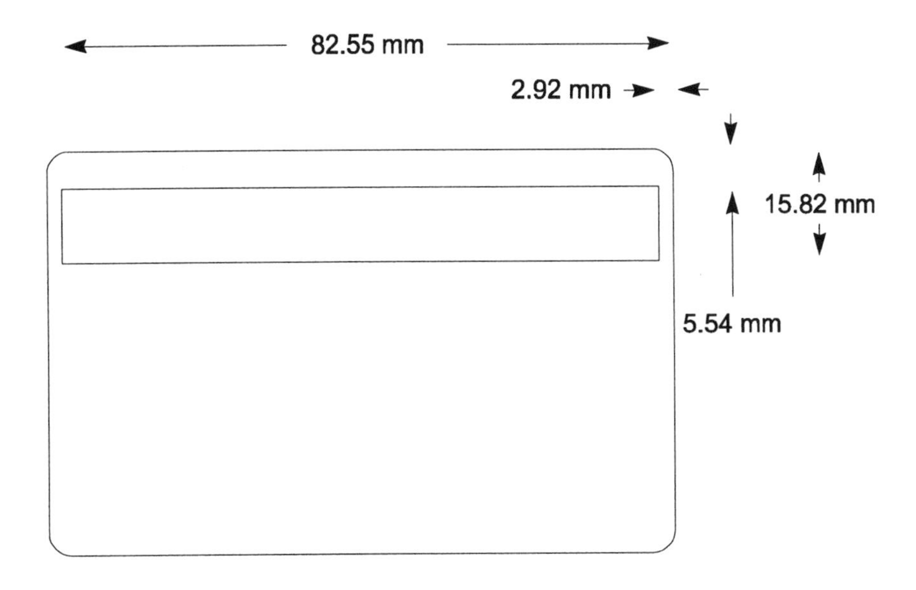

Figure 2.2 Position of magnetic stripe on the ID-1 card

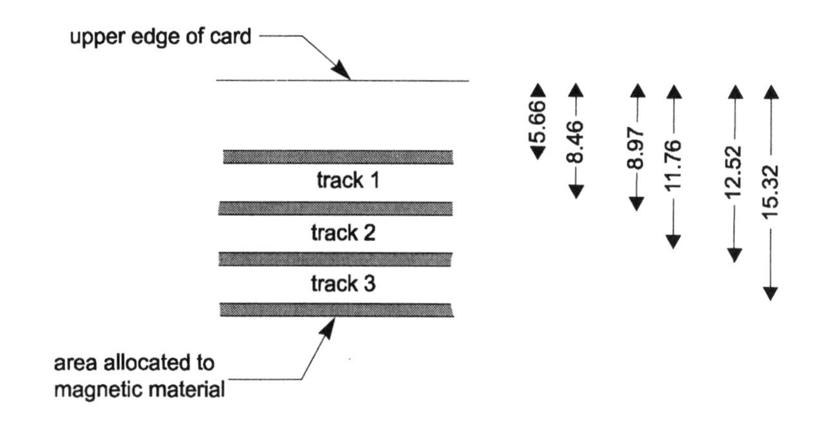

Figure 2.3 Position of separate tracks on the ID-1 card (in mm)

The main drawback of the magnetic stripe technique consists in the considerable ease with which the stored data may be altered. Whilst tampering of the embossed marks requires at least some mechanical skill and can also be easily detected by the trained eye, it is straightforward to change the data coded on the magnetic stripe with a standard read/write device, and difficult to prove at a later stage. Furthermore, this type of card is often used in automatic machines where visual inspection is impossible, as in cash dispensers. The potential criminal, having obtained valid card data, can use simple duplicates of cards at such unattended machines without having to forge the visual security features designed to prove the card's authenticity.

Magnetic card manufacturers have developed various techniques for protecting the data on the magnetic stripe against forgery and duplication. Thus, for example, German Eurocheque cards contain an invisible and unalterable coding in the body of the card, which makes it impossible to amend the data on the magnetic stripe or even to duplicate it. However, this and other procedures require a special sensor in the terminal, which considerably increases costs, which is why none of these procedures have so far succeeded in becoming internationally established.

2.3 SMART CARDS

The Smart Card is the youngest and cleverest member of the identification card family in the ID-1 format. It is characterized by an integrated circuit incorporated in the card, which contains elements used for data transmission, storage and processing. Data transfer can take place either via the contacts on the card's surface, or without contact through electromagnetic fields.

The Smart Card offers a range of advantages compared with the magnetic stripe.

The maximum storage capacity of a Smart Card is of a far greater order of magnitude than that of a magnetic one. Circuits with over 20 kbytes of memory are already available, and this figure is expected to multiply with each new chip generation. Only the optical memory cards described in the next chapter have a greater capacity.

However, one of the most important advantages of Smart Cards consists in the fact that their stored data can be protected against unauthorized access and tampering. As access to data takes place only via a serial interface supervised by the operating system and by a security logic system, it is possible to write confidential data to the card which can never be read from outside. This secret data can then only ever be processed internally by the chip's arithmetic unit. In principle, the memory functions of writing, erasing and reading can be controlled both by hardware and by software, and be linked to specific conditions. This allows the construction of numerous security mechanisms, which can also be tailored to the special demands of the relevant application.

Together with the ability to compute cryptographic algorithms, the Smart Card makes it possible to implement a convenient security module which can be carried about at all times, for example in one's briefcase.

Further advantages of the Smart Card are its high degree of reliability and long life compared with magnetic cards, whose circulation is generally limited to one or two years.

The fundamental characteristics and functions of Smart Cards are laid down in the 7816 series of ISO standards. These standards will be examined in detail in the following chapters.

Due to the differences in functionality but also in price, Smart Cards are divided into memory cards and microprocessor cards.

2.3.1 Memory cards

The architecture of a memory card is presented as a block diagram in Figure 2.4.

The data required for the application is stored in the memory, mostly on EEPROM. Access to memory is controlled by the security logic, which in the simplest case consists only of write or erase protection for the memory or its individual domains. However, there are also memory chips with a more complex security logic, which can also carry out simple coding. The data is transmitted to and from the card via the I/O port. Part 3 of ISO 7816 defines a special synchronous transmission protocol, which allows a particularly simple and economical chip implementation. However, some Smart Cards use the I^2C bus which is common in serial access memories.

Memory card functions are mostly optimized for a particular application. Although this severely restricts their flexibility, it makes them particularly good value for money. Typical memory card applications are pre-paid telephone cards and health insurance cards.

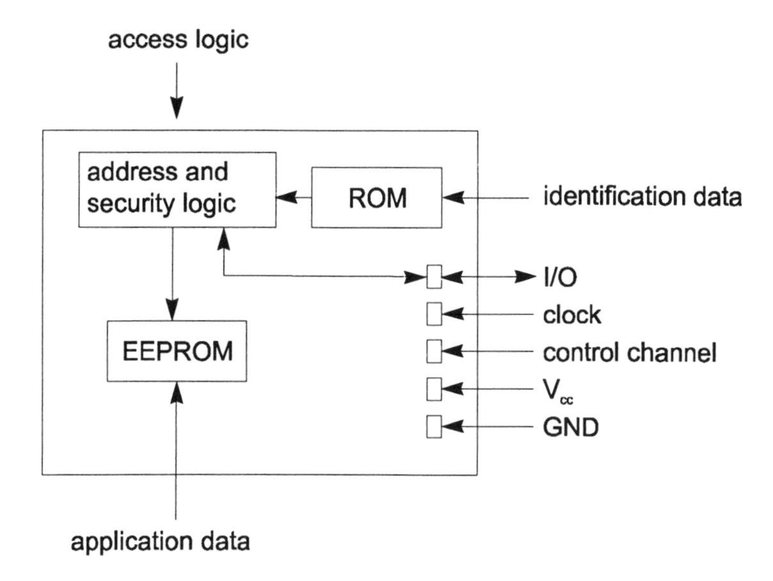

Figure 2.4 Typical architecture of a memory card with security logic

2.3.2 Microprocessor cards

The heart of the chip in a microprocessor card is – as the name implies – a processor, which as a rule is surrounded by four additional functional blocks: the mask-ROM, the EEPROM, the RAM and the I/O port.

The mask-ROM contains the chip's operating system, and is etched during manufacture. The ROM's contents are theoretically identical for all the chips in one production run, and cannot be changed during the chip's lifetime.

The EEPROM is the chip's non-volatile memory, to and from which data and also program codes may be written and read under the operating system's control.

The RAM is the processor's working memory. This area is volatile, and all the data stored in it is lost when the chip's supply voltage is switched off.

The serial I/O interface usually consists only of a single register, through which the data is transferred bit by bit.

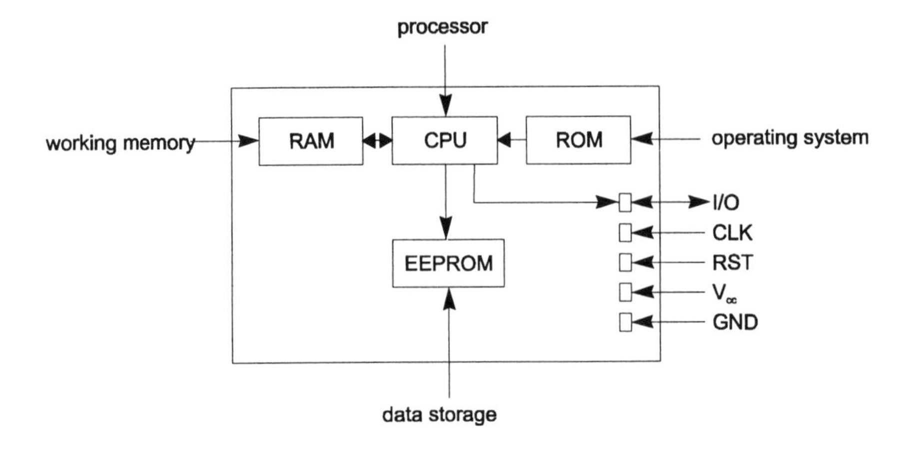

Figure 2.5 Typical architecture of microprocessor cards

Microprocessor cards are very flexible in their application. In the simplest case they contain a program optimized for a single application, and thus can only be used for this application.

However, modern Smart Card operating systems make it possible to integrate several different applications in a single card. In this case, the ROM contains only the operating system's basic instructions, whilst the program's application-specific part is only loaded into the EEPROM after the card's manufacture.

2.3.3 Contactless Smart Cards

Where Smart Cards contain contacts, these follow the eight-pin layout specified in ISO standard 7816 part 1. As a result of experience gained by manufacturers in recent years, the reliability of Smart Cards with contacts has consistently improved. For instance, the failure rate of telephone cards over their expected lifetime of one year is currently well under 1 in a 1000. Nevertheless, as has always been the case, contacts are one of the most frequent causes of failure in electromechanical systems; in other words errors can result from dirt or contact wear. When used in mobile instruments, vibrations can lead to short-term breaks in contact. Since the contacts which are located on the card's surface are connected directly to the inputs of the integrated circuit built into the card, there is the danger that electrostatic discharges – and several thousand volts are by no means rare – could weaken or even destroy the circuit.

These technical problems are elegantly circumvented by the contactless Smart Card.

In addition to these technical advantages, contactless technology also offers the issuer and user a range of interesting new possibilities during use. Thus, for example, contactless cards need not necessarily be inserted into a card reader, since there are systems which work at a distance of up to one metre. This is a great advantage in access-control systems where a door or turnstile needs to be opened, since the person's access authorization can be checked without the card having to be taken out of one's pocket and inserted into a reader. One extensive field of application for this technology is in local public transport, where the largest quantity of personal data must be captured in the shortest possible time.

However, contactless technology is also advantageous in systems which require deliberate insertion of the card into the reader: the card's orientation can be arbitrary, in contrast to magnetic cards or those with metallic contacts which only function in a particular position. This makes handling considerably easier, and thus increases customer acceptance.

A further interesting variation on the use of contactless cards is where the card is not inserted in a slot, but is simply placed on a marked location on the card-reader's surface. In addition to simplicity of use, this solution is also attractive in that it significantly reduces the risk of vandalism (e.g. through depositing chewing gum or superglue in the slot).

As far as marketing is concerned, contactless technology offers the benefit that no technical elements are visible on the card's surface, so that the card's visual design is not constrained by magnetic stripes or electric pins.

Why is it, then, that despite all these advantages, the use of contactless technology has not yet been adopted more comprehensively?

Certainly, one important reason is that this technology has not yet achieved the same stage of development as its rival with the built-in contacts. The electronics required in the card contain analogue and digital elements, which creates extra problems during integration. Furthermore, additional com-ponents such as transmission loops or capacitor plates normally have to be integrated into the card. As a result, the production of contactless cards is still more expensive. Since insufficient experience has been gained through mass production, the reliability of this technology has not yet been thoroughly proved. Nevertheless, we can expect a larger volume of contactless Smart Cards, particularly those which can be used over a distance of up to 30 centimetres, to be employed in public transport systems in the near future. Once these mass applications have proved successful, many others will surely follow. In this respect, the piloting application of contactless cards in local public transport plays a similar role to that of contact-based cards in public telephone systems.

2.4 OPTICAL MEMORY CARDS

In applications where the storage capacity of Smart Cards is insufficient, it is possible to use optical cards carrying many megabytes of data. However, using currently available technology, the cards can only be written once and the data cannot be erased.

Standards ISO/IEC 11693 and 11694 define the optical memory cards' physical characteristics and linear data-recording technology.

Combining the optical card's high storage capacity with the Smart Card's intelligence results in interesting new features: the data may be written optically in encrypted form, and the key stored securely in the chip's secret memory. Thus the optically stored data is protected from unauthorized access. Figure 2.6 shows a typical layout of a Smart Card with contacts and a magnetic stripe. It can be seen that the available optical storage area is restricted by the chip's contacts, which of course reduces its storage capacity. The magnetic stripe is located on the card's reverse.

The read and write devices for optical storage cards are currently still very expensive, which so far has seriously restricted their applicability.

Optical cards may be used in the health sector for recording patient data, where the high capacity available enables even the storage of X-ray images.

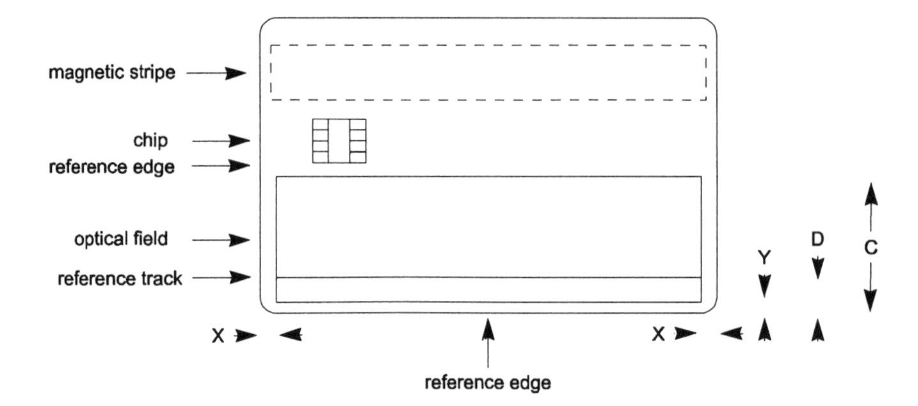

Figure 2.6 Layout of optical storage area on an ID-1 card, in accordance with ISO/IEC 11'694-2. C: 9.5 mm to 49.2 mm, D: 5.8 mm ± 0.7 mm; X: with PWM: maximum 3 mm; with PPM: maximum 1 mm; Y: with PWM: Y<D and minimum 1 mm; with PPM: maximum 4.5 mm (PWM: pulse width modulation; PPM: pulse position modulation)

Physical and electrical properties

3

The Smart Card's body inherits its fundamental properties from its predecessors, the familiar embossed cards. Technically speaking, Smart Cards are simply plastic cards, which are personalized by being imprinted with a variety of user features such as name and customer number.

In later implementations, the cards were provided with a magnetic stripe which allowed simple automated processing. When the idea of implanting a chip in the card was mooted, the original cards were used as a template, and a microprocessor was added. Many standards relating to the card's physical properties are not even specific to Smart Cards, but apply equally to magnetic and to embossed cards as well.

3.1 PHYSICAL PROPERTIES

The most obvious feature of the Smart Card is its format. The next distinguishing feature is the presence or absence of a contact area; sometimes there is no visible electric interface at all (a contactless Smart Card). A magnetic stripe, embossing and a hologram may be the next features to catch the eye. All these features and functional components form part of a Smart Card's physical properties.

In reality, a large part of the physical characteristics are of a purely mechanical nature, such as size and resistance to bending or shearing. These are familiar to the user from personal experience. In practice, however, typical physical properties such as susceptibility to temperature and light, and resistance to moisture are also important.

The interaction between the body of the card and the implanted chip also needs to be considered, since only the combination of the two can create a functional card. For instance, a card which is suitable for use at very high ambient temperatures is of little use if its microprocessor does not share this property. Both components must satisfy all the requirements separately as well as together, otherwise high failure rates can be expected during use.

3.1.1 Formats

Small cards sharing the typical Smart Card dimensions of 85.6 mm by 54 mm have been in use for a very long time. Almost all Smart Cards are produced in this, the best-known format. It is designated ID-1, and its size is specified in International Standard ISO 7810. As can be seen clearly from the abbreviation "ID", which stands for identification card, this first standard, created in 1985, had nothing to do with Smart Cards as we know them today. It only describes an embossed plastic card carrying a magnetic stripe, designed for personal identification. A chip built into the card was never considered. It was only several years later that further standards defined the presence of a chip and its position on the card.

With today's variety of cards, which are used for all kinds of purposes and come in a wide range of dimensions, it is often difficult to determine whether a card really is an ID-1 Smart Card or not. Besides the implanted chip, one of the best identifying features is the thickness of the card. If it measures 0.76 mm and contains a microprocessor, then it may be considered to be a Smart Card in the real sense of the word.

The conventional ID-1 format has the advantage of being very convenient to handle. The card's format is so specified that it is not too large to carry about in a wallet, but not so small that it is easily lost. In addition, the card's flexibility makes it very practical.

Nevertheless, this format is often incompatible with the requirements of modern miniaturization. Some portable phones weigh only 200 g and are not much bigger than a packet of tissues. Hence it has become necessary to define an additional, smaller format than ID-1, which takes into account the requirements of small handsets. The card can be very small, since usually it will only be inserted in the instrument once and remain there for good. The ID-000 format was defined for these circumstances, and goes by the more descriptive name of "plug-in" card. Currently this format is only implemented in the field of GSM telephone instruments, in which there is very little physical space and where the card does not need to be replaced very frequently.

However, since the ID-000 is hardly convenient to handle, either to manufacture or to use, it has led to a further format, known as ID-00 or "Mini-card", whose dimensions are approximately halfway between ID-1 and ID-000. This type of card is more convenient to handle and is also cheaper to produce, since it can be printed more efficiently. Nevertheless, the ID-00 definition is fairly new, and this format has not yet become established either nationally or internationally.

Format definitions in their respective standards have been optimized in terms of dimensions. Thus, the ID-1 card's height and width must be such that without the rounded corners, it fits between two concentric and symmetrically positioned rectangles with the following dimensions:

- External rectangle: Width 85.72 mm (= 3.375")
  Height 54.03 mm (= 2.127")

- Internal rectangle: Width 85.46 mm (= 3.365")
  Height 53.92 mm (= 2.123")

The thickness must be 0.76 mm (= 0.03") with a tolerance of ±0.08 mm (= ±0.003"). The corner radii and the card's thickness are defined conventionally. Using these definitions, an ID-1 card's dimensions can be represented by Figure 3.2.

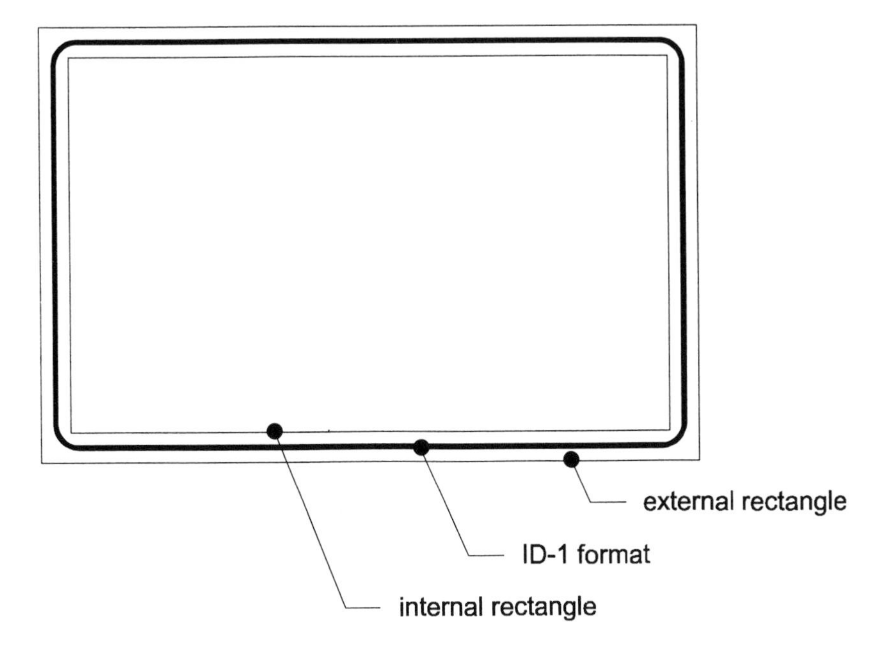

Figure 3.1 Definition of dimensions of ID-1 format

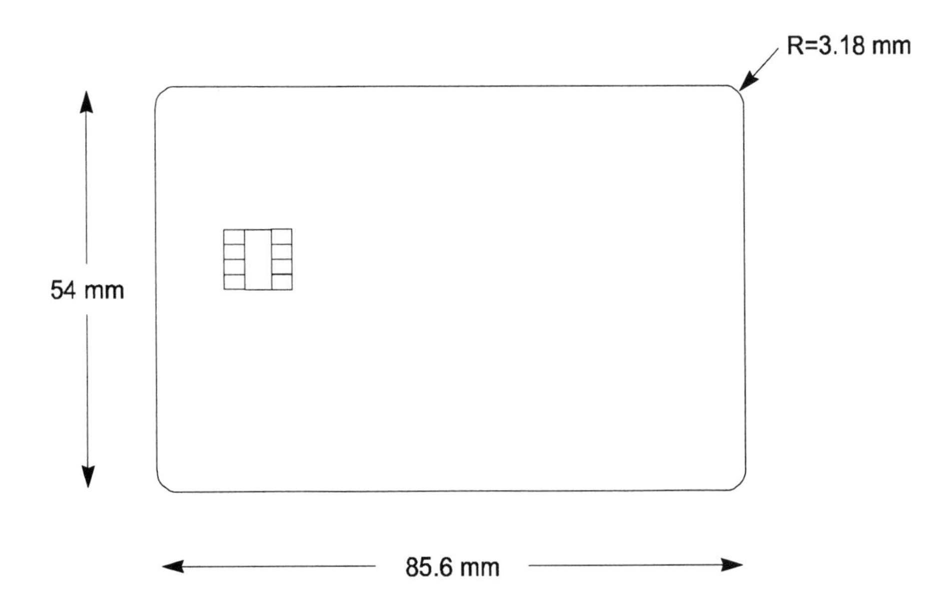

Figure 3.2 The ID-1 format. Thickness: 0.76 mm ± 0.08 mm; corner radius: 3.18 mm ± 0.30 mm (the individual measurements represent the dimensions without allowances)

The ID-000 format is also defined using two concentric rectangles. Since this format originated in Europe (based on the GSM mobile radiotelephone system), the basic dimensions are metric. The two measurements of the two rectangles are:

- External rectangle: Width 25.10 mm
  Height 15.10 mm
- Internal rectangle: Width 24.90 mm
  Height 14.90 mm

The Plug-in's bottom right-hand corner is cut off at an angle of 45°, in order to facilitate the card's correct insertion into the card reader.

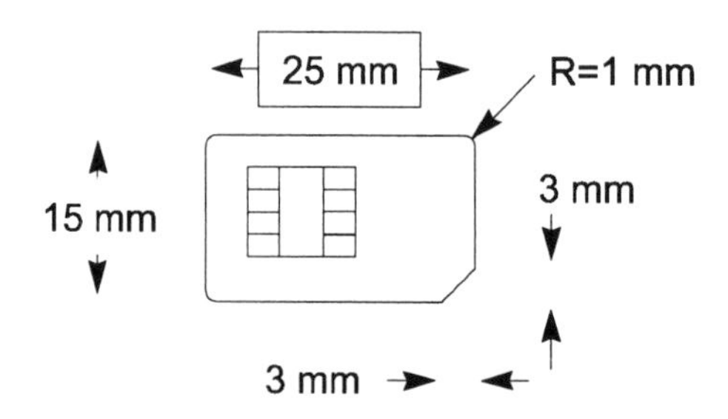

Figure 3.3 The ID-000 format. Thickness: 0.76 mm ± 0.08 mm; corner radius: 1 mm ± 0.10 mm; corner: 3 mm ± 0.03 mm (the individual measurements represent the dimensions without allowances)

The ID-00 format is also based on metric measurements, and its maximum and minimum dimensions are defined by two concentric rectangles which measure:

- External rectangle: Width 66.10 mm
  Height 33.10 mm
- Internal rectangle: Width 65.90 mm
  Height 32.90

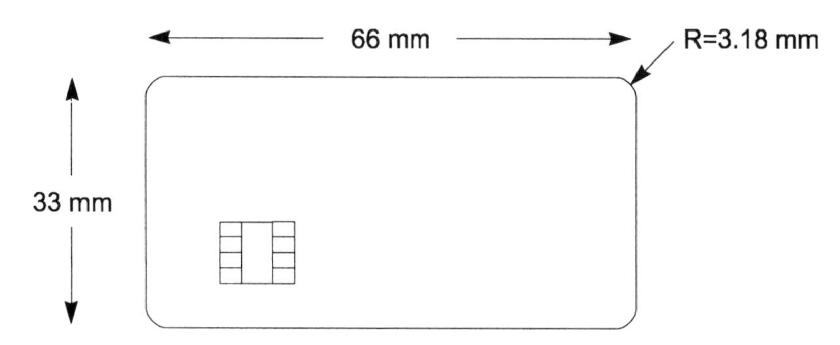

Figure 3.4 The ID-00 format. Thickness: 0.76 mm ± 0.08 mm; corner radius: 3.18 mm ± 0.30 mm (the individual measurements represent the dimensions without allowances)

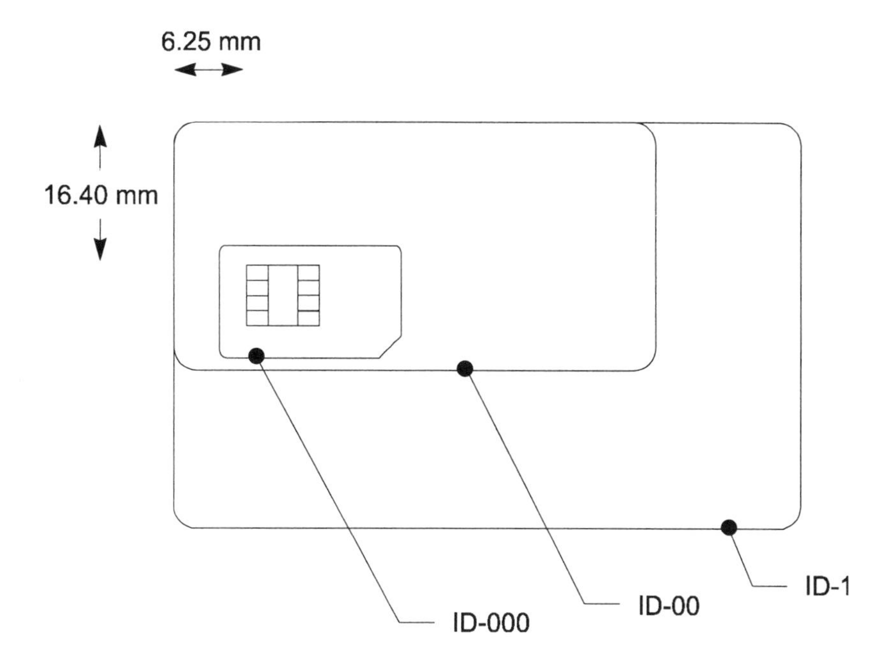

Figure 3.5 Relationship between the different card formats ID-1, ID-00 and ID-000

The three formats, ID-1, ID-00 and ID-000, can all be pressed out of the same blank. This is of particular interest for the manufacturer, since it allows the production process to be optimized and made more economical.

Thus, for example, a manufacturer may only produce cards in a single format (preferably ID-1), then embed the modules in them and personalize them completely. Depending on the exact application for which these Smart Cards are designed, they may be modified to the desired format in a single subsequent production step.

Alternatively, this may be done later by the client, as is already the case to some extent in the GSM field. The client receives a pre-punched ID-1 card, so that an ID-000 card is obtained by breaking off three segments. In another procedure, the ID-000 is completely pressed out of the ID-1 and attached to the remaining card with a single-sided tape on the reverse (the side without the contacts). The customer can now manufacture his own card, using whichever format is suitable for his telephone handset, and production and delivery by the manufacturer are standardized.

3.1.2 Cards with contacts

The main difference between a Smart Card and all other types of cards is the microprocessor implanted in it. Where the power supply and the data transfer require a physical contact, this is done via an electrical terminal which consists of the six or eight gold-plated contacts present on all conventional Smart Cards. The position of these contacts on the card and their dimensions are laid down in ISO 7816-2, dated 1988.

Figure 3.6 Example of a GSM card in ID-1 format in which the client obtains an ID-000 card by breaking off three segments of a pre-punched ID-1 card (Giesecke und Devrient)

A national AFNOR standard has existed in France since long before ISO 7816-2. It specifies a somewhat higher position for the contact field than in the ISO standard. The former is still listed in ISO as a "transitional contacts location", but the standard recommends that it be dropped in the future. However, many French cards still use the original position for the contacts, so that it is unlikely to be superseded in the immediate future.

The contact field is located in the card's upper left corner, as shown in Figure 3.7.

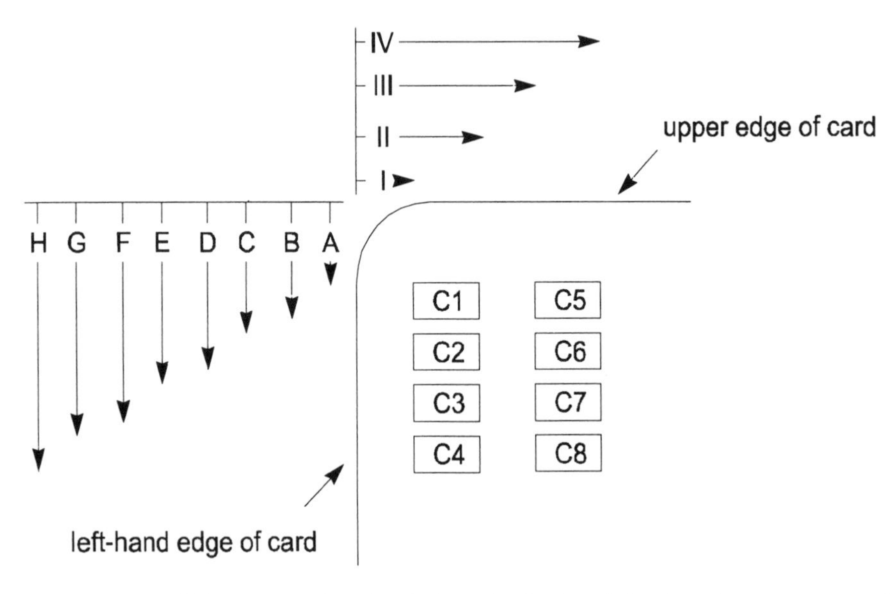

Figure 3.7 Contact locations in relation to body of card (the locations are not to scale)

I maximum 10,25 mm A maximum
19,23 mm
II minimum 12,25 mm B minimum
20,93 mm
III maximum 17,87 mm C maximum
21,77 mm
IV minimum 19,87 mm D minimum
23,47mm
E maximum
24,31mm
F minimum
26,01 mm
G maximum
26,85 mm
H minimum
28,55 mm

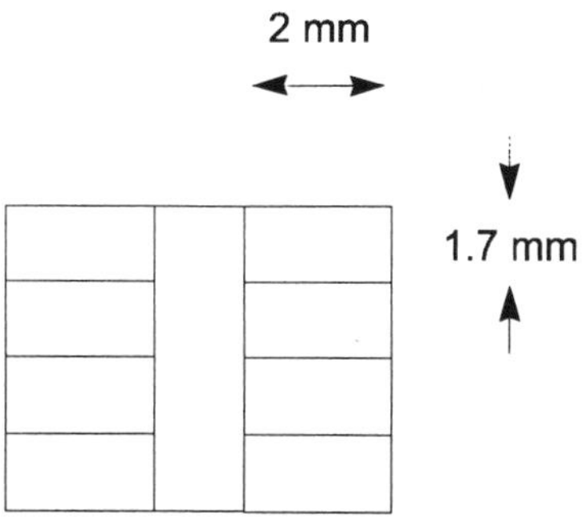

2 mm

1.7 mm

Figure 3.8 Minimum size of contacts in accordance with ISO 7816-2

The contacts should be not less than 1.7 mm high, and not less than 2 mm wide. Their maximum size is not specified. However, it is limited by the necessity of electrically isolating the contacts from each other.

The module's location on the card is fixed by the standards. Similarly, the magnetic stripe and the embossed field are specified precisely (ISO 7811). All three elements may coexist on one card. However, the following relationships must be observed: where a card carries only a chip and an embossed field, these may be located on the same or opposite faces. If there is also a magnetic stripe, it must always be on the opposite face to the embossing.

3.1.3 Cards without contacts

Cards which do not contain contacts use a technology which allows the transmission of data over short distances without ohmic coupling. The technical principles are not new, and have been known and widely used for many years in animal implants and in the transponders involved in starter locking in motor vehicles. The potentially large number of technical options is severely restricted in applications using ID-1 formatted Smart Cards – to which discussion will be restricted here – since all functional elements must be incorporated into a card measuring only 0.76 mm in thickness. Thus, for example, no flexible batteries are available in this thickness to supply power to the electronics.

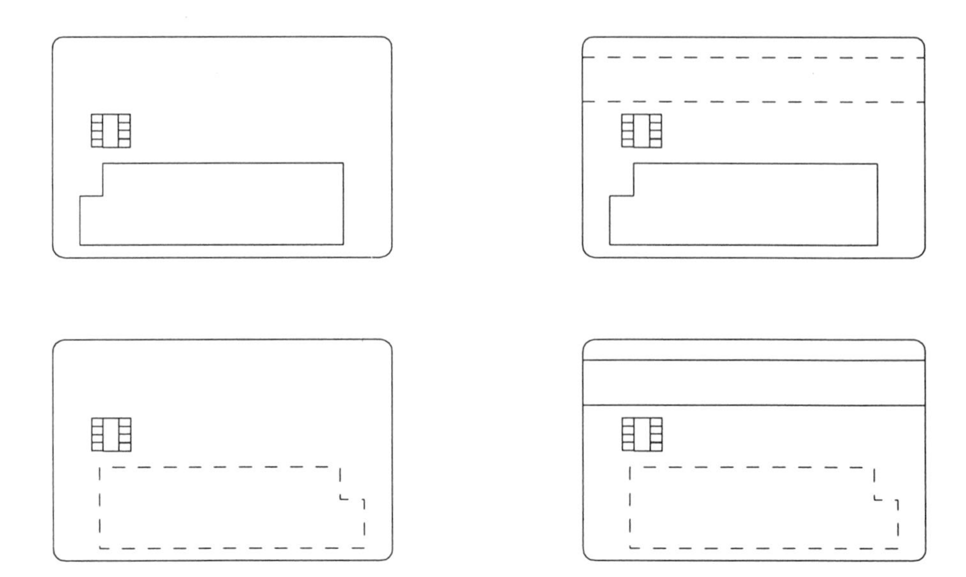

Figure 3.9 Variations in the arrangement of card elements in accordance with ISO 7816-2: chip, embossed field and magnetic stripe

Figure 3.10 Example of a card with chip, magnetic stripe, signature field and embossing (Giesecke und Devrient)

Four problems must therefore be solved if a contactless card is required:

- Power supply to the integrated circuit
- Transmission of the clock signal
- Data transmission to the Smart Card
- Data transmission from the Smart Card

Various concepts have been developed to solve these problems, some of them tailor-made for special applications. It is therefore very important to know whether the operational distance between terminal and card is a few millimetres or about a metre. The development of optimized solutions for specific applications tends, unfortunately, to lead to incompatibility between them.

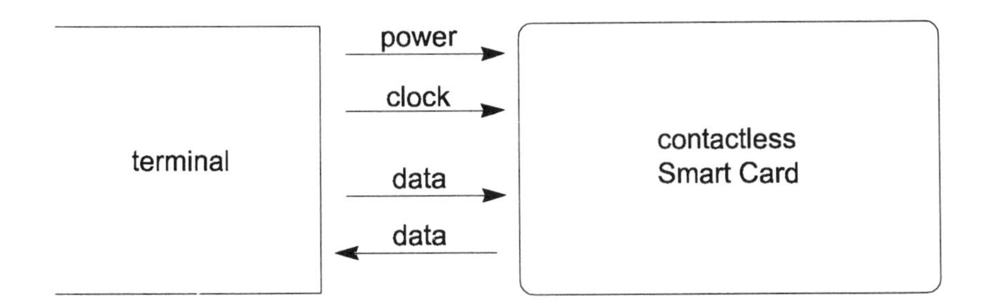

Figure 3.11 Requirements for power and data transfer between terminal and contactless Smart Card

The most frequently used techniques are microwave transmission, optical transmission, capacitive coupling and inductive coupling. Capacitive and inductive couplings are most suited to the Smart Card's flat format. Since a detailed description of all available techniques would go beyond the physical scope of this book, capacitive and inductive couplings only will be examined, not least because so far they are the only techniques to have been considered in the ISO/IEC 10536 and 14443 standardization series.

In capacitive coupling, conductive surfaces are incorporated into the card which act as capacitor plates and have a useful capacitance of a few tens of picofarads. As a rule, this is insufficient for power transmission. Hence this method is usually implemented only for data transmission, whilst power transfer is carried out inductively. Data transmission makes use of a differential procedure using a symmetric pair of coupled surfaces, where the range is limited to about 1 mm.

Inductive transmission is currently the most widespread method, used in both data and power transmission. A variety of requirements and boundary conditions, for example postal regulations, have led to diverse implementations.

In applications such as access control, it is usually sufficient for the data stored in the card only to be read, allowing for technically simple solutions. Due to the low power requirement of a few tens of microwatts, the range of these cards extends to a metre or so. The storage capacity is usually only a few hundred bits. Where the data needs to be written as well, the power requirement increases to over 100 μW. The consequence is that the range in writing mode is limited to about 10 cm, as the writing device's radiated power may not be increased arbitrarily due to postal regulations. Microprocessor cards without contacts have an even higher power requirement, reaching about 100 mW. As a result, the distance to the terminal is limited to a few millimetres.

Regardless of range and power requirements, all inductively coupled cards operate on the same principle.

One or several loops are integrated into the card and act as coupling elements. Energy transmission follows the principles of weakly coupled transformers. The carrier frequency is in the range 100 – 300 kHz or a few MHz.

Data transmission from terminal to card relies on various methods such as frequency, phase or amplitude modulation. In the opposite direction, from Smart Card to terminal, a type of amplitude modulation is used in which a change in the card's resistance is controlled by the data signal.

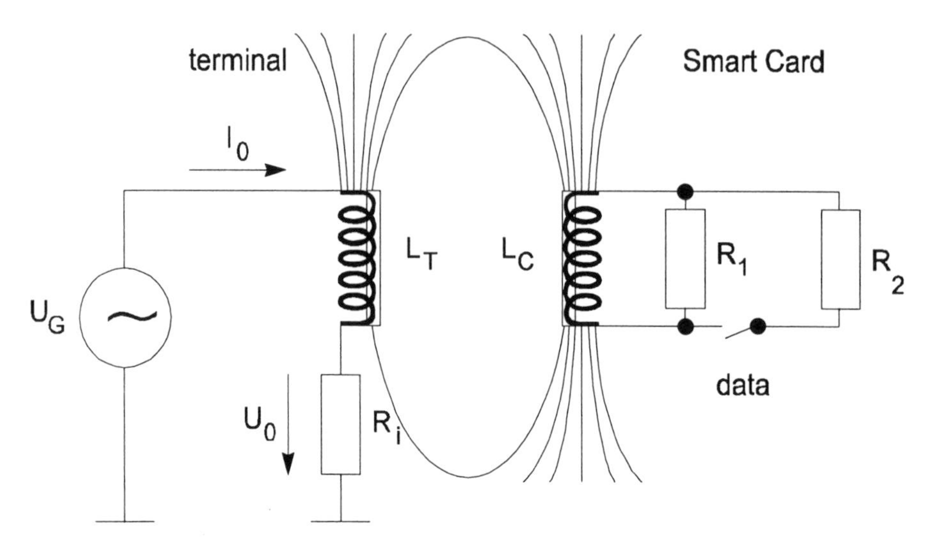

Figure 3.12 Example of data-transmission principle in a contactless Smart Card

When resistor R_2 is switched on by the data signal, the current passing through the coupling loop in the terminal rises, and thus also the potential drop in the AC generator's internal resistance R_i. This increase in potential can be detected in the terminal and interpreted as a data signal.

Due to the many different methods used by individual manufacturers, the standardization process launched by ISO/IEC in 1988 has been as difficult and time-consuming as expected. The working party charged with this task was required to standardize a contactless card which would be largely compatible with the other standards set for ID cards. That means that a contactless card may also contain other functional elements such as magnetic stripes, embossing and chip contacts. This allows implementation of contactless technology simultaneously with existing systems using other methods.

3.1.3.1 Close coupling card

As already explained, the technical possibilities for contactless power and data transfer depend fundamentally on the distance from card to terminal in the read and write modes. In order to structure the wide variety of options, ISO/IEC standards distinguish between Close Coupling and Remote Coupling cards, where the exact boundary between the two types is hard to define. In ISO/IEC standard 10536 for Close Coupling Cards, the application is described as "slot or surface operation", which refers to the fact that during operation the card must be inserted into a slot or placed on a defined surface which forms part of the terminal. ISO/IEC standard 10536, entitled "Identification Cards – Contactless Integrated Circuit Cards" consists of four parts:

- Part 1: Physical characteristics
- Part 2: Dimension and location of coupling areas
- Part 3: Electronic signals and reset procedures

- Part 4: Answer to reset and transmission protocols.

Parts 1 to 3 have by now become international standards, whereas Part 4 is still in preparation. The following were important preconditions for this standardization:

- Extensive compatibility with ISO 7816
- Operation at any orientation between card and reader
- Transmission carrier frequency between 3 and 5 MHz
- Bidirectional data transmission with inductive or capacitive coupling
- Card power requirement below 150 mW (sufficient for operating microprocessors).

Part 1 of the above-mentioned standard defines the physical features. Essentially, it lays down the same requirements which apply to Smart Cards with contacts, in particular with regard to bending and shearing. One difference relates to resistance to electrostatic discharges. Since in a contactless card, no conductive connections are needed between the card's surface and the integrated circuits inside the card, such a card is largely insensitive to damage by electrostatic discharge. Therefore, the standard specifies a test voltage of 10 kV, compared with 1.5 kV in cards with contacts.

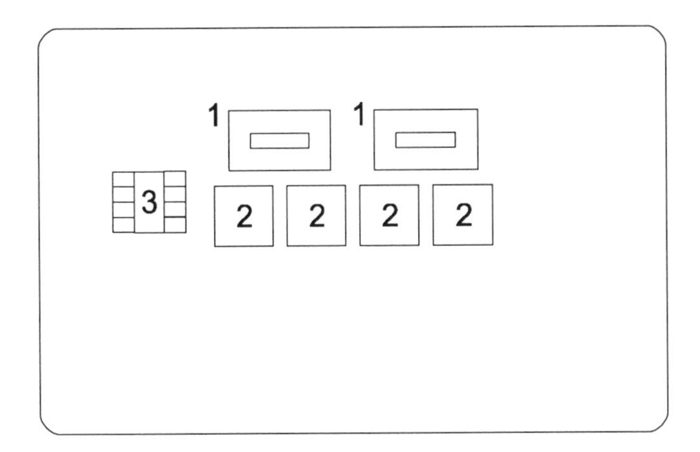

Figure 3.13 Arrangement of coupling elements in Smart Cards: 1 – coupling loops in the Smart Card; 2 – capacitive coupling areas in the Smart Card; 3 – contact areas of the chip

Part 2 of ISO/IEC standard 10536 specifies the location and dimensions of the coupling elements. Since it has proved impossible to agree on a single method, both capacitive and inductive elements are defined in such a way as to ensure that both techniques can be simultaneously implemented in the card and the terminal. An example is shown in Figures 3.13 and 3.14.

The arrangement was chosen so that the orientation could be arbitrary, given appropriate control circuitry in the terminal.

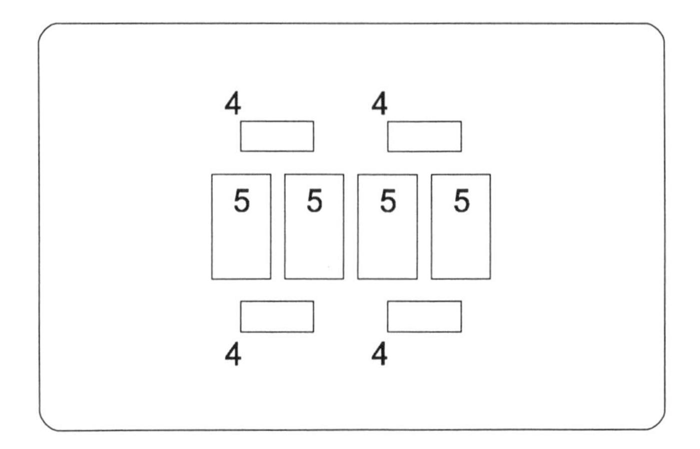

Figure 3.14 Arrangement of coupling elements in Smart Card terminals: 4 – coupling loops in the terminal; 5 – capacitive coupling areas in the terminal

Part 3 of ISO/IEC standard 10536 was adopted in 1995. It is the standard's most important section so far, and describes the modulation procedure for both inductive and capacitive data transmission, since no single method has been agreed upon. Hence, a terminal which meets the standard must support both methods, and these may also be implemented in a single card.

Power transmission

Power transmission takes place via an alternating sinusoidal magnetic field with a frequency of 4.9152 MHz, which intersects with one or several of the inductive coupling surfaces, depending on the number of coupling loops in the card. The terminal must generate all four fields.

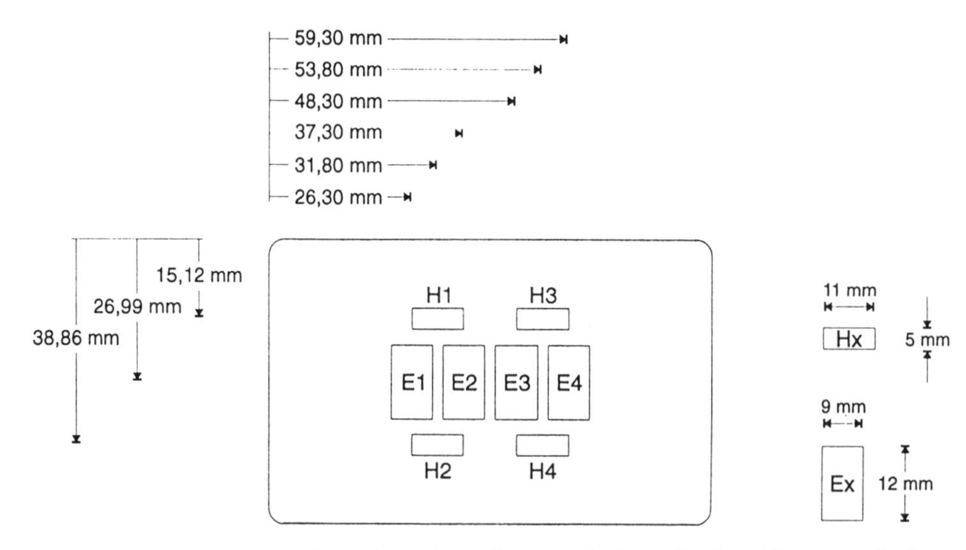

Figure 3.15 Location and dimensions of coupling areas in Smart Cards and in the terminal

The alternating magnetic fields F1 and F2 through surfaces H1 and H2 have a phase difference of 180°, and similarly, fields F3 and F4 through surfaces H3 and H4. Fields F1 and F3, as well as F2 and F4, are at a phase difference of 90°. Each individual magnetic field is powerful enough to transmit at least 150 mW to the card. However, the card must not receive more than 200 mW in total. This complicated definition of the magnetic fields is necessary in order to allow invariant inductive data transmission in four positions, as explained below.

Inductive data transmission

Different modulation methods are used for inductive data transmission in the two directions.

Transmission from card to terminal

For transmitting data from card to terminal, an auxiliary carrier with a frequency of 307.2 kHz is first generated via impedance modulation (see Figure 3.12), with the resistance varying by at least 10%. Data modulation then follows through a 180° phase switching of the auxiliary carrier, resulting in two phases which can be interpreted as logic 0 and 1. The initial condition following establishment of the magnetic field is defined as logic 1, with this initial condition (time interval t_3 in Figure 3.19) remaining stable for at least 2 ms. Each subsequent phase shift of the auxiliary carrier indicates a reversal of the logic state, resulting in NRZ (non-return to zero) coding. Transmission speed for ATR is 9600 bit/s.

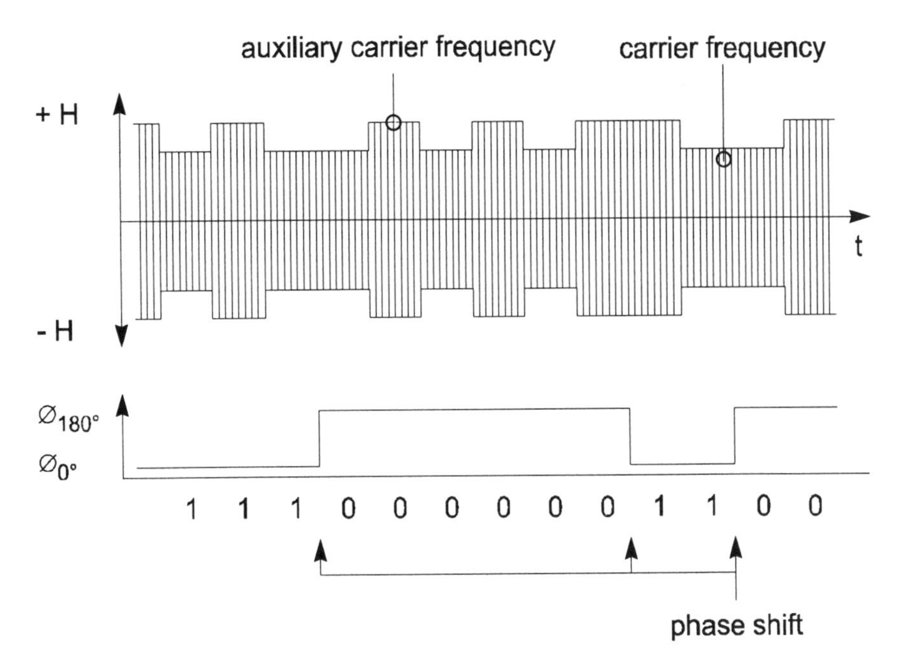

Figure 3.16 Representation of the principle of phase modulation in data transfer with a Smart Card (top: alternating magnetic field; bottom: phase duration). Carrier frequency: 4.9152 MHz; auxiliary carrier: 307.2 kHz

Transmission from terminal to card

For transmitting data from the terminal to the card, the four alternating magnetic fields F1 to F4 which penetrate the four coupling surfaces H1 to H4 are phase-modulated (PSK, phase shift keying). Specifically, the phases of all fields shift simultaneously by 90°. Thus, two phases A and A´ are defined. Depending on the card's orientation in relation to the terminal, this results (as seen from the card's perspective) in two different phase constellations, shown in Figures 3.17 and 3.18.

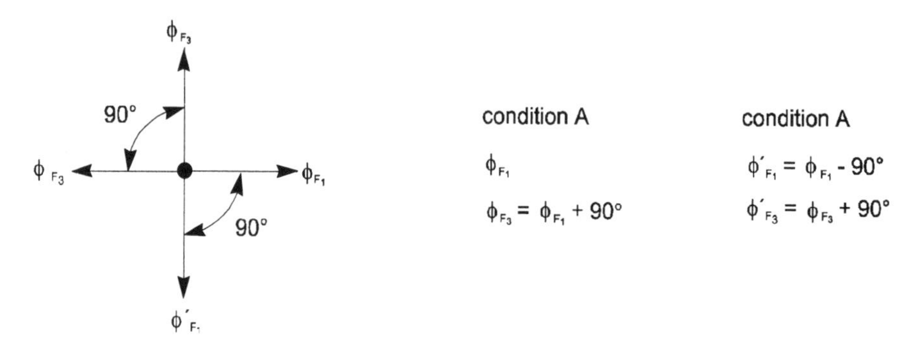

Figure 3.17 First variation of phase modulation in data transfer using a Smart Card (the arrows represent phase shift)

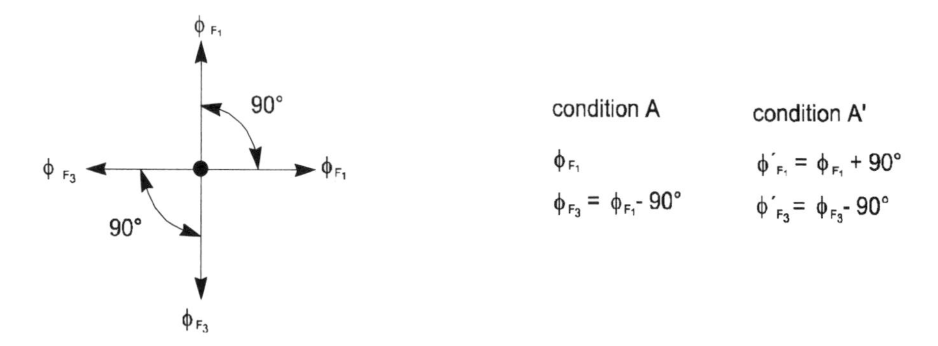

Figure 3.18 Second variation of phase modulation in data transfer using a Smart Card (the arrows represent phase shift)

Since the card needs to function in all four possible positions relative to the terminal, the initial condition (time intervals t_2 and t_3 in Figure 3.19) is interpreted as logic 1, regardless of the actual situation. Each subsequent phase shift is interpreted as a reversal of the logic state, again resulting in NRZ coding.

Capacitive data transmission

For capacitive data transmission from card to terminal, one pair of the coupling surfaces is used, either E1 and E2 or E3 and E4 (Figure 3.15), depending on the card's orientation in relation to the terminal. The other pair may be used for data transmission in the opposite direction. Since the card sends the ATR over one of the two surface pairs, the terminal can detect the card's orientation. The maximum potential difference between a surface pair is restricted to 10 V, but it must be large enough for the receiver's minimum potential difference threshold to be exceeded by ±300 mV. A differential NRZ coding is used for data transmission. The sender switches the voltage between surfaces E1 and E2 or between E3 and E4. Once again, the logic 1 status is fixed in the time interval t_3 (Figure 3.19). Each subsequent voltage reversal indicates a switching of the logic state.

Initial status and answer to reset

For the terminal to be able to determine unambiguously the method of data transmission and the card's position at the start of a data exchange, time intervals must be precisely defined for the start of power and data transmissions. Figure 3.19 shows the conditions and values for the reset interval t_0, the power increase interval t_1, the preparation interval t_2, the interval for stable logic state t_3 and the interval for answer to reset t_4.

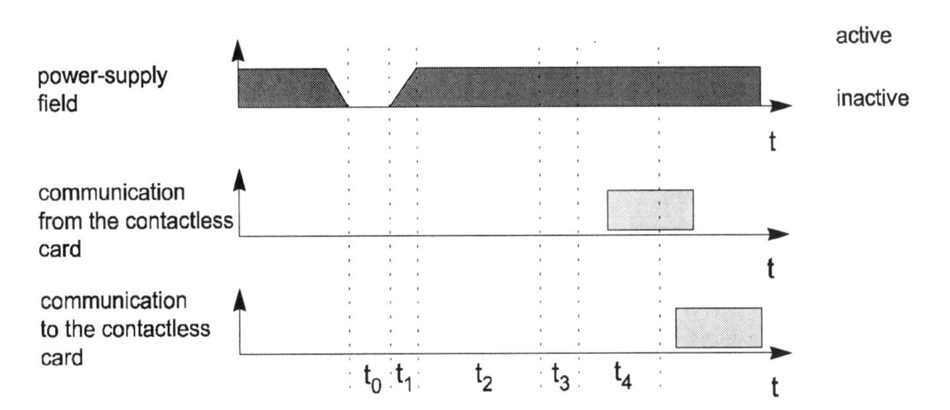

Figure 3.19 Progress of data transfer in Smart Card, in accordance with ISO/IEC 10536-3: $t_0 \geq 8$ ms; $t_1 \leq 0.2$ ms; $t_2 = 8$ ms; $t_3 = 2$ ms; $t_4 \leq 30$ ms

Minimum reset recovery time interval t_0

When a reset needs to be generated by switching power-transmitting fields on and off, the interval during which no power is transmitted must be no shorter than 8 ms.

Maximum power rise time interval t_1

The rise time interval of the power transmitting field established by the terminal must not exceed 0.2 ms.

Preparation time interval t_2

The card's preparation time interval during which it reaches a stable condition is 8 ms.

Stable logic state interval t_3

Before answer to reset, the logic state is held at 1 for an interval of 2 ms. During this time, the card and terminal are set to logic 1 for inductive data transmission.

Maximum response time for ATR t_4

The card must start ATR transmission before 30 ms have elapsed.

The card can show in ATR that the conditions for subsequent operation must be modified in respect to power level, data transmission speed or field frequency. The free interval produced thereby can be used in accordance with the application's requirements. For example, substantially higher transmission speeds may be selected for time-critical applications.

The schedule to ISO/IEC standard 10536-3 describes detailed testing methods and formats which make unambiguous measurement of the fields at the card's surface possible, as well as those of the terminal.

Part 4 of the standard is currently in draft form. Beyond the protocols already known from ISO/IEC 7816, a new full–duplex protocol is being defined which is also likely to be of interest in the context of cards with contacts.

3.1.3.2 Remote coupling cards

The term "cards with remote coupling" covers Smart Cards which can transmit data to the terminal over a distance of several centimetres, up to about 1 metre. This possibility is of great interest for all applications where data needs to be exchanged between card and terminal, without the user having to handle the card and insert it in the terminal. Examples are:

- Access control
- Vehicle identification
- Electronic driving licence
- Ski pass
- Airline ticket
- Electronic payment
- Luggage identification

The diversity of applications suggests the existence of a variety of possible technical implementations, making standardization a complex and also, unfortunately, a tedious task. The ISO working group has been dealing with this standardization since 1994. The planned standard is designated ISO/IEC 14443 Identification Cards – Remote Coupling Communications Cards. The projected standard will be unable to cover all possible applications, so the following key points have been defined:

- Identification, i.e. the reading of general data from a card
- Access control, i.e. using the card to authorize access to a protected area
- Electronic payment, i.e. the purchase of goods or services against payment.

Definition of the technical solution is still in its early stages, and the eventual approach has not yet been decided upon.

In recent years, a high level of demand has been identified for contactless cards, for applications in the field of local public transport. Industry has developed systems which have already been tested successfully in pilot studies. These new systems have distinct advantages for both operators and users. Besides cashless operation, the operator benefits from higher flexibility in the tariff structure and from the ability to capture and evaluate data relating to customer behaviour. Another advantage is a higher turnstile throughput and considerably better protection against vandalism. For the user, extensive automation makes fare payment considerably simpler, doing away with awkward manipulation of coins at ticket dispensers.

Since international standardization for these applications is still in its infancy and quick results are not expected, it appears that *de facto* industry standards will dominate the marketplace over the next few years. One example is the Mifare system, developed by the Austrian firm Mikron, which is also supported by Philips and Siemens. Mifare was developed specifically for use in local public transport. It operates with inductive transmission between card and terminal, at a frequency of 13.56 MHz and data transmission speed of over 100 kbit/s. At 10 cm, the range is sufficient to save the passenger the tedious removal of the card from his wallet. Since in this mode of operation it must be assumed that several cards would simultaneously be present within the terminal's range, a fast anti-collision algorithm ensures that the data carried on the different cards does not interfere with each other. The architecture of the card makes it possible to run 16 independent functions on one card. Data access is protected through an encryption procedure. In use, different data, such as travel information (journey start and end), can be captured, and the fare paid electronically. The application described in Section 13.1 and used by the German airline Lufthansa also employs the Mifare system.

3.1.4 Security features

Since Smart Cards are mostly used as authorization for certain operations or to identify the holder, security features are often present on the card's surface, in addition to the implanted chip. In contrast to the microprocessor, the security of these features is based not on cryptographic procedures but to some extent on concealment. It is also underpinned by using techniques involving very high cost or considerable expertise. Since normally human beings and not machines check whether cards are genuine, many security features are based on visual identification characteristics. However, some security features rely on a modified Smart Card microprocessor, and thus are only machine-readable.

Signature strip

A rather simple method of identifying a card's owner is the signature strip firmly fixed to the card, found on all credit cards. It cannot be changed once written to and is thus permanent. Any new strip glued on top is immediately obvious due to a very fine colour print. The irreversible attachment of the printed paper strip can be achieved through a hot adhesive process. In another approach, the signature strip is part of the card's top layer and is laminated into it during manufacture.

Security patterns

A somewhat more expensive process involves foils inserted under the card's top transparent layer and printed with coloured patterns. The name "guilloche" refers to the closed and interwoven areas of fine lines which are generally curved or oval, of the type used in various banknotes and share certificates. At the moment, these patterns can only be produced by printing, and are also impossible to forge.

Microscript

Another security process based on fine printed structures is known as microscript lines. The text is only visible to the naked eye as a line, can only be read with a magnifier and is also impossible to copy.

Holograms

Holograms integrated into the card for security purposes have become well known to all users. These are pressed into the card in a process involving hot embossing, and cannot be separated from it without damaging the card. Security is based first and foremost on the fact that only a couple of firms around the world make these holograms, and they are not widely available.

Multiple laser image

The multiple laser image (MLI) is similar to a hologram. It is a type of 3-D image which is very similar to simple holograms. The main difference is in the representation of individualized information in each card. For example, the card holder's name may be made into an image. MLI is produced by blackening of embossed synthetic foils with a laser beam.

Lasergravure

Lasergravure is the name given to the blackening of special synthetic layers by burning them with a laser. Unlike embossing, this is a particularly secure method of writing individual data, such as name and card number. The two available variants are known as vector and scanning gravure. In the former, the laser beam follows a continuous and uninterrupted path, highly suitable for representing line markings, with the advantage of high speed. In contrast, scanning lasergravure involves blackening a large number of adjacent points, a process similar to ink-jet or dot-matrix printing. The main application of this method is in the printing of pictures on the cards. However, though it has the advantage of high definition, it suffers from being very time-consuming. Lasergravure of a typical passport picture of normal quality, for instance, takes approximately 10 seconds.

Embossing

Another alternative used in recording user data on the card's surface involves embossing alphanumeric marks. Since embossing can be very easily tampered with, a further security feature which is sometimes used is to allow some of the marks to overlap the hologram.

The widest range of visual security features were developed in the time period between the mass introduction of cards and that of Smart Cards. During this time they were the only means of checking if cards were genuine. As a consequence of the microprocessors implanted in new cards and the cryptographic procedures made possible thereby, those methods have to some extent been superseded. Nevertheless, they are still of great importance where cards need to be checked by people rather than by machines, since the former can never gain access directly to the chip without some electronic means.

The above list of security features describes only the most important and best known measures briefly. There are many other types, such as invisible codes only readable in IR or UV light, magnetically readable codes and special printing processes in iridescent colours. Though these are of great interest, they must be omitted here for reasons of space.

In the future, security features will be present not only on the card, but even in the chip itself. Some chips may be treated similarly to high-security banknote paper. This is also a vital precondition for the printing of genuine banknotes. Special chips and modified hardware are necessary in order to introduce similar features into the IC industry. The terminal will have to be capable of measuring any modifications to the hardware and using the results to evaluate a chip's authenticity.

A fast cryptographic algorithm may be implemented on a given chip as an additional feature. The time required for the computation of a particular value would be so short, thanks to the hardware solution, that software emulation on another chip would not be possible in such a short time. Thus, a terminal would have to distinguish between chips purely by time-interval measurement. Numerous chips are currently on the market which contain this hardware feature or a similar device. They are consequently no longer freely obtainable, just like banknote paper. Such features are only suitable for very large applications, due to the high costs involved in developing special hardware. In addition, the monopoly almost necessarily acquired by the chip, with no secondary suppliers available, is unacceptable to many card issuers.

3.2 THE BODY OF THE CARD

The materials, construction and production of the body of the card are effectively determined by the card's functional elements, as well as by the stresses to which it is subjected during use. Typical functional elements include:

- Magnetic track
- Signature strip
- Embossing
- Imprinting of personal data via laser beam (text, photo, fingerprint)

- Hologram
- Security printing
- Invisible authentication features (e.g. fluorescence)
- Chip with contacts or other coupling elements

Clearly, even a relatively small card, only 0.76 mm thick, must at times contain a large number of functional elements. This places extreme demands on the quality of the material used to produce the card as well as on the manufacturing process.

The minimum requirements as far as card robustness is concerned are laid down in ISO standards 7810, 7813 and 7816 part 1. The requirements relate essentially to the following areas:

- UV radiation
- X-ray radiation
- The card's surface profile
- Mechanical robustness of the card and contacts
- Electromagnetic susceptibility
- Electrostatic discharges
- Temperature resistance

ISO standard ISO/IEC 10373 specifies the test methods for many of these requirements, which allow users and card manufacturers to perform objective tests of card quality. For Smart Cards, the bending and shearing tests are of particular importance in this context. This is because the chip, with its physical properties (brittleness and fragility), is a fragile foreign object within the elastic card; it must be protected against mechanical strains which arise when the card is subjected to bending and shearing stresses, through special structural features. Chapter 9 contains a detailed list of tests and the methods used to perform them.

3.2.1 Smart Card material

The material originally adopted for ID cards and still widely used is amorphous thermoplastic PVC (polyvinyl chloride). It is the most economical of all the available materials, easy to process and suitable for a wide range of applications. This material is used in credit cards worldwide. The drawbacks are its limited lifespan due to physical deterioration, as well as poor resistance to extremes of temperature. In production, PVC can only be processed into cards as a foil, since injection moulding is not possible. PVC is considered to be damaging to the environment due to its chlorine content, and the starting compound, vinyl chloride, is listed as a carcinogen.

In order to bypass these disadvantages, cards have, for some time, been produced from ABS (acrylonitrile-butadiene-styrol). This is also an amorphous thermoplastic which is distinguished by its strength and resistance to extremes of temperature. Hence this synthetic is often used in mobile telephone cards, which are subjected to very high temperatures. ABS can be processed both as a foil and by injection moulding. The material's drawbacks are its limited acceptance of colour as well as sensitivity to weather conditions. No environmental risks are known in connection with ABS, though the starting compound, benzene, is a carcinogen.

A PVC substitute which has been in use for some time in the field of packaging is known as either polyethylene terephthalate (PET), or more commonly as polyester. This thermoplastic is used in Smart Cards, both in its amorphous form (A-PET) and in the partly crystalline form (PETP). Both are suitable for processing as foils, and also for injection moulding. However, PETP cannot yet be laminated, which makes additional processing necessary.

The fourth frequently used card material is polycarbonate (PC). It is used in applications in which high strength and long life are required, such as ID cards. Polycarbonate's main disadvantages are its susceptibility to scratches, as well as its very high cost compared with all other materials used in this field.

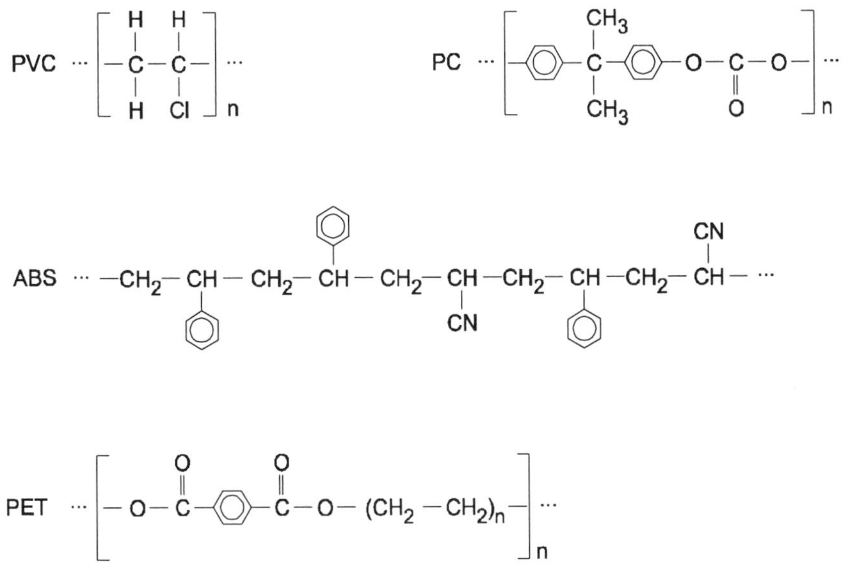

Figure 3.20 Structural formulae for important materials used in manufacturing body of card

3.2.2 Production methods

Smart Cards may be manufactured in several ways. In most production sequences, the chip is first mounted on a flexible film which carries the (usually gold-coated) contacts. The individual chip modules are pressed out of the finished film and inserted in the card. In this method, there is no direct attachment of the chip to the card. The advantage of this approach is that bending strains generated when the card is mechanically stressed are largely kept away from the chip.

Currently a distinction may be made between two fundamental techniques for mounting the chip on the carrier film, the TAB technique (tape automated bonding) and the wire bond technique.

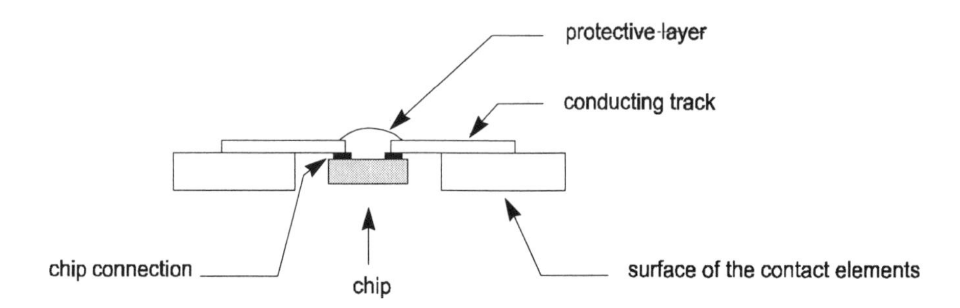

Figure 3.21 Cross-section of chip module using TAB technology

Figure 3.21 represents a chip module in the TAB technique. The special feature here is that metallic "bumps" are first galvanized to the chip's connecting pads. The carrier film's conducting tracks are then soldered to these bumps. Mechanically, this soldering is strong enough not to require additional attachment of the chip, which simply hangs from the conducting tracks. The chip's active surface is further protected against environmental damage by a covering layer. The advantage of the TAB method lies in the high mechanical strength of the chip connections, and in the module's flatness when installed. However, this advantage must be weighed against higher costs than the wire bond module.

Figure 3.22 Example of module using TAB technology

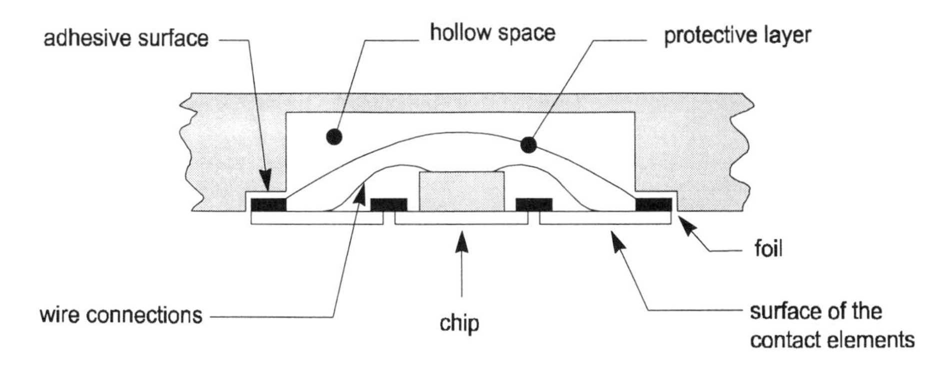

Figure 3.23 Cross-section of chip module using wire-bond technique

The most common method used nowadays is the wire bond technique, shown in Figure 3.23.

The carrier is again a plastic foil, the gold-coated contacts being galvanized to its front surface. Holes are punched out of the carrier film for taking up the chip and connecting wires. At this stage, the chip is placed into the punched section from the back, and attached to the conducting track (die-bonding). Then the chip's terminals are connected to the back of the contacts with thin wires (a few microns). Finally, the chip and connecting wires are protected against environmental damage by a cast layer of material.

The advantage of this method derives from the fact that it relies largely on processes common in the semiconductor industry for packaging chips in standardized housings. Less specialized expertise is required than in the TAB method, and as a result it is more economical.

The drawback is that the module's height, length and width are significantly larger than with the TAB module, since not only the chip but also the bonded wires need to be protected by the covering layer. This increases the problems associated with the chip module's installation into the card.

Three separate procedures are now established for the installation of the module: lamination, insertion into milled hollows and mounting in pre-moulded cards.

Lamination relies on the conventional card-manufacturing process. Most cards are produced by the lamination of the cover foil and the inlet foil. Before lamination, appropriate holes are punched in the foils and the chip module is inserted in them.

During lamination, the module is welded to the card. This approach achieves high levels of bonding between the module and the body of the card. In practice, the chip can no longer be removed from the card without destroying the latter.

In the second process, a hole is milled in the finished card and the chip module is glued into it.

This approach is useful when very complex cards need to be produced, containing many functional elements. Faulty cards can be rejected before the expensive chips are installed in them.

Cards containing only a few functional elements and with relatively modest demands in terms of printing quality, can be most economically produced using the injection moulding process. Here the entire card, including the hole for the chip, is manufactured as one unit and the chip module glued in it. The drawback of this approach is the fact that these cards can only accept a limited number of additional functional elements, and can only be printed individually.

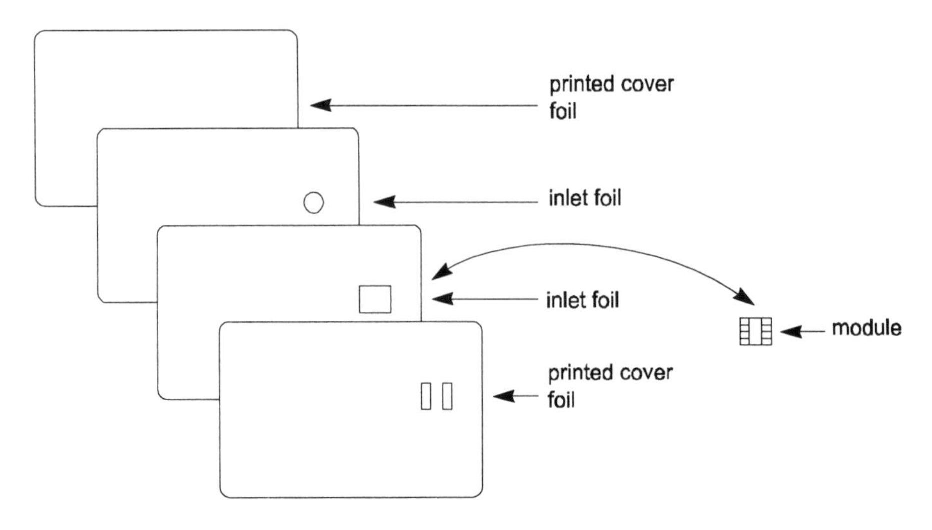

Figure 3.24 Installation of module using lamination procedure

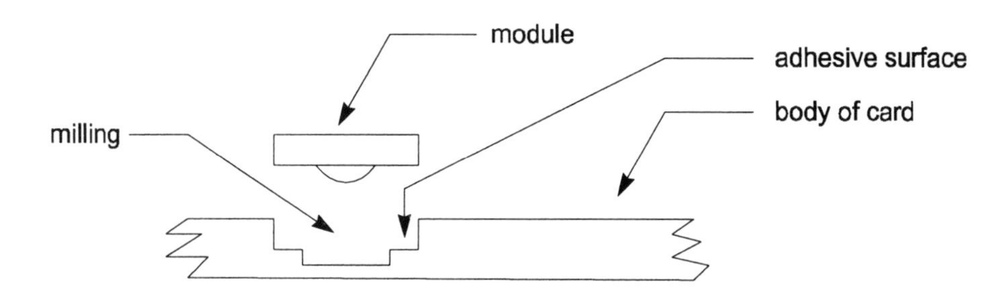

Figure 3.25 Installation of chip module into card body through holes milled into the surface

More recently, processes have been developed which allow the card to be manufactured in a single production step. A printed foil, the chip module and a label are inserted automatically into a form, and injected in one go. This and similar processes reduce the number of production steps. It is to be expected that further optimization of this process will lead to even lower Smart Card prices in the future.

3.3 ELECTRICAL PROPERTIES

A Smart Card's electrical properties depend solely on the incorporated microprocessor, since it is the card's only circuit component. In the early days of Smart Card technology, the only crucial factor was for the microprocessor to be functional, and less attention was paid to overall electrical properties. The applications were closed almost without exception, using a single type of card and a terminal specifically developed for it. The Smart Card's electrical properties were only relevant insofar as the terminal had to be laid out to accept the type of card in use. This situation is in the process of changing. With current large-scale applications, in which different Smart Cards must coexist with many different terminals, it is critical to ensure that all the cards in use are electrically identical, or at least behave uniformly within clearly defined electrical domains.

The large application which determined these specifications was the GSM mobile telephone network. This system, in which different terminals originating from the most diverse manufacturers had to work with many types of cards, clearly sets the boundary conditions. The large volume of cards in use, too, meant that the electrical properties defined in GSM 11.11 became the model for all semiconductor manufacturers. All new microprocessor products destined for Smart Cards throughout the world meet the electrical specifications laid down in GSM 11.11, without which they cannot be sold.

The other fundamental standard in this field, and the one with the widest international acceptance, is ISO/IEC 7816-3. It defines a very wide range of electrical conditions for Smart Cards, such as the booting and shutdown sequence.

3.3.1 Circuitry

Most Smart Cards have eight contact fields on the front face. These form the electrical interface between the terminal and the card's microprocessor. All electrical signals are carried across these contacts. However, in accordance with ISO/IEC 7816-2, two of the eight fields (C4 and C8) are reserved for future functions yet to be defined, and for reasons of compatibility should not as yet be used. One of these two contacts is planned as a second I/O interface, so that at some point Smart Cards will support full duplex data transmission. For this reason, some recent Smart Card modules have only six contact fields, which slightly reduces production costs. However, their functionality is identical to that of modules with eight contacts.

The contacts are numbered sequentially from top left to bottom right, as in conventional semiconductor components. The eight defined contacts are designated and electrically specified as shown in Figure 3.27, in line with ISO 7816-2.

V_{cc}		GND
RST		V_{pp}
CLK		I/O
RFU		RFU

V_{cc}		GND
RST		V_{pp}
CLK		I/O

C1		C5
C2		C6
C3		C7
C4		C8

Figure 3.26 Electrical properties and numbering of contact fields in a Smart Card in accordance with ISO 7816-2

Contact	Description	Function
C1	V_{cc}	Supply voltage
C2	RST	Reset
C3	CLK	Frequency
C4	RFU	Reserved for future functions and presently not in use
C5	GND	Mass
C6	V_{PP}	External voltage for programming (generally no longer used)
C7	I/O	Entrance/exit for serial communication
C8	RFU	Reserved for future functions and presently not in use

Figure 3.27 Description and function of contacts in accordance with ISO 7816-2

A few years ago it was still necessary to supply EEPROMs with external voltage for programming and erasing, since the microprocessors then in use did not have a charge pump. Hence, contact field C6 was reserved for this purpose. However, current state-of-the-art technology makes it possible to generate voltage directly on the chip using a charge pump, so this field is no longer used. Nevertheless, it cannot be employed for other functions, as this would conflict with ISO standards. Thus, every Smart Card carries a contact field with no actual function, but which must still be present. Since the programming voltage contact lies between two others, which are necessary for the card to function, it could not be eliminated in any case, thus reducing the disadvantage somewhat.

3.3.2 Supply voltage

A Smart Card's supply voltage lies at 5 volts, with a maximum deviation of ±10%. This voltage, which is conventional for TTL supply circuits, is standard for all cards currently on the market.

In mobile telephony, the weight reduction in end-user sets demanded by the market is pressing for a conversion from 6 volt or 4.5 volt sets to 3 volt types. As these days all mobile phone components are available in 3 volt technology, the Smart Card is the only one which still requires 5 volts. Hence, mobile phones need a voltage converter for the card's power supply, which is expensive and increases costs unnecessarily. As a result, future developments will include Smart Cards with a voltage range of 3 to 5 volts and ±10% tolerance, resulting in an effective voltage range of 2.7 to 5.5 volts.

Theoretically, it would also be possible to develop cards specially for the 3 volt range. This would suffer from the drawback of losing compatibility with the millions of existing 5 volt cards. In the worst case, use of a 3 volt card in a 5 volt terminal would destroy the card's chip. This would require users to be careful never to insert cards into a terminal with the wrong voltage, which one can certainly not expect them to be able to do. The advantages of simple and straightforward use would be lost.

In principle, the extended voltage range poses a problem neither for the processor nor for the memory components. However, there is also an EEPROM integrated into the microprocessor. It is precisely this EEPROM, and its charge pump, which constitutes the greatest obstacle on the road to 3 volt technology. It is technically difficult, and in fact has only recently become possible, to install EEPROMs and their charge pumps on chips working in the 2.7 to 5.5 volt range.

On the other hand, this wide voltage range will become mandatory for all microprocessors in the near future.

3.3.3 Supply current

The card's microprocessor derives its supply voltage through contact C1. According to the GSM 11.11 standard, the current may not exceed 10 mA. Although the ISO standard, in its current version, still specifies a figure of 200 mA, this is out of date and is likely to be modified in future, in the light of modern technology.

With a 5 volt supply voltage, and assuming a current consumption of 10 mA, the card's power input is 50 mW. This amount is so low that one need not be concerned about the chip heating up during operation, even when this energy is distributed over an area of only 25 square millimetres.

A microprocessor's current uptake is directly proportional to the frequency of the applied clock, which means that one may quote either the maximum possible current or the current as a function of the clock. The current also depends on temperature, though the relationship is not as strong.

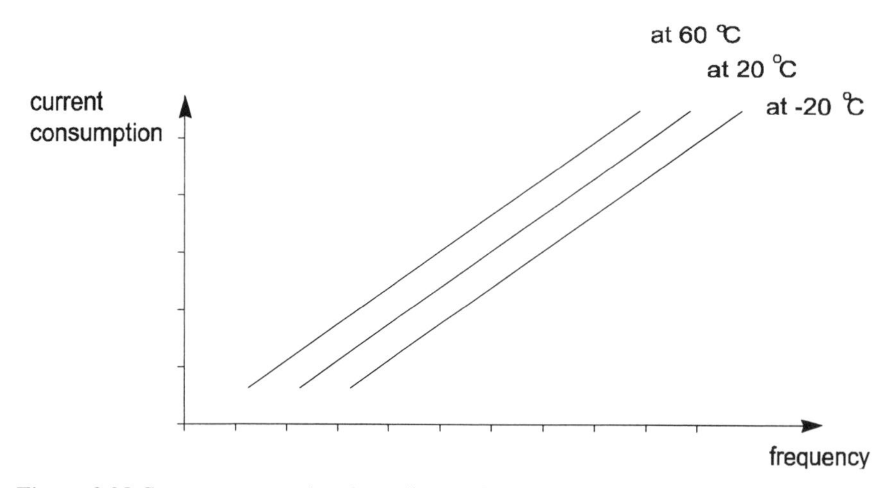

Figure 3.28 Current consumption depending on frequency of applied clock and temperature (micro-controller not in the sleep mode)

Another important point in relation to the supply current has caused headaches for several terminal manufacturers who have chosen to ignore it. All microprocessors in use are based on CMOS technology. Under certain circumstances, transistors can give rise to high cross-currents. These result in spikes in the nanosecond domain, reaching several times the nominal current. These spikes can also result from the EEPROM's charging pump being switched on. If this high current cannot be provided by the terminal during this short interval, the supply voltage will drop below the permitted value. This can lead to a write error in the EEPROM, or trigger the chip's low-voltage detector.

That is the reason why each valid standard and specification contains a comment concerning these possible spikes. They require the current supply to be capable of absorbing spikes with a maximum duration of 400 ns and a maximum amplitude of 200 mA. Assuming a triangular spike, this corresponds to 40 nAs of charge. The problem can be solved in a straightforward fashion by installing a 100 nF ceramic capacitor between the earth and the supply voltage, very close to the card's contacts.

In future, the GSM standard plans to reduce the maximum permitted current to a value of 1 mA, in order to extend the life of mobile phone batteries as far as possible. With the fast advances being made in semiconductor technology, this should soon present little difficulty. It is only the current need for EEPROM writing that still constitutes an obstacle, but with increasing miniaturization of EEPROM cells and the resulting decrease in the charges in the cells, the figure of 1 mA should be achievable within a few years.

Almost all microprocessors have a special current-saving mode, which deactivates all parts of the chip other than the I/O interrupt. When the microprocessor is in this state, also called sleep mode, current consumption drops substantially as most sections of the chip are disconnected from the supply voltage. In principle, only the I/O interface's interrupt logic, as well as the RAM, need to be powered, the other components can be switched off. The RAM's contents need to be conserved during this time in order to store the current data. The processor is also powered in practice, but the ROM and EEPROM are disconnected. This mode is of particular interest in battery-driven mobile phones. Therefore GSM 11.11 specifies the maximum current permitted in sleep mode, namely 200 μA with a 1 MHz clock at 25°C. There are plans to reduce the maximum figure in the future to 10 μA.

3.3.4 Clock supply

Smart Card processors do not possess internal clock generation. It must therefore be provided externally, and acts as the reference for data transmission speeds. The pulse ratio must be in the range of 40% to 60%, in accordance with GSM 11.11.

The clock provided at the contact field is not necessarily identical with the one available to the processor internally. In a few microprocessors, an optional divider may be added between the external and internal clocks. This often has the value 2, so that the external clock is twice as fast as the internal one. The reasons for this are connected on the one hand with the chip's hardware, and on the other with the option this provides of using the oscillators already present in the terminal for clock generation.

Most Smart Card microprocessors allow the clock supply to be switched off when the CPU is in sleep mode. Switching off, in this context, means freezing the clock supply at a certain level. Depending on the semiconductor manufacturer's preferences, the "off" level may be high or low.

The current drawn by the Smart Card from the clock's supply line is only a few microamps, and switching it off may appear at first sight rather strange. Nevertheless, the amount of current saved within the terminal may be substantial, and a particular application may make it worthwhile.

3.3.5 Data transmission

Where errors occur during data transmission, it may happen that terminal and card are attempting to send simultaneously. This results in data collision on the connecting I/O channel. Quite apart from the problem this causes at the application level, it can lead to physical currents along the I/O channel which at certain magnitudes can destroy the interface components.

In order to prevent damage to the semiconductors should such an eventuality arise, the I/O conductor in the terminal is set to a +5 V level with a 20 kΩ pull-up resistance. With the additional convention of never sending an active 5 volt level, the problem of the two parties sending inconsistent voltages to each other in an error situation is avoided. If, during communication, the I/O conductor needs to be set to a level of +5 V, the relevant party switches its output to a high ohmic state (tri-state), and the conductor is automatically pulled up to +5 V by the pull-up resistor.

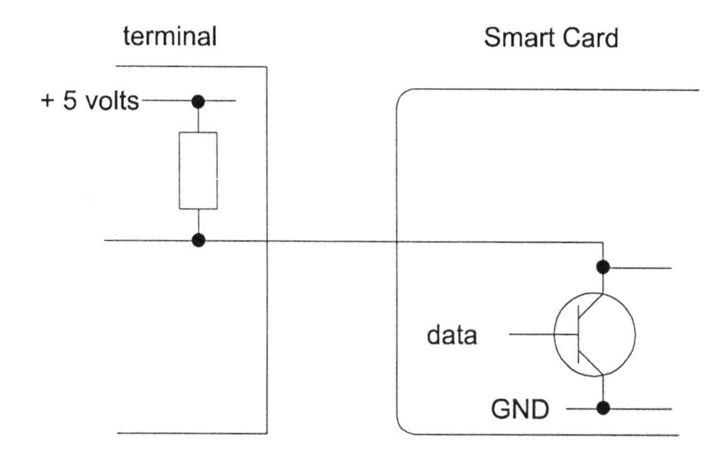

Figure 3.29 Use of the I/O channel between terminal and Smart Card

3.3.6 Booting/shutdown sequence

All Smart Card microprocessors are protected against charges and potentials at the contacts. Nevertheless, a precisely defined booting and shutdown sequence is prescribed, and must be strictly adhered to. This is also reflected in the appropriate part of ISO/IEC 7816-3. The sequences define the electrical part of booting and shutdown, and have nothing to do with the sequence of mechanical contacts, which in fact is not stipulated.

As shown in Figure 3.30, an earth must be connected before the supply voltage. Supply voltage connection is followed by the clock connection. If, for example, an attempt is made to connect the clock before the supply voltage, the microprocessor will attempt to draw its entire power over the clock supply line. This can damage the chip irreversibly, causing complete functional failure. A faulty shutdown sequence may also have similar effects on the microprocessor.

When the microprocessor is operating, it can be reset via the appropriate wire. This requires first a low level on this wire, and resetting is initiated with the rising slope. Such resetting during operation is called a warm reset, as in other computer systems. Correspondingly, a cold reset is one established by ISO-compatible switching on of all supply connections.

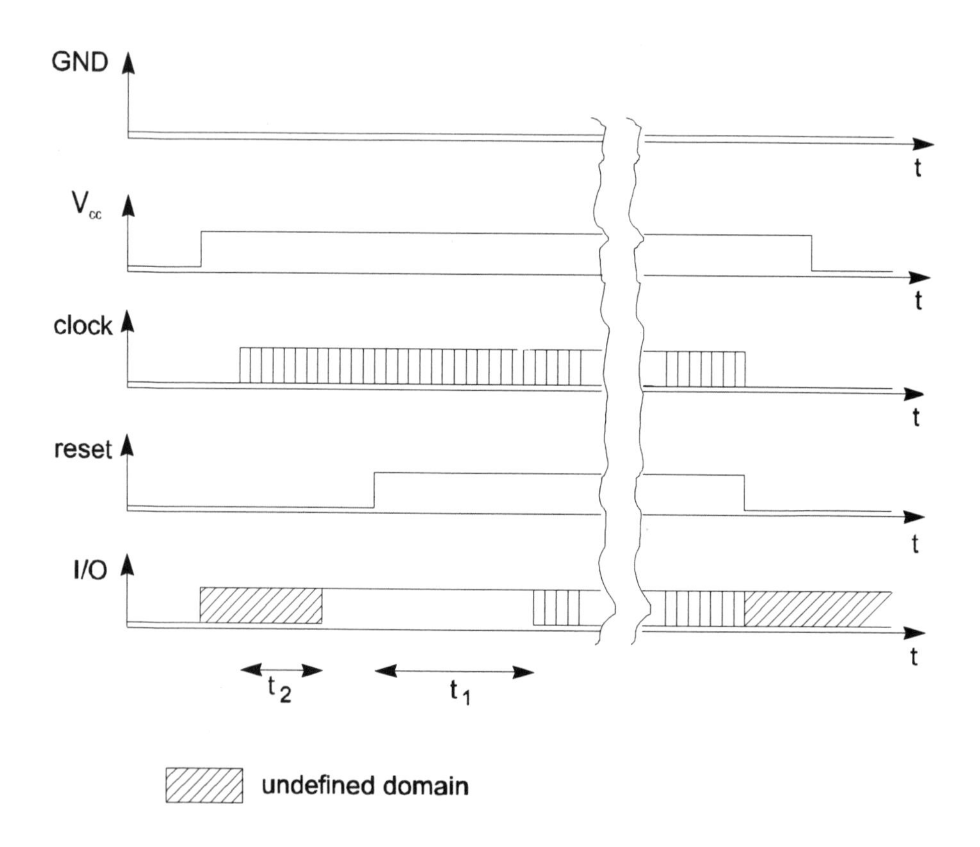

Figure 3.30 Booting/shutdown sequence in a Smart Card in accordance with ISO/IEC 7816-3; $(400/f) \leq t_1 \leq (40000/f)$, $t_2 \leq (200/f)$

3.4 SMART CARD MICROPROCESSORS

From an IT perspective, a Smart Card's central component is the microprocessor embedded under the contacts field. It controls, initiates and monitors all activities. The microprocessors specially designed and developed for this purpose are complete computers in themselves. They are supplied with a processor, a memory and an interface to the outside world.

The most important functional elements of a typical Smart Card microprocessor are the CPU, the address and data bus and the three types of memory (RAM, ROM, EEPROM). In addition, the chip contains a simple interface module which is responsible for serial communication with the outside world. However, this interface must not be imagined to be a complex functional unit on the chip, which can send and receive independently. In this context, a serial interface only means an address which can be accessed by the CPU and which is connected to the contacts field via an I/O wire.

In addition, some manufacturers also provide special arithmetic units on the chip which function as a kind of mathematical coprocessor. However, the functions available to these units only extend to exponential and modular operations on integers. Both operations are vital to public key encryption procedures, like the RSA algorithm.

The microprocessors used in Smart Cards are not standard components which are widely available. They are modules developed specially for this purpose, and are not implemented in other applications. There are several important reasons for this, as follows.

Manufacturing costs

The area required by the microprocessor on the silicon wafer is one of the decisive factors affecting manufacturing costs. Large chips result in a more difficult, and hence more expensive, packaging in the module. Great efforts are therefore made to reduce chip size as far as possible.

Furthermore, many standard modules available on the market contain technical functions not needed in Smart Cards. Since these functions occupy additional areas of silicon, chips designed for Smart Cards can be optimized by deleting them. Although the manufacturing costs per chip are only reduced minimally as a result, the bulk savings are significant, which justifies these modifications to the chip design.

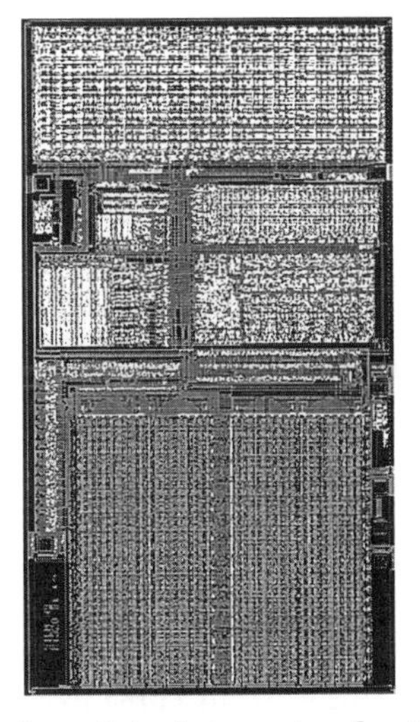

Figure 3.31 Comparison of size between two functionally identical Smart Card microprocessors before and after reduction of chip surface (shrink procedure). Left: SLE 44C80 in 1 μm technique using 21.7 mm^2 surface. Right: SLE 44C80S in 0.8 μm technique using 10 mm^2 surface (Siemens)

Figure 3.32 Photograph of Smart Card microprocessor P 83 C 852 with the functional elements ROM, EEPROM, CPU and RAM (from top left to bottom right).
The chip has a surface area of 22.3 mm^2 and contains 183 000 transistors (Philips)

Functionality

Due to the requirement that all the functional elements of a computer be integrated on a single silicon chip, the range of acceptable semiconductor modules is extremely restricted. With the specification of a minimal chip area, 5 volt power supply and a serial interface on the chip, effectively all standard modules are ruled out. Additionally, the chip must contain a memory (EEPROM) which can be erased and written to but does not require a permanent power supply for data storage.

Security

Since Smart Cards are used in security-related fields, making both passive and active security features necessary on the chip, the requirement of a chip specially developed for this purpose becomes unavoidable.

Chip area

The area occupied by the microprocessor on the chip strongly affects the chip's fragility. The larger the area, the easier it is for a crack to occur when the card is subjected to bending or shearing forces. Consider a phone card carried in a wallet. The bending loads on the card and the chip embedded in it are enormous. Even the finest hairline cracks in the chip are sufficient to make it useless. Therefore, most card manufacturers place an upper limit of about 25 square millimetres on chip area, in as square a layout as possible, so as to minimize the risk of fracture.

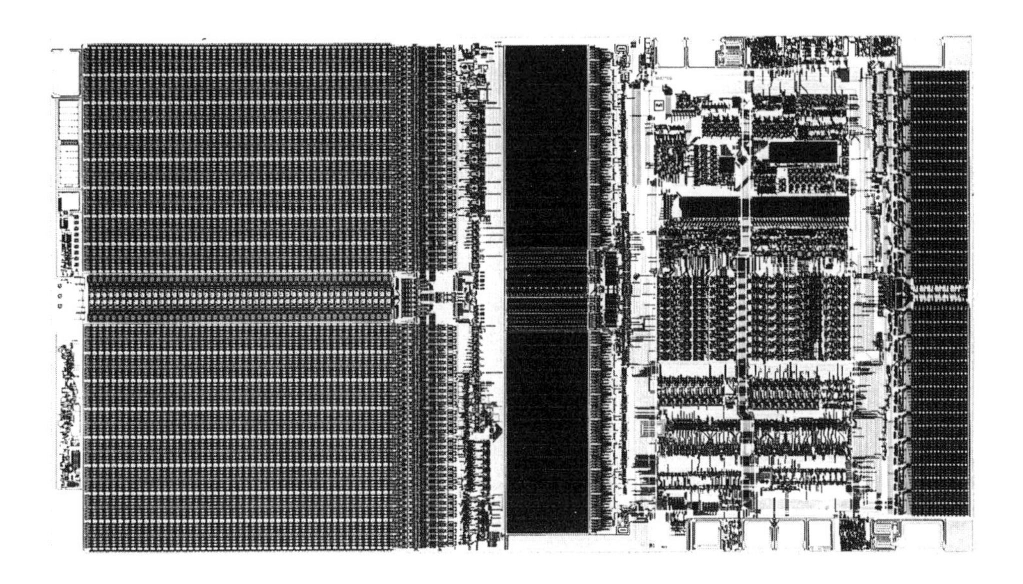

Figure 3.33 Photograph of Smart Card microprocessor ST 16 623 with the functional elements EEPROM, ROM, module and RAM (from left to right) (SGS-Thomson)

Availability

The security policy of many card manufacturers is to ensure that the microprocessors are not available to all and sundry on the open market. This makes reverse engineering of the chip's hardware considerably more difficult, as normally the competition would have no access to it.

These are the arguments for using only specially developed microprocessors, none of them freely available on the market. In the early days of Smart Card technology, repeated attempts were made to meet the five requirements listed above with standard modules. However, eventually it became clear that this was impossible. Unfortunately, such narrow specialization, using only a few types of microprocessors and hence manufacturers, suffers from the disadvantage that card manufacturers have little freedom to manoeuvre. In the event of production bottlenecks at a semiconductor manufacturer, it is impossible to quickly switch over to alternative modules.

3.4.1 Processor types

The CPUs used in Smart Cards are not special developments, but proven modules used in other areas over long periods. It is not, in fact, the practice to develop processors for special applications, since generally this is too expensive. Also, in such a case one would end up with a completely unknown device, for which the operating system manufacturers do not have suitable function libraries.

Additionally, Smart Card processors must be extremely reliable, and therefore it is better to trust older, proven processors rather than the latest developments dreamt up by semiconductor manufacturers, with little practical exposure. For the same reasons, the very safety-conscious aerospace industry only utilizes modules which are one or two generations behind state-of-the-art technology.

Since Intel brought out the first 4004 microprocessor integrating 2300 transistors in 1971, developments in this field have made huge strides. More recent products make this clear, such as the Pentium processor containing 3.1 million transistors. In the area of Smart Cards, however, as already explained, the processors used are not those which are the most technologically advanced. A total of 200 000 transistors on the chip is considered high in this context.

Because the size of addressable memory in a Smart Card is between 6 kB and a maximum of 30 kB, the use of processors with 8-bit memories does not represent an excessive restriction. The processors themselves are based on a CISC (complex instruction set computer) architecture. Hence they require several clock-cycles per machine instruction, and usually have a very extensive instruction set. The address range of the 8-bit processors is always 16 bits, with which a maximum of 65 536 bytes can be addressed. The processors' instruction sets use either the Motorola 6805 or the Intel 8051 architecture. The existing instructions are in part supplemented with additional ones by the relevant semiconductor manufacturer. These mostly relate to additional options for 16-bit memory addressing, which only exist in the most rudimentary form in the two instruction sets acting as the template.

One exception, a Smart Card processor which deviates from the two architectures already discussed, is the Hitachi H8 chip. It uses a 16-bit processor based on an architecture and instruction set similar to RISC (reduced instruction set computer).

3.4.2 Memory types

Other than the CPU, the various memories are the most important components of a microprocessor. They serve to store program code and data. Since Smart Card microprocessors must be complete computers, the memory is typically divided into RAM, ROM and EEPROM. The split depends very strongly on the chip's eventual field of application. However, efforts are always made to keep the RAM and EEPROM as small as possible, since they require the most space per bit. In multi-application Smart Cards, the most frequently used chips are those in which the ROM is roughly twice the size of the EEPROM, in order to be able to store its very complex operating system. In single-application Smart Cards, the microprocessor used has an EEPROM which is only slightly larger than the application's data. Thus all the application's variables, including some parts of the operating system, are stored in the EEPROM, which is very area-intensive and hence expensive, but in this way its use is optimized.

The integration of three different types of semiconductor memory on a single piece of silicon is technically complex. It requires a considerable number of production steps and exposure masks. The space required by individual memory types is also very different, due to the different construction and mode of operation. Thus, a RAM cell needs about four times more space than an EEPROM one, which in turn needs four times as much as a ROM cell. This is why Smart Card microprocessors are provided with so little RAM: 256-byte RAM is considered large. This is obvious if one considers that the same amount of space can carry 1024 bytes of EEPROM or 4096 bytes of ROM.

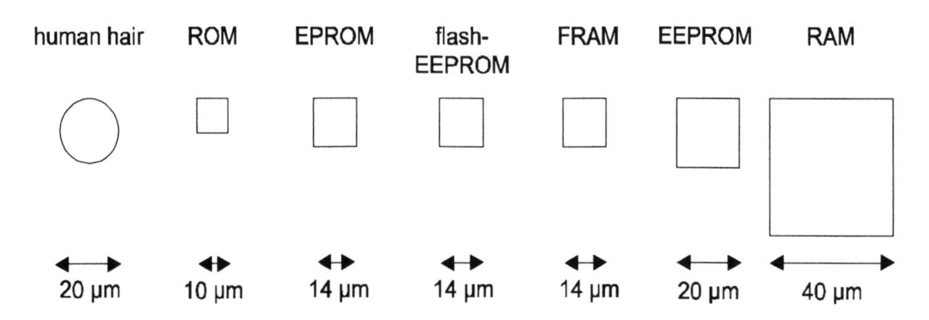

Figure 3.34 Comparison of space requirements per 1 bit depending on memory type. Sizes are approximate and refer to 0.8 µm technology. Comparative widths: DNA double helix: ≈ 0.1 µm; first transistor in 1959: 764 µm

Memory type	Number of write/erase cycles	Write-time per memory cell	Average cell-size in 0.8 µm technology
Volatile memory			
RAM	∞	≈ 70 ns	≈ 1700 μm^2
Secure memory			
EEPROM	10000 – 1000000	3 – 10 ms	≈ 400 μm^2
EPROM	1(as erasing with UV light is not possible)	≈ 50 ms	≈ 200 μm^2
Flash-EEPROM	≈ 100000	≈ 10 µs	≈ 200 μm^2
FRAM	≈ 10^{10}	≈ 100 ns	≈ 200 μm^2
PROM	1	≈ 100 ms	—
ROM	0	—	≈ 100 μm^2

Figure 3.35 Comparison of memory types used in Smart Card microprocessors (Flash-EEPROM and FRAM are still in the development stage)

A new storage technology has been discussed recently. This is the so-called Flash-EEPROM cell, which allows much faster write and erase access than conventional EEPROM. Depending on the precise implementation, its size is roughly half that of currently used EEPROM cells.

ROM (read only memory)

As the name implies, this type of memory cannot be written to. No voltage is needed to hold the data in memory, since they are represented in the chip by hard-wiring. A Smart Card's ROM contains most of the operating system's routines, as well as various testing and diagnostic functions. These programs are built into the chip during production. The program is made into a so-called ROM-mask. This is used to place the program in the chip by lithography. Only the data, which is the same for all the chips, can be introduced into the ROM during manufacture.

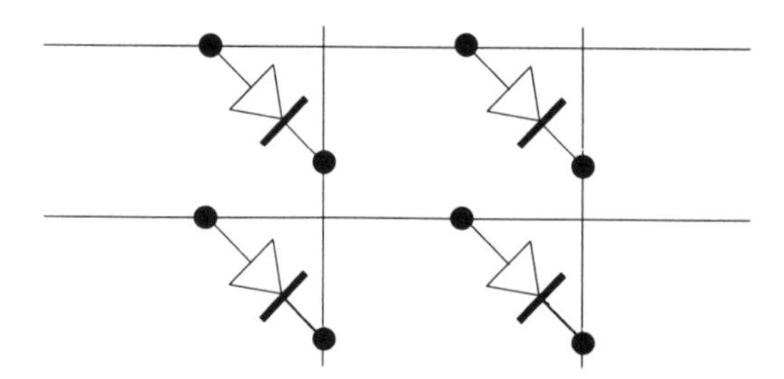

Figure 3.36 Basic functional structure of a ROM

PROM (programmable read only memory)

PROMs are not used in Smart Card microprocessors, though this would offer several important advantages. In contrast to ROMs, they need not be programmed during manufacture but can have the data written to them shortly before chip installation. Nor would voltage be needed for data storage. The reason for not using this type of memory lies mainly in the fact that programming requires access to the address, data and control buses. But since the ability to access the buses would allow data not only to be written but also to be read, PROMs must be strictly ruled out in Smart Cards, in view of the confidential data stored in the memory.

EPROM (erasable programmable read only memory)

EPROMs often found application in the early years of Smart Card technology, since at the time this was the only type of memory in which data could remain stored without voltage and also be written (though only once per bit). But since EPROMs could only be erased with UV radiation, these memories could not be erased. This is why they are no longer of practical significance in new applications.

The only sensible use is in irreversible storage of a chip number during semiconductor production, but currently this can be achieved through a special non-erasable EEPROM.

EEPROM (electrical erasable programmable read only memory)

This type is more complex than ROMs and RAMs, and is used in Smart Cards for all data and programs which need at some stage to be modified or erased. Functionally, an EEPROM corresponds to a PC hard disk, since data remain in memory in the absence of power and can be modified as necessary.

An EEPROM cell is, in principle, a tiny capacitor which may be charged or discharged. The charging status can be interrogated by a type of sensor. A charged capacitor represents a logic 1, and vice versa. Hence, in order to store one byte one needs exactly eight of these small capacitors, plus the appropriate sensing circuitry.

In order to understand how an EEPROM cell operates, it is useful to visualize the semiconductor device as shown in cross-section in Figure 3.37. The actual construction is somewhat more complicated, but this schematic diagram explains the working principle.

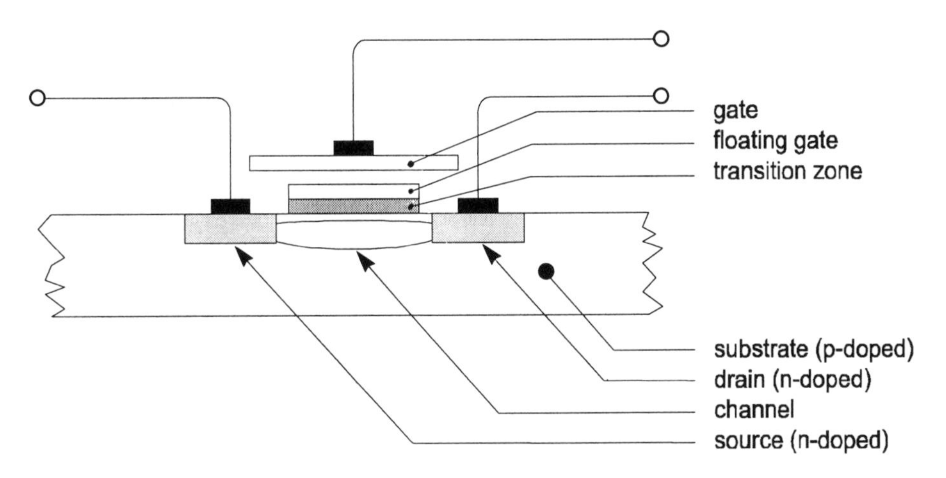

Figure 3.37 Cross-section of structure of semiconductor of EEPROM cell

An EEPROM cell, in its simplest form, is basically a modified field-effect transistor (MOSFET), mounted on a silicon layer. This layer carries a source and a drain. Between these is a control gate, and applying a potential to the latter controls the current between source and drain. As long as no potential is applied to the control gate, no current can flow since two diodes (n-p and p-n) are located in between. With a positive potential applied to the gate, electrons are attracted from the substrate thus forming an electrically conducting channel between source and drain, the FET is conducting and the current flows.

In an EEPROM cell, an additional so-called floating gate lies between the control gate and the substrate. It is not connected to any external voltage source, and the gap between it and the substrate is very small, of the order of 10 nm. The floating gate can exploit the tunnel effect (the Fowler-Nordheim effect) to be charged or discharged via the substrate. This effect causes charge carriers to penetrate thin oxide layers. This requires a sufficiently large potential difference at the thin insulating layer, the tunnel oxide layer. The flow of current between source and drain can be controlled by the floating gate's charge. This means that depending on the current, a logic 0 or 1 can be interpreted at this gate.

A high positive potential is applied to the control gate in order to charge the floating gate. This causes a high potential difference between the substrate and the floating gate, followed by electrons tunnelling through the oxide layer to the floating gate. The current is measured in picoamps. Now the floating gate is negatively charged and causes a high threshold voltage between source and drain, i.e. the field effect transistor is in the non-conducting state: no current can flow between source and drain. Thus the storage of electrons in the floating gate is comparable to the storage of information.

The voltage needed to charge the EEPROM cell is about 17 V at the control gate, but through coupling this figure drops to about 12 V at the floating gate. However, since Smart Card microprocessors are only supplied with 3 to 5 volts, a so-called charging pump is needed. In principle, this is a cascading potential-doubling circuit. It generates an output voltage of about 25 V from the low input voltage, which after stabilization is approximately equal to the required 17 V. Depending on construction details, charging an EEPROM cell requires between 3 and 10 ms per memory section (1 byte to 32 bytes).

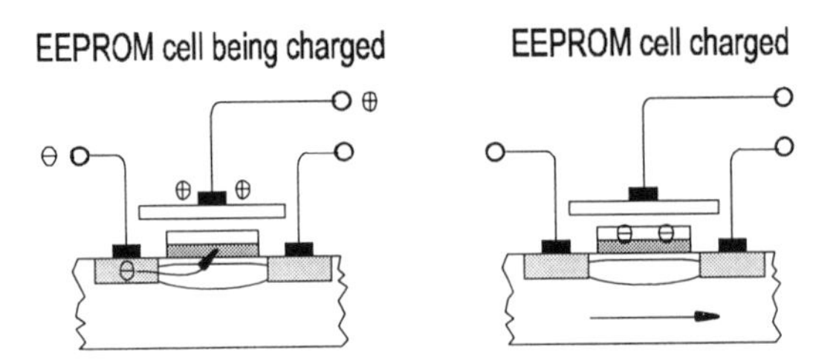

Figure 3.38 Loading an EEPROM cell

To erase an EEPROM cell, a negative potential is applied to the control gate which causes the electrons to leave the floating gate and return to the substrate. The EEPROM cell is now discharged and the threshold voltage between source and drain is low as the FET is conducting.

The floating gate can also be discharged by heating or by powerful radiation (such as X-rays or UV radiation) and then returns to the so-called ground state. This situation is of fundamental significance for the design of Smart Card operating systems, since otherwise it would be possible to penetrate security devices by deliberate alteration of environmental conditions. Depending on the technical implementation of an EEPROM cell, the ground state can correspond to logic 0 or 1. This is specific to each Smart Card microprocessor, and may need to be checked with the manufacturer.

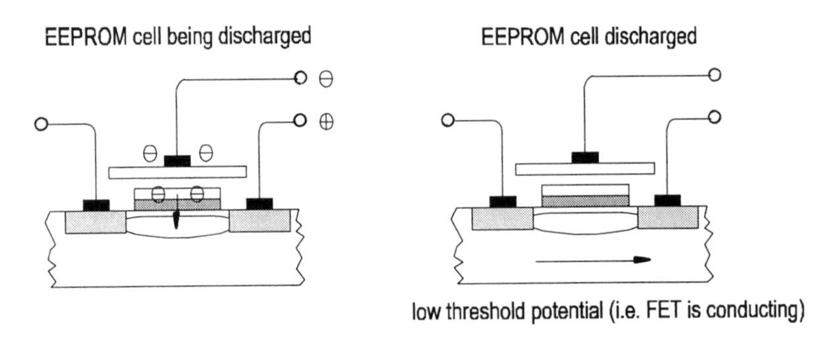

Figure 3.39 Downloading an EEPROM cell

EEPROMs are one of the few semiconductor memories which do not allow unrestricted access. It is a type of memory which can be read any number of times, but can only be accessed, and thus programmed, a limited number of times. The reason for this constraint can be found in its construction. Its life-expectancy depends strongly on the type, thickness and quality of the tunnel oxide layer between the floating gate and the substrate. Since this layer must be produced at a very early phase of the semiconductor's manufacture, it must be subjected to heavy thermal loads during later production steps. These loads may result in damage to the oxide layer, which in turn affects the EEPROM cell's life. During production, and during each writing cycle, the tunnel oxide layer absorbs electrons which are not re-released. These absorbed electrons are located in the channel between the source and the drain, and once they reach a certain number can exercise a stronger effect on the threshold potential than the floating gate's charge. When this occurs, the EEPROM cell has reached the end of its life. It can be written to, but the floating gate's charge only affects the characteristics of the channel between source and drain to a minimal degree, as a result of which the threshold potential always retains the same value. The number of write/erase cycles this makes possible varies greatly, depending on construction details. Typical values range from 10 000 to 1 000 000 cycles over the entire temperature and supply voltage range. However, at optimized power supply and at room temperature, the number of possible cycles is 10 to 50 times higher.

Once an EEPROM cell is near the end of its life, data is only stored for a short time. This time interval can vary from hours through to minutes or even seconds. The more "used up" the EEPROM is, i.e. the more electrons have been absorbed by the tunnel oxide layer, the shorter the storage interval.

A charged floating gate loses its charge with time, due to insulation losses and quantum mechanical effects. The time interval for this phenomenon to become noticeable can vary between 10 and 100 years. In this context, it is also interesting to note that a charged floating gate carries between 100 000 and 1 000 000 electrons, depending on the implementation. Nowadays, all semiconductor manufacturers guarantee data storage for 10 years. In order to reduce this effect, the contents of EEPROM cells can be refreshed periodically through reprogramming. However, this is only practicable where data needs to be stored over long periods.

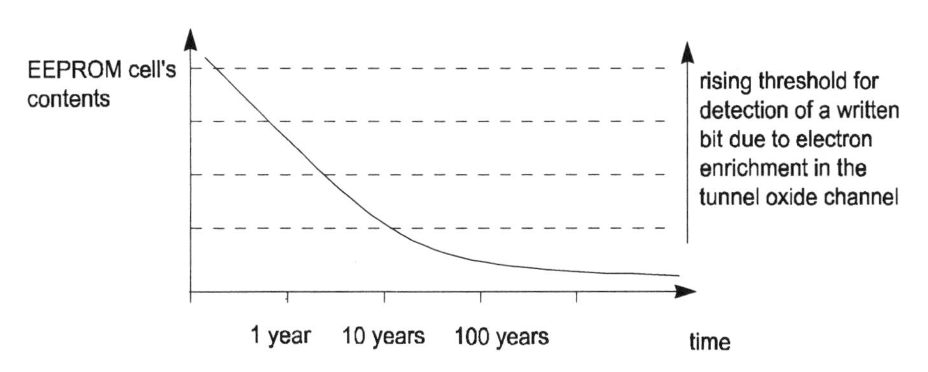

Figure 3.40 Modification of downloading curve in EEPROM cell depending on number of programming–erasing cycles carried out

Flash-EEPROM (flash electrical erasable programmable read only memory)

Flash-EEPROMs have been available for some years as discrete components, similar to EEPROMs in their functions and semiconductor construction. The fundamental difference between them and EEPROMs is in the writing process, which occurs not through the tunnel effect but by so-called hot electron injection. Fast electrons are produced by a high potential difference between source and drain, of which some penetrate the tunnel oxide layer through a positively charged control gate, and are stored in the floating gate. Writing time is reduced by this effect to about 10 μs, a great advance compared with 3-10 ms for previous EEPROMs. In addition, the required programming voltage is only 12 V, compared with 17 V for EEPROMs. Currently there are no Smart Card microprocessors with flash-EEPROMs, but various semiconductor manufacturers have announced that such components will become available in the next few years.

FRAM (ferroelectric random access memory)

FRAMs are a new semiconductor development. Despite the name, a FRAM is not a volatile memory like RAM, but retains its contents without a power supply. The properties of ferroelectric materials are used to store data. Construction is similar to EEPROM cells, but a ferroelectric substance is present between the control and floating gates.

This type of storage would be ideal for Smart Cards, since it exhibits all the desirable properties of an optimized memory. Only 5 V are needed for programming, programming time is around 100 ns, and the number of possible programming cycles is of the order of 1 billion. Integration density is similar to that of flash-EEPROMs. Nevertheless, FRAM memories suffer from two disadvantages. The first is that the number of read cycles is limited, therefore a kind of refresh cycle is necessary. The second drawback is even more significant: no efforts worth mentioning are currently being made in the field of Smart Card microprocessors to implement this new technology. But perhaps this situation will change in a few years' time, since FRAM technology possesses all the features needed to completely supplant EEPROMs, which at the moment hold almost exclusive sway.

RAM (random access memory)

The RAM is a Smart Card's memory in which data is stored and can be modified during one session. Access is possible an unlimited number of times, and none of the restrictions found with EEPROMs apply. It needs a power supply for its operation. Once this is switched off, or fails temporarily, the RAM's contents are no longer defined.

A RAM cell consists of numerous transistors, so connected that they operate as a bistable circuit. The switching state represents the contents of one bit in RAM. The RAM used in Smart Cards is static (SRAM): its contents need not be refreshed periodically. As such, it does not depend on an external clock, in contrast to dynamic RAM (DRAM). Static RAM is important since it must be possible to stop the clock supply to Smart Cards, which with dynamic RAM would lead to a loss of the stored information.

3.4.3 Supplementary hardware

Some Smart Card requirements cannot be completely met by software, so a number of Smart Card manufacturers offer additional functions in the form of non-standard hardware on the chip.

Coprocessors

There are specially developed arithmetic units for computations relating to public key algorithms, carried on the chip in addition to the microprocessor's usual functional units.

These units are only capable of the few calculations necessary for this type of algorithm. This includes exponentiation and modulo arithmetic of large numbers. The speed of those units which are optimized exclusively for these two operations is achieved by very wide architecture, up to 140 bits. In their special field of application they are superior even to a very fast PC, by a factor of at least 6.

The arithmetic unit is called up by the processor, which supplies either the data directly or pointers to the data, and starts the processing with the appropriate instruction. Once the task is completed and the result stored in RAM, the processor takes over control of the chip once again.

Random number generator

Random numbers are often needed to authenticate the Smart Card and the terminal. For security reasons, these should be genuine random numbers and not pseudo-random, as is common in Smart Cards. Some microprocessors contain a genuine random number generator in the chip.

However, the quality of this generator must not be susceptible to external physical factors such as temperature or supply voltage. It may generate the random numbers on the chip with the assistance of these external factors, but this must happen in such a way that a deliberate modification of one or several of these parameters cannot lead to predictability in the random numbers generated.

Since this is very difficult to achieve purely with the semiconductor, a different approach is taken. The random number generators use the processor's various logic states, such as pulse or register contents, and pass these on to a feedback shift register, which is in circuit with a clock; the latter is also generated by various parameters. When the CPU reads the random number register, it obtains a relatively useful random number which cannot be determined from the outside. The quality of the genuine random numbers so obtained can be further improved by additional procedures and algorithms.

Error detection in EEPROM

The fundamental limit on a Smart Card's life is determined by the EEPROM and the technically restricted number of its possible write/erase cycles. One way to minimize this restriction utilizes various software error correction codes for certain heavily used EEPROM areas. Alternatively, the error correction codes can be implemented as hardware circuitry on the chip. This detects EEPROM errors in a software-transparent fashion, and if the errors are not too extensive they can be corrected.

Naturally, an additional EEPROM is needed to store the code. However, since good error correction codes require relatively high storage space, one is confronted by a strategic decision. If the error correction is good, it requires additional space which may use up to 50% of the protected memory. But the memory storing the error correction procedure cannot be used for any other purpose. If a lower quality of error correction is used, then the additional usage of memory is lower, but the benefit very questionable.

There are many chips on the market with EEPROM error detection and correction in hardware implementation, but which also require half of the protected memory. The end-result is that the EEPROM available to the user is not that large, but the EEPROM's lifespan is increased many times beyond conventional values.

Hardware-supported data transmission

The only communication between a Smart Card and the outside world takes place over a bidirectional serial interface. Up to now, data reception and transmission over this interface have been controlled exclusively by the operating system's software, without any hardware support. However, this makes extreme demands on the software. It may even give rise to additional software errors, but the main problem is that the speed of software-supported data transmission is limited, since the processor's speed itself is highly constrained.

If a higher communication speed is needed, it is necessary to use either internal clock doubling or a UART: universal asynchronous receiver transmitter. As the name suggests, this module allows data reception and transmission independently of the processor. It is not restricted by the processor's speed, nor does it need software for communicating at byte level. Of course, the higher levels of the relevant communications protocol still need to be present in the Smart Card as software, but the lowest level is implemented as hardware in the UART module.

This is likely to become the standard in future Smart Card microprocessors, but at the moment there are only very few chips which allow hardware-supported communications. The technical capability has long been around, but calculation demonstrates that a send-and-receive routine implemented as software in ROM requires less physical storage space on the semiconductor than a functionally comparable UART module. Since the chip's surface area exerts a decisive influence on the price of Smart Card microprocessors, up to now almost all semiconductor manufacturers have decided against the latter. With increasing circuit integration density, however, this balance may change rapidly.

Internal clock doubling

The computational demands on Smart Cards are constantly increasing. This certainly applies in the area of cryptographic algorithms. In order to meet these requirements, it is possible simply to provide a higher clock rate in specially designed microprocessors. Computational power increases linearly with clock rate; a doubling of the latter would result in a doubling of the processor's output. However, for reasons of compatibility it is generally counter-productive to increase clock frequency beyond about 5 MHz.

It was proposed to get around this restriction by using clock doubling in the chip, so that with the external clock remaining unchanged, the internal one would increase. This could be implemented, for instance, with a PLL (phase-locked loop) circuit, which has long been standard technology. Thus, a Smart Card connected to an external 3.5 MHz clock, for example, would be driven internally at 28 MHz. This would be of considerable benefit for the computation times of complex cryptographic algorithms.

Nevertheless, processor speed is not the only bottleneck in a Smart Card. Data transmission speed and the EEPROM's write/erase time, specified by the relevant standard, would not benefit from such a solution, thus severely limiting the advantages. It can still be a great advantage for some applications, however, to use an internally increased clock rate, all the more so when one realizes that the circuitry required on the card is minimal.

The fact cannot be ignored, though, that clock doubling considerably increases the current used by a microprocessor. To some extent, these relationships are even linear: e.g., quadrupling the clock quadruples current consumption. But it is precisely with battery-operated terminals that a higher current is undesirable.

Hardware-supported memory management

In the latest Smart Card operating systems it is possible to load an executable machine code into the card[1]. It can be called up with a special instruction, and perform a cryptographic function only known to the card issuer. However, once an executable program is loaded into the card, it is no longer possible in principle to prevent it from containing a function for reading secret data from memory. Operating system manufacturers have been very careful to maintain the confidentiality of system structures and program codes; the same applies to secret keys or algorithms forming part of various applications on the card. The accessibility of such confidential information would have fatal consequences for an application provider. One technical solution is to have the new program tested by an independent organization. However, even this cannot guarantee complete security, since a different program to the one tested may then be loaded. Another possibility is to keep the program so secret that nobody other than the application provider may know it.

An acceptable way out of this impasse consists in providing the Smart Card's microprocessor with a so-called memory management unit, or MMU for short. This controls the actual application's memory boundaries in parallel with program execution. The permitted memory area is fixed by an operating system routine before the application is called up, and can no longer be changed during program execution within the application. This ensures that an application is completely self-contained, and cannot access memory areas forbidden to it.

MMUs have been utilized in many fields for years, though only very few Smart Card microprocessors have them. This additional hardware will greatly increase in importance in the future, since it is the only practical way to safely separate several applications which are sharing a single Smart Card.

Chip hardware extensions

Where the chip hardware needs to be extended for some reason, this leads to considerable expenditure in development time and cost by the manufacturer. Hardware that is tailor-made to fit an individual customer's requirements can only be created in two ways, by constructing the new hardware using current generation semiconductors or by using a two-chip solution, with all its drawbacks.

An acceptable way around this involves a compromise solution. The principle used is a mixture of both of the above solutions. The chip containing the new hardware module is directly glued to the existing chip, and electrically connected by wires. This solution benefits from the fact that most Smart Card microprocessors possess not one but several I/O conductors which can be used to communicate with other chips. The resulting thickness, using this sandwich construction, differs only insignificantly from that of normal chips, since the silicon carrier can be ground down to make it thinner. A sandwich chip can thus be built into a standard module without further expenditure.

The procedure described is ideal for implementing individual customer requirements for additional hardware without expensive modifications. Existing chips can be combined with

---

[1] See also 5.10 Code programmed in circuit.

new modules which execute fast cryptographic computations (such as DEA), or possess a special serial interface for testing security features in other chips. It is also possible to introduce into the card a special ASIC with a secret cryptographic algorithm. The procedure is not cost-effective for large numbers, in the million plus range, since here it pays to develop special chips. However, for small to medium runs, sandwich-chips are a very effective solution for prototype series or for special applications, such as security modules for terminals.

Fundamentals of information technology

"Smart Cards are small computers in credit card format without a man–machine interface". This statement is entirely fitting. The specific properties of Smart Cards, compared with all other types of cards, are determined by the microprocessor integrated into the card.

Functionally, the plastic card is mainly a carrier for the microprocessor. Of course, further elements may be present in addition, but they are not essential for the card's actual operation. To understand the characteristics of these small computers and the data-processing mechanisms based on them, we need to look at some aspects of information theory.

4.1 ERROR DETECTION AND ERROR CORRECTION CODES

When transmitting or storing data, it should be possible to detect any modification of this data. In particular, stored programs must be secured against corruption, since a single altered bit of program code may destroy the program or change its execution to such an extent that the required functions are no longer available. The EEPROM memory used in Smart Cards is especially sensitive to external effects such as heat, voltage fluctuations and suchlike. Therefore, those sections which perform a security-related function must be protected, undesirable modifications identified by the operating system and their negative effects nullified. In particular, this applies to data such as program code, keys, access conditions, pointer structures and the like, which is very sensitive to modification. Error detection codes (generally known as EDC for short) are used for this purpose, with varying degrees of detection probability which depend on the particular code. A more extensive tool is known as ECC, for error correction codes. They allow not only the recognition of errors in the tested data, but also error correction to a limited degree.

The principle of operation of all these codes is the allocation of a checksum to the protected data. Normally it is stored adjacent to the latter. It is computed with a widely known algorithm and not a secret one. Where necessary, the data can be checked for alteration by using the EDC thus obtained. This is done by comparing the stored checksum with one calculated afresh.

It is a feature of error detection and correction that it uses the most diverse mathematical procedures. In some cases, the higher-value bits are better protected in order to reduce as much as possible any adverse effects. In most cases, however, the cost of such algorithms by far exceeds that of program code. Hence it is more common to use procedures in which error detection does not distinguish between higher and lower parts of a byte, and which operate on the byte as a unit.

EDCs and ECCs are very similar to message authentication codes (MAC) and cryptographic checksums (CCS). However, there is a fundamental difference. EDC and ECC checksums can be calculated and checked by anyone. In contrast, MAC and CCS require a secret key as they are designed to protect against tampering with the data and not against inadvertent corruption.

The most widely known error detection code is, probably, the use of a parity bit, attached to each byte as part of many transmission procedures and also in some memory modules. Before computing the parity, one must determine whether even or odd parity is to be used. With even parity, the parity bit is selected so that the overall sum of all the bits in the data byte and in the parity bit is an even number. With odd parity, the bits must add up to an odd number.

This parity calculation mechanism ensures detection of one incorrect bit per byte. On the other hand, correction of the error is impossible since the parity bit provides no information about the position of the altered bit.

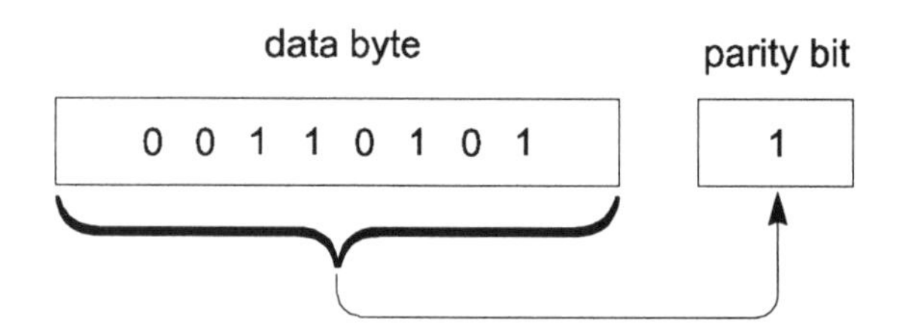

Figure 4.1 Example of error detection using an additional parity bit (odd parity)

Should two bits in one byte happen to be simultaneously wrong, parity remains unchanged and hence no error detection occurs. Another drawback of parity-based error detection is the relatively large overhead of one parity bit per eight data bits. This represents an additional memory requirement of 12.5%. Furthermore, it is very difficult to work with additional parity bits when the memory is organized in bytes, since it would require a considerable programming effort. This is why parity bits are not used for error recognition in Smart Card memories. XOR or CRC checksums are better suited to this task.

4.1.1 XOR checksums

XOR checksums, also known as longitudinal redundancy check (LRC) due to the method used, can be computed very simply and very quickly: both important criteria for error detection codes designed to be used in Smart Cards. In addition, the algorithm can be implemented extremely easily. Besides the protection of memory-held data, XOR checksums are typically used in data transmission (ATR and transmission protocol T=1).

Calculation of an XOR checksum is carried out by consecutive logical XOR concatenation of all data bytes. That is, data byte 1 is XOR-ed with data byte 2. The result is XOR-ed with byte 3 and so on.

If the checksum is placed after the tested data and a checksum is again calculated, using both the data and the first checksum, the result is '00'. This is the simplest way to check whether the data and the checksum still retain their original values and thus are uncorrupted.

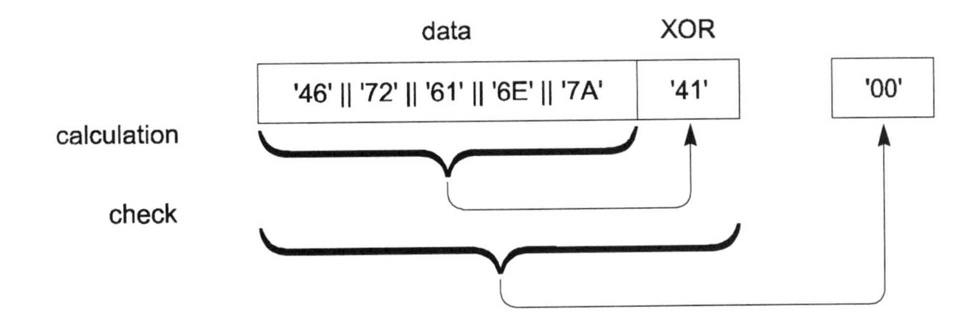

Figure 4.2 Calculation and checking of an XOR checksum

The great advantages of XOR checksums are their fast computation and the simplicity of the algorithm. This algorithm is so simple to construct that its program code in Assembler is only between 10 and 20 bytes long. One of the reasons for this is that the XOR operation is available in all processors as a direct machine instruction. In addition, as demanded by various ISO standards (data transmission with T=1), almost every Smart Card operating system must implement an algorithm for XOR computation, so that it can be used for other tasks without further overheads.

Unfortunately, XOR checksums also suffer from several serious drawbacks, which considerably limits their practical application. They are not very certain in principle. For example, they do not allow detection of the interchange of two bytes within the overall data. Also, multiple errors may occur at the same position in several bytes and cancel each other out. The consequence of all this is that XOR checksums are mainly used in the area of data transmission, and have only restricted application for checking the consistency of memory contents.

4.1.2 CRC checksums

The CRC procedure (cyclic redundancy check) also derives from the field of data communications, but is rather more useful than the one based on XOR logic. Nevertheless, CRC checksums too are only an error detection code, and cannot be used for error correction. The procedure has been in use for quite some time in transmission protocols such as X/Z modem and Kermit, and is widespread as a hardware implementation in hard disk controllers. It is based on CCITT Recommendation V.41.

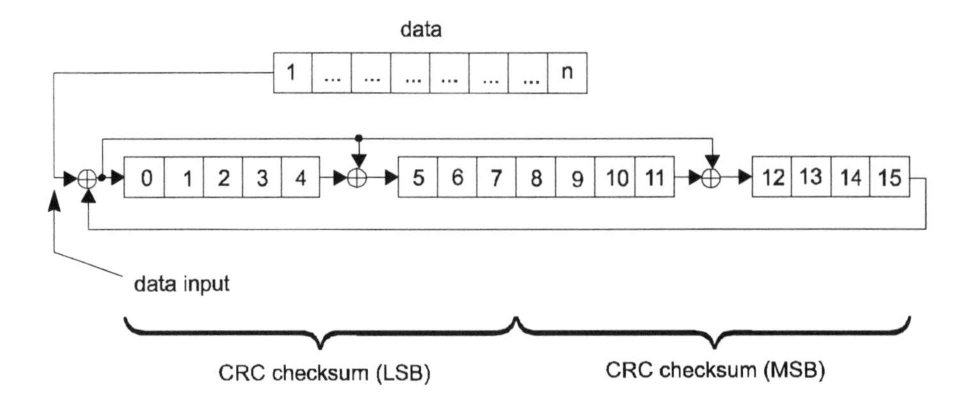

Figure 4.3 Calculation of CRC checksum with generating polynomial $G(x) = x^{16}+x^{12}+x^{5}+1$ (data and CRC register are shown in bits)

CRC checksums are generated by a 16-bit, cyclic feedback shift register. The feedback is controlled by a generating polynomial. Considered mathematically, the tested data is recorded as a large number and divided by the generating polynomial. The residual represents the checksum. The procedure should only be used up to 4 kbytes of data, as beyond that the ability to detect errors drops sharply. However, this restriction can easily be circumvented by dividing the data into blocks not exceeding 4 kbytes.

The following polynomials are the ones normally used:

- **CRC CCITT V.41** $G(x) = x^{16} + x^{12} + x^{5} + 1$
- **CRC-16** $G(x) = x^{16} + x^{15} + x^{2} + 1$
- **CRC-12** $G(x) = x^{12} + x^{11} + x^{3} + x^{2} + x + 1$

Thus, with CRC checksums it is always necessary to know the generating polynomial as well as the shift register's details, otherwise the calculation cannot be done correctly. The shift register's starting value is 0 in the overwhelming majority of cases (e.g. ISO 3309), but several transmission procedures (such as CCITT Recommendation X.25) use the value 1 for all bits.

When calculating the CRC checksum as shown in Figure 4.3, the following steps are executed: first the 16-bit register is set to its starting value. Then all data bits, beginning with the lowest, are fed into the feedback shift register one after the other. The feedback, and the polynomial division, are based on logic XOR concatenation between the bits. After all data has been fed into the register, the calculation is complete and the contents of the 16 bits represents the desired CRC checksum.

CRC checking of plaintext is achieved by computing a new checksum and comparing the new outcome with the original one. If the two are identical then we may conclude that the plaintext and the checksum have not been changed.

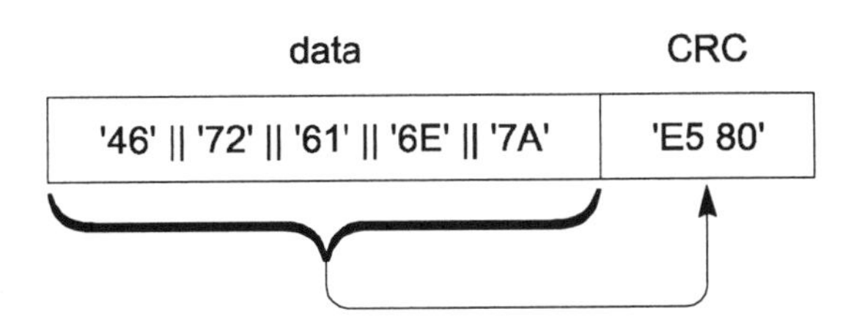

Figure 4.4 Principle of calculating CRC-checksum

(generating polynomial $G(x) = x^{16}+x^{12}+x^{5}+1$, starting value '0000')

The great advantage of CRC checksums is the certainty of detection even with multiple errors, which can only be achieved by very few procedures. In addition, and in contrast with the XOR procedure, it is possible to detect interchanged data bytes with CRC, since byte order plays a crucial role in checksum formation through the feedback shift register. However, it is very difficult to specify exact detection probabilities, since these depend strongly on error position.

The CRC algorithm is relatively simple, and thus the size of implemented code is appropriate to the needs of small Smart Card memories. The greatest drawback is the slowness of the calculation. Computational speed drops considerably due to the fact that the algorithm requires data to be shifted bit by bit. CRC checksums were originally designed for hardware implementation, which fact militates strongly against implementation by software. The throughput of a CRC routine is lower than that of XOR by a factor of around 200. A typical figure would be 0.2 ms/byte at 3.5 MHz clock frequency. Accordingly, calculation of a CRC checksum for a 10 kbyte Smart Card ROM would require 2 seconds.

4.1.3 Error correction by multiple storage

Error correction codes must be used where it is necessary not just to detect alterations to stored data, but also to correct those which are erroneous. However, since computation of these codes is expensive in terms of program code, using them to protect Smart Card memory poses certain problems. Furthermore, the algorithms are usually designed only to correct low error rates. Because EEPROM memory in Smart Cards is constructed in sections, and errors mostly cause a whole section to fail, only procedures capable of correcting clustered errors would make sense. Error correction, therefore, takes a different route.

Technically, the simplest solution is multiple data storage on physically separate sides of the memory, with a majority vote during the reading phase. The usual approach is triple storage, with a 2-out-of-3 decision. A less space-intensive variant of this procedure is double data storage with additional protection by an EDC checksum. Occurrence of an error can be recognized by checking both EDC values. At the same time it is possible to determine in which of the two separated memory sites the error is present. The other site must then contain the true data, which is used to restore the faulty ones.

The additional storage capacity made necessary by these error correction procedures is considerable, but for smaller quantities of data is still well within reasonable limits. The great advantage is that no costly and code-intensive algorithms are required.

Nevertheless, several comments need to be made in respect of the implementation of error correction codes in Smart Cards. At first sight it seems tempting to use the procedure for the removal of errors in EEPROM. But this supposed data security is fraught with several severe drawbacks. The storage space needed for it is huge. The time needed for writing data into memory also increases considerably, since they must be stored more than once. Algorithms capable of correcting clustered errors of the order of magnitude typical of section-based EEPROMs are very costly, and also require a great deal of storage space for EDC code. However, the fundamental disadvantage is worse still. Even when using error correction algorithms, errors may still be present since the correction only works properly up to a certain number of errors. When an operating system corrects memory errors automatically, it is never certain, in principle, whether the correction was carried out as intended.

Imagine automatic error correction applied to the balance in an electronic purse. The customer can never be sure what happens to the credited amounts in the event of error. The balance may be corrected properly, but there is a given probability that it is still too high or too low after correction. In this context it must be remembered that Smart Cards are cheap mass-produced articles, which can simply be exchanged when faulty. As a rule, if the data is incorrect, a supervisory system accessible to humans must decide what to do. For example, on the first occasion of an error appearing in a Smart Card-based purse, one would surely refund the customer's losses. However, if the error keeps cropping up regularly, the goodwill offered to this person would be sharply restricted, since the recurrence may suggest fraud by attempting to tamper with the EEPROM. This situation cannot be regulated by an error correction code in the card: the system administrator must intervene.

4.2 ENCRYPTION

Besides their function as data storage devices, Smart Cards also act as proof of authorization and as encryption modules. The result, even in the early days of Smart Card development, was that cryptology became of central importance. Currently the processes and methods of this science are thoroughly entrenched in Smart Card technology.

Cryptology may be divided into two areas of activity, cryptography and cryptoanalysis. Cryptography is the science dealing with data encryption and decryption methods. The science of cryptoanalysis attempts to crack existing cryptographic systems.

As far as Smart Cards are concerned, the main objective and the main area of interest is the practical application of existing cryptographic procedures and methods. Hence this chapter will deal more with the practical aspects of cryptology, rather than the theoretical ones. Nevertheless, we cannot ignore either the application of the method or the fundamental aspects of the theoretical background.

The aim of cryptology, on one hand, is to maintain the secrecy of information, and on the other, to ensure its authenticity. Both aims are independent of each other, and make different demands on the underlying system. Secrecy means that only the chosen recipient or recipients can decode the contents of a message. Where the recipient can ensure that the received message has not been altered during transmission, this is termed authentication.

The cryptographic notation used in this book is listed in Figure 4.6.

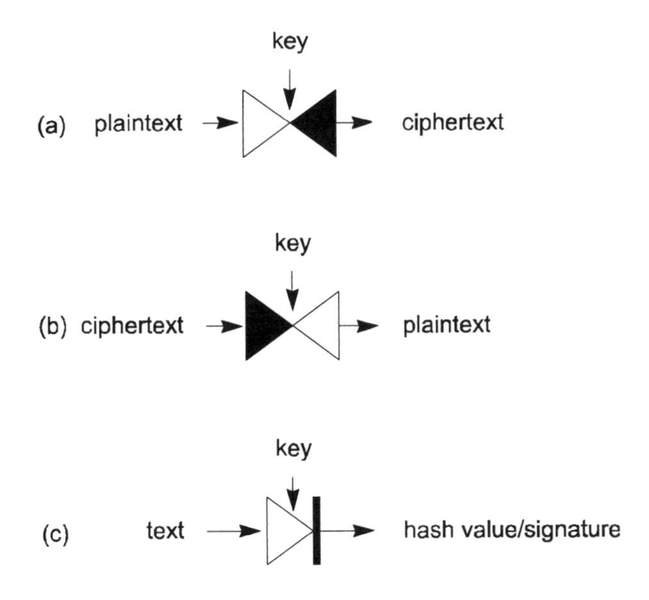

Figure 4.5 Codes used in encryption algorithms:
(a) encryption;
(b) decryption;
(c) one-way function/hash value/signature

The following terminology and principles are fundamental to cryptology, and are necessary for understanding its methods.

In its simplest form, encryption recognizes three types of data. Non-encrypted data is termed plaintext. In contrast, encrypted data is known as ciphertext. One or more keys are needed for encryption and for decryption, this being the third type of data. These three types are processed by an encryption algorithm. The algorithms currently used in Smart Cards are generally block-oriented. This means that plain- and ciphertext can only be processed in packets of fixed length (e.g. 8 bytes in DEA).

As a rule, modern cryptographic algorithms are based on the Kerckhoff principle. This idea, named after Auguste Kerckhoff (1835–1903), states that the entire security of an algorithm should only rely on the confidentiality of the key and not on that of the algorithm. The effect of this generally accepted but frequently ignored principle is that many cryptographic algorithms used in the civilian domain have been published, and to some extent are even standardized.

The opposite approach to that of Kerckhoff relies on the principle of security through confidentiality. Here, the system's security relies on the fact that a supposed opponent does not know how the system works. This principle is very old and is still often used today. However, it is inadvisable to develop a cryptographic system, or any other system, solely based on it. So far, all systems using it have been broken, mostly in a very short time. In our

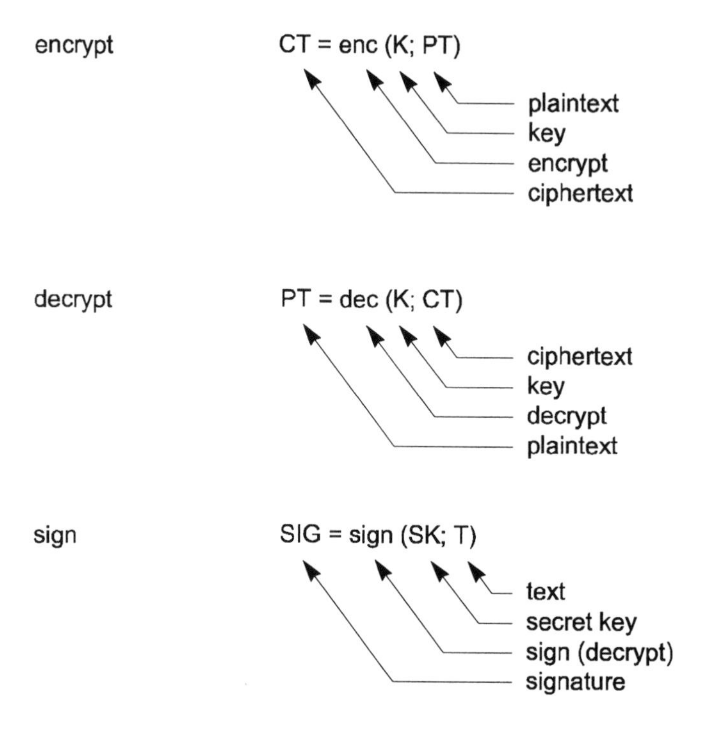

Figure 4.6 Cryptographic notation

information-based society it is generally no longer possible to keep the technical details of a system secret over any length of time, and this is precisely what the principle requires.

Nevertheless, confidentiality allows the occurrence of accidental disclosures of information to be reduced to a large extent. Therefore this principle is used repeatedly in parallel with Kerckhoff's. In many large systems it is built in as an additional security stage. Since, in practice, the security of modern and published algorithms only relies on the computational powers of computers, additional confidentiality achieves a considerable improvement in the degree of protection.

Should one rely purely on the protection offered by computational powers not being available to the supposed opponent, then in certain circumstances one may very quickly be overtaken by rapidly advancing technical developments. Statements such as "one would need a thousand years to crack this cryptographic system" are unreliable, since they are based on the computational powers and algorithms available today and cannot take into account future developments, which by their nature are unknowable. The power of processors doubles roughly every 18 months, hence over the past 25 years it has increased by a factor of about 25 000.

In recent years, the increasing networking of computers has created a further possibility of concerted and efficient attacks on keys and cryptographic systems. Consider the challenge made on the Internet to crack a DEA key, passed on to millions of users whose numbers snowball step by step. If only 1% of current subscribers[1] participate in such an action, the potential attacker would have at his disposal a parallel computer made up of 300 000 individual machines.

Crypto-algorithms are divided into symmetric and asymmetric ones. This distinction relates to the key used. Symmetric means that encryption and decryption algorithms use the same key. In contrast, the asymmetric algorithms postulated in 1976 by Whitfield Diffie and Martin E. Hellman employ a different key for encryption and for decryption.

A term which is often mentioned in connection with crypto-algorithms is that of the magnitude of cipherspace. This denotes the number of possible keys for a particular algorithm. A large cipherspace is one of the many criteria needed to ensure a secure algorithm.

One requirement for the technical implementation of crypto-algorithms in Smart Cards which has only recently moved into the foreground, is that of freedom from noise. In this context, the term means that the algorithm's execution time must be independent of the key, plaintext and ciphertext. Should this requirement not be met, then the result may well be that the secret key can be computed within a very short time and the whole cryptographic system broken.

Cryptology distinguishes strictly between a system's or algorithm's theoretical and practical security. A system is theoretically secure when an attacker has unlimited time and means at his disposal and even then is unable to breach the system. For instance, this means that a system may no longer be termed theoretically secure when an attacker would need hundreds of years and many supercomputers to breach it. If the attacker has only limited time and means at his disposal and cannot manage to breach the system with their assistance, the system is termed secure in practice.

A cryptographic system can secure message secrecy and/or authenticity. Breaching the system means that secrecy and/or authenticity are no longer ensured against attack. When an attacker discovers the secret key to an encryption algorithm, he is able to decipher the protected data, read it and alter it if necessary.

There are various options for cracking a key. In "ciphertext only attack", the opponent only knows the ciphertext and uses it to discover the key or the plaintext. A "known plaintext attack" is more likely to be successful, and means that the hacker knows several plaintext/ciphertext pairs and uses them to discover the key. In "chosen plaintext attack" and "chosen ciphertext attack", the attacker can generate his own text, either plaintext or ciphertext; this line of attack is the most likely to succeed, since the secret key can be discovered by trial and error.

Discovering the key purely through brute force attack is, of course, the most trivial. It uses maximum computational power to discover the key. Using a known plaintext/ciphertext pair, all possible keys are attempted. It goes without saying that this usually requires the power of supercomputers. Considered statistically, half of all possible keys need to be attempted, on average, to find the correct one. A large cipherspace makes this line of attack considerably more difficult.

---

[1] It is estimated that there were about 30 million Internet subscribers in early 1996. The growth—rate has not yet reached saturation, and currently is still rising exponentially. Hence it is likely that 80 million computers will have access to this network in 1997.

4.2.1 Symmetric crypto-algorithms

Symmetric crypto-algorithms are based on the principle of using the same key for encryption and for decryption, hence the term "symmetric". The best known and most widely used example is the data encryption algorithm, or DEA for short. This algorithm was developed by IBM in cooperation with the US National Bureau of Standards (NBS), and published in 1977 as a US standard (FIPS Pub 46). The standard describing DEA is frequently referred to as DES (data encryption standard). Of course, since this algorithm is designed according to the Kerckhoff principle, it could be published without a loss of security. So far, however, not all development criteria have been released, which is repeatedly leading to theories concerning the possibility of attack and traps present in the system. Nevertheless, all attempts to break this algorithm have up to now failed.

The DEA design contains two important encryption principles, confusion and diffusion as described by C. Shannon. The principle of confusion states that the ciphertext's statistics should relate to the plaintext's statistics in such a complex way that an attacker cannot derive any benefit from it. The second principle, diffusion, states that each bit of plaintext and of the key should affect as many bits of the ciphertext as possible.

DEA is a symmetric block-encryption algorithm. It does not lead to any expansion of the ciphertext, which means that plaintext and ciphertext blocks are of equal length. Block length is 64 bits (= 8 bytes), just like the key.

The key, though, contains 8 parity bits, which reduces the available space. If the 64 bits in the key are numbered consecutively from left (msb) to right (lsb), bits 8, 16, 24, ... , 64 represent the parity bits. Due to the 8 parity bits, DEA's cipherspace is 2^{56}. This gives $2^{56} \approx 7.2 \times 10^{16}$ possible keys. At first sight, this cipherspace, with its 7.2×10^{16} possibilities of choosing a key, may appear very large, but in fact this is DEA's main weakness[2]. In view of the steadily increasing computational power of modern computers, a cipherspace of this magnitude is regarded as the lower limit for the security of a crypto-algorithm. Where the

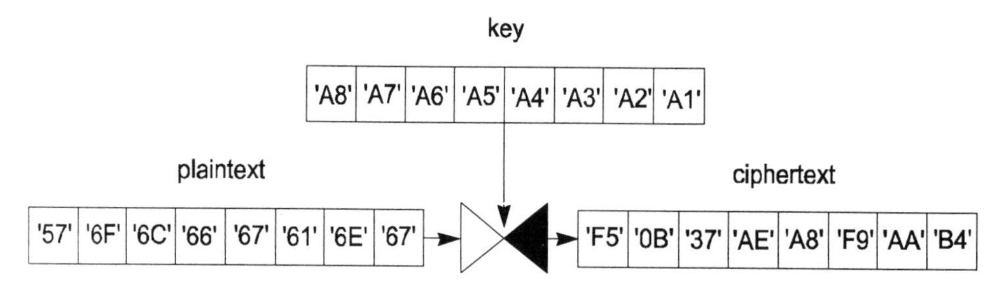

Figure 4.7 Function of the DES algorithm in encryption

---

[2] In order to gain an idea of large numbers, consider the following comparison. The earth's mass is about 5.974×10^{27} grams. Using this very rough approximation, the number of electrons, protons and neutrons making up the earth would be about 10^{52}. If only one elementary particle is needed to store a bit, the total mass of the earth could store a maximum of 10^{52} bits.

cipherspace is too small, and with a pair of plaintext/ciphertext at one's disposal, all possibilities can simply be attempted until the secret key is found.

If a plaintext/ciphertext pair is obtained by tapping into a communication between the terminal and the Smart Card, the following brute force attack can be carried out. The plaintext is encrypted with all possible keys. By comparing the result with the ciphertext already known, it is possible to determine whether the right key has been found. This procedure is eminently suitable for parallel processing. Each individual computer attempts a small part of the cipherspace. The following example makes clear the time needed for such a brute force attack. The currently fastest DEA modules[3] need 64 ns for complete block encryption. If 10 000 such computation modules are connected in parallel, they can each independently check a small part of the cipherspace. Assuming that on average, only half the cipherspace must be searched to find the right key, the following calculation can be made:

$$\frac{1}{2}\times\left(\frac{2^{56}\quad 64\,\text{nsec}}{10\,000}\right)\approx 64\,\text{h}$$

It is beyond the scope of this book to describe the implementation of DEA in detail. The reader is referred to FIPS Publication 46 or to C. H. Meyer and S. M. Matyas: *Cryptography* (Wiley, New York, 1982). However, one point is significant. The DEA encryption algorithm was designed so as to be capable of easy implementation as hardware circuitry. Unfortunately, however, no Smart Card microprocessors exist as yet which contain a DEA module. Hence it must be realized in software. As a result, its size even in highly optimized versions is about 1 kbyte of Assembler code. As a necessary consequence, the speed of computation is rather low.

Figure 4.8 shows a comparison of typical Smart Card encryption and decryption times compared with a hardware module and a PC. Times can vary with implementation, and quote only pure computation time for a DEA encryption or decryption of an 8-byte block, assuming all registers are pre-loaded.

Realization	Speed
Smart Card with 3.5 MHz	17.0 ms / 8 byte block
Smart Card with 4.9 MHz	12.0 ms / 8 byte block
PC (80486, 33 MHz)	30 μs / 8 byte block
PC (Pentium, 90 MHz)	16 μs / 8 byte block
DEA hardware circuitry	64 ns / 8 byte block

Figure 4.8 Comparison of DEA encryption and decryption times

[3] A gate array made by DEC using gallium arsenide technology for ECB and CBC modes.

Selection of a key for the DEA algorithm can be carried out with the aid of a random number generator. It produces an 8-byte random number to act as a key, which only requires an additional parity calculation.

There are many other symmetric crypto-algorithms besides DEA. The only one we will mention here is known as IDEA (international data encryption algorithm), developed by Xuejia Lai and James L. Massey. Since its publication in 1990 as PES (proposed encryption standard) it has been improved once, in 1992, and at that time was known provisionally as IPES (improved proposed encryption standard). However, nowadays it is generally known as IDEA. This algorithm's complete development criteria and internal structure have been published, so it meets the Kerckhoff principle.

Like DEA, IDEA is a block-oriented crypto-algorithm and also uses 8-byte plaintext/ciphertext blocks. However, in contrast to DEA, the key length is 16 bytes, making available a considerably larger cipherspace of $2^{128} \approx 3.4 \times 10^{38}$. Because of this construction, it is compatible with DEA in all respects other than the increased key length. There is also full compatibility with triple-DEA systems, which also use 16-byte keys, so that a change of algorithm does not affect the length of the key or of input/output data. Naturally, compatibility in this context does not mean that data encrypted with DEA can be decrypted by IDEA. IDEA is generally considered a very good cryptoalgorithm, and its use is widespread in data security[4]. It is as yet very difficult to predict its security over the longer term, as this algorithm has not yet been in the public domain for long enough.

Very few Smart Card IDEA implementations are in existence. The program's storage requirement is around 1000 bytes. Typical encryption and decryption times are slightly lower than with DEA. However, the development of IDEA was based on the assumption that a 16-bit processor is used. Since, generally speaking, Smart Cards use 8-bit processors, the speed advantage compared with DEA is not as big as was expected. Figure 4.9 compares IDEA operation under various conditions, with an 8-byte block and assuming precalculated keys.

Realization	Speed
Smart Card with 3.5 MHz	12.3 ms / 8 byte block
Smart Card with 4.9 MHz	8.8 ms / 8 byte block
PC (80386, 33 MHz)	70 µs / 8 byte block
IDEA-hardware circuitry	360 ns / 8 byte block

Figure 4.9 Typical encryption and decryption times with IDEA

Types of operation for block encryption algorithms

Like all block encryption algorithms, DEA can be run in four distinct ways, standardized in ISO 8372. Two of these are particularly suitable for sequential texts with no block structure. The other two (ECB and CBC mode) are based on 8-byte blocks. These two block-oriented modes, more than any others, are used in the field of Smart Cards.

---

[4] Currently, IDEA has probably achieved its widest exposure through Philip Zimmermann's public domain program for data communications security, PGP — Pretty Good Privacy.

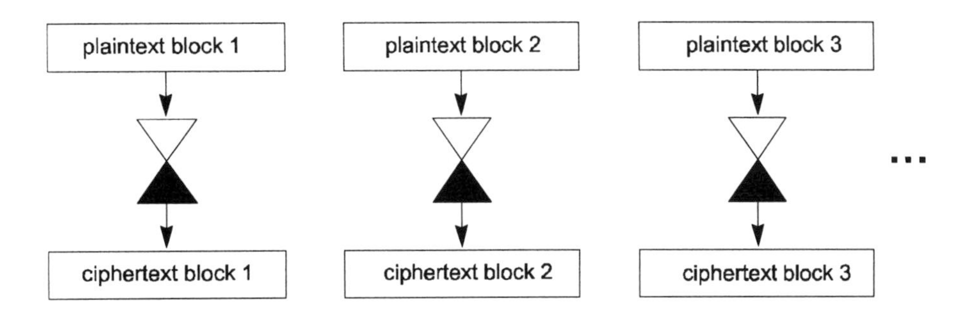

Figure 4.10 ECB operation for block encryption algorithms (analog decryption)

DEA's basic mode of operation is designated ECB (electronic code book). Eight-byte plaintext blocks are encrypted independently of each other with the same key. This is DEA in its purest form, without additional features.

The second block-oriented mode of operation is designated CBC (cipher block chaining). A data string consisting of many blocks is so concatenated via an XOR operation during encryption, that later blocks become dependent on earlier ones. This can reliably show up any interchange, addition or deletion of encrypted blocks, which is impossible with ECB.

When plaintext blocks are appropriately constructed (pre-placed transmission sequence counter or transmission sequence counter in the initialization vector), such concatenation ensures that even identical plaintext blocks give rise to different ciphertext ones. This makes cryptoanalysis of captured data considerably harder, since codebook analysis becomes impossible to perform.

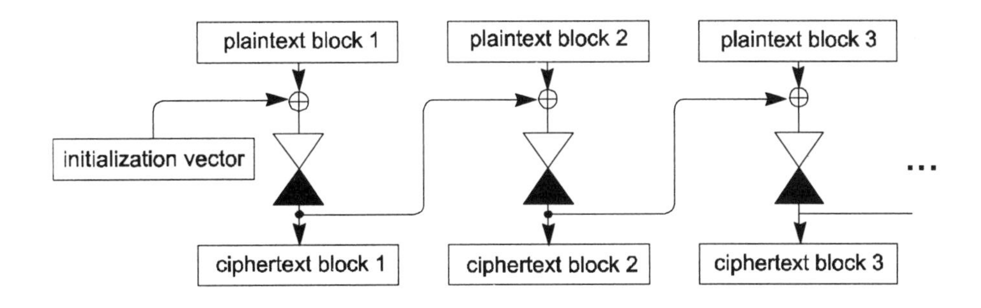

Figure 4.11 CBC operation for block encryption algorithms (analog decryption)

The first plaintext block is XOR-ed with an initialization vector (often known as IV), and then encrypted. The result is the ciphertext, which is again XOR-ed with the next plaintext block. All subsequent blocks are treated similarly.

As a rule, the initialization vector is the null vector. However, in some systems a session-specific random number is written to it, in lieu of using temporary keys. Naturally, this number must be known during decryption.

Triple-DES

In addition to the four operation modes used by block encryption algorithms, there is one further variant whose application is designed to increase security. However, in practice it is only used in DEA, due to the small cipherspace. But in principle it can be used by all block encryption routines which do not form a group. Should an algorithm have this property, then dual encryption with two different keys would not lead to an increase in cryptographic security, since the result would be identical to encryption with a third key. This means that dual encryption with an algorithm which does not possess the group property does not increase the size of the cipherspace, since an attacker would only need to discover the third key in order to obtain the same outcome as the previous encryption with two different keys.

- *ciphertext* := enc (*ciphertext 2*, (enc (*ciphertext 1*, *plaintext*))

But DES is a group, so that in principle one could carry out dual encryption with two different keys, and there would be no third key, the use of which in a simple encryption would produce the same outcome.

However, in 1981 Ralph C. Merkle and Martin E. Hellman published a method of attack called "meet-in-the-middle", which can be used very successfully against any dual encryption with block encryption algorithms. It depends on knowing a number of plaintext/ciphertext pairs. The principle of operation relies on computing all possible encryptions of the plaintext with the first of the two keys, followed by decrypting the known outcome (the ciphertext), with every possible second key. Now the list produced in the first stage is compared with the results of the second stage. As soon as an agreement is found, both keys have been discovered with some degree of certainty. To increase the likelihood of having found the right keys, a comparison is made with further known plaintext/ciphertext pairs. It can be seen from the above that the effort is only insignificantly greater than with a conventional attack, in which the entire cipherspace must also be examined. Hence one does not use cascaded dual encryption with DEA.

Instead, a procedure known as Triple-DES is used. Three DEA operations are connected in a chain, with alternating encryption and decryption. Decryption of the triple-encrypted block proceeds by reversing the order of operations, i.e. decryption–encryption–decryption. Where all three keys are identical, the alternating encryption and decryption gives the same outcome as a simple encryption. That is the reason why the method does not use three encryption operations in the chain.

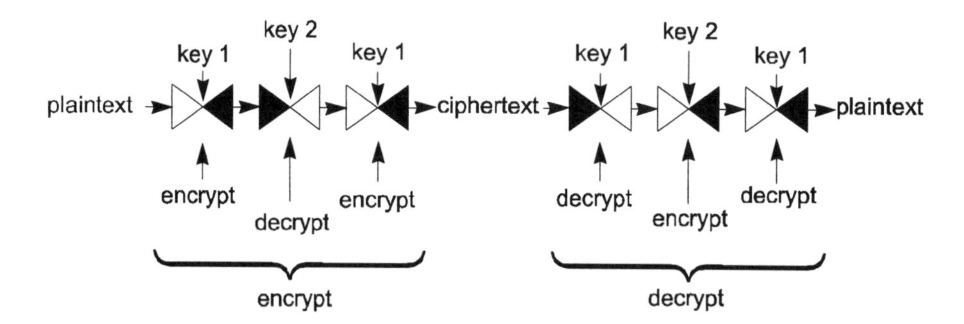

Figure 4.12 Theoretical encryption operation in triple-DES procedure

Triple-DES is considerably more secure than consecutive dual encryption with different keys, since the meet-in-the-middle attack would not work here. Instead of an 8-byte key, three are needed of which the first and third are equal. Therefore the key length is 16 bytes. Thus, this procedure is compatible with normal DEA and requires no more effort other than the double-sized key. It is precisely this which is the main argument for using Triple-DES in Smart Cards. Due to the clearly higher level of security compared with that of a single encryption, it is used particularly for key derivation or for protecting very sensitive data, for example in the transmission of keys.

Name	Type	Length of plaintext	Length of cyphertext	Key length
DEA	symmetrical	8 bytes	8 bytes	8 bytes
IDEA	symmetrical	8 bytes	8 bytes	16 bytes
Triple-DES	symmetrical	8 bytes	8 bytes	16 bytes
RSA	asymmetrical	512 bits = 64 bytes 768 bits = 96 bytes	512 bits = 64 bytes 768 bits = 96 bytes	512 bits = 64 bytes 768 bits = 96 bytes
DSS (512 bits)	asymmetrical	20 bytes	20 bytes	(64 + 20) bytes

Figure 4.13 Crypto-algorithms used in Smart Cards

4.2.2 Asymmetric crypto-algorithms

In 1976, Whitfield Diffie and Martin E. Hellman described the possibility of developing an encryption algorithm based on two different keys[5]. One key should be public, the other secret.

This would make it possible to encrypt a message with the public key, and only the secret key's owner would be able to decrypt it. This would do away with the problem of replacing and distributing secret symmetric keys; and new procedures, such as digital signatures which can be verified by anyone, would be possible for the first time.

Figure 4.14 Encrypting and decrypting using public key algorithms

[5] Whitfield Diffie and Martin E. Hellmann: *New Directions in Cryptography.*

RSA algorithms

Two years later, Ronald L. Rivest, Adi Shamir and Leonard Adleman proposed an algorithm which meets the above criteria[6]. The algorithm, named RSA after its three inventors, is the best known and most versatile asymmetric crypto-algorithm currently in use. Its very simple principle of operation (Figure 4.15) is based on the arithmetic of large integers. The keys are generated from two large prime numbers.

$$\begin{aligned} \text{Encrypt}: &\quad y = x^{e} \bmod n \\ \text{Decrypt}: &\quad x = y^{d} \bmod n \\ \text{with} &\quad n = p \times q \end{aligned}$$

Figure 4.15 Encrypting and decrypting using RSA algorithms
x: plaintext e: public key
y: cyphertext d: secret key
n: public modulus
p, q: secret prime numbers

Before encryption, the plaintext block must be padded to the appropriate length. This can vary according to the key length used. Encryption itself takes place by exponentiation of the plaintext, followed by a modulo operation. The result of this calculation forms the ciphertext. It can only be decrypted if the secret key is known, and is performed analogously to encryption.

Thus, the algorithm's security is based on the problem of factoring large numbers. It is very simple to compute the public modulus from the two prime numbers by multiplication, but very difficult to factor it into its two prime components since no efficient algorithms exist for this purpose.

Key generation for the RSA algorithm follows a simple scheme, which we illustrate here with a small example:

1. Choose two prime numbers p and q	$p = 3, q = 11$
2. Calculate the public modulus	$n = p \times q = 33$
3. Calculate an auxiliary variable z for key generation	$z = (p-1) \times (q-1)$
4. Calculate the public key e with the following properties: $e < z$ and $\gcd(z,e) = 1$, i.e. the greatest common divisor of z and e is 1. Since there are many numbers with this property, one of them is chosen	$e = 7$
5. Calculate the secret key d with the property that $(d \times e) \bmod z = 1$	$d = 3$

[6] Ronald L. Rivest, Adi Shamir and Leonard Adleman: A Method for Obtaining Digital Signatures and Public-Key Cryptosystems.

This concludes calculation of the key. The public and secret keys can now be tested on a further example, for encrypting and decrypting with the RSA algorithm:

1. Use the number 4 as plaintext x (with $x < n$)	$x = 4$
2. Encrypt	$y = 4^7 \bmod 33 = 16$
3. Calculate, producing a ciphertext y with the value 16	$y = 16$
4. Decrypt	$x = 16^3 \bmod 33 = 4$

The result of decrypting the ciphertext is, as expected, the plaintext.

A Smart Card's RAM would not be sufficient for the exponentiation of the large numbers required for encryption and decryption, as they reach large magnitudes before calculation of the modulus. Therefore, the method uses so-called modulo exponentiation, in which the intermediate results are never larger than the modulus. For example, in calculating x^2 mod n, the calculation does not involve $(x \times x) \bmod n$, since the intermediate result (i.e. $x \times x$) would become an unnecessarily large number before its size is reduced by taking the modulus. Instead, one calculates $((x \bmod n) \times (x \bmod n)) \bmod n$, which mathematically makes no difference to the result. The advantage of this method is that it requires considerably fewer steps and less storage space, since the intermediate results are reduced in length. Another procedure to speed up the RSA algorithm consists of using the Chinese remainder theorem[7].

The secret key should be as large as possible, to make attacks more difficult. Public and secret keys may be of different length, which is the usual arrangement, since the time needed for calculation in checking a digital signature is significantly reduced when the public key is as small as possible. Fermat's fourth number is often employed as the public key. This prime number has the value $(2^{16})+1=65\,537$, and due to its small size is very suitable for fast checking of digital signatures. The Fermat number $(2^1)+1=3$ is also used in this context.

Should an attacker succeed in splitting the public modulus into its two prime factors, the entire key computation could be executed. With a small value, e.g. 33, this is quite straightforward, but no fast algorithm exists as yet for this purpose in the case of large numbers. If the two prime factors can be found, the system is breached since the secret key is revealed[8].

We have mentioned the requirement that RSA keys be sufficiently large. Currently, 512-bit keys (= 64 bytes) are regarded as large enough. Nevertheless, 768 bits (= 96 bytes) and 1024 bits (= 128 bytes) are also used. The computational overheads for encryption and decryption shoot up with increasing key length, not linearly but almost exponentially.

---

[7]For further information on both methods, see Gustavus J. Simmons: *Contemporary Cryptology*, IEEE Press, New York, 1992.

[8] In summer 1994, 1600 computers linked via the Internet managed to crack a 426-bit key in 8 months. The task required 5000 MIPS years. In comparison, a 100 MHz Pentium represents 50 MIPS.

It is one of the strengths of the RSA algorithm, though, that it is not bound to one particular key length, in the way that DEA is. If one requires higher security, it is possible to use longer key lengths without changing the algorithm. It is still necessary to keep an eye on the time and memory space needed for the calculation, since even 512-bit keys are still regarded today as secure. With current factorization algorithms, one may take it as a rule of thumb that an increase of 15 bits in key length leads to a doubling in the effort needed for factorization.

Key length	Maximum number of decimal places	Time and memory space	Number of prime numbers in time and memory space
8 bits = 1 byte	3	256	54
512 bits = 64 bytes	155	1.3×10^{154}	$\approx 3.8 \times 10^{151}$
768 bits = 96 bytes	232	1.6×10^{231}	$\approx 1.5 \times 10^{228}$
1024 bits = 128 bytes	309	1.8×10^{308}	$\approx 2.5 \times 10^{305}$

Figure 4.16 Typical RSA key lengths

Figure 4.16 shows that even for a 512-bit key, the number of primes is so large that no collision ever occurs between two different key pairs[9].

It is normally impossible to carry out an RSA calculation within a period of a few minutes on Smart Card microprocessors, with their 8-bit wide CPUs. However, there now exist microprocessors with additional arithmetic units developed specially for fast exponentiation. With these it is possible to perform RSA calculations within an acceptable time and with reasonable software overheads. Program code for such a hardware-supported 512-bit RSA algorithm takes up in the region of 300 bytes. With 768 bits and a 1024-bit key, roughly 1 kbyte of Assembler code is needed in the card.

RSA is rarely used for data encryption, despite its very good security, due to the long execution times required. The main area of application is in digital signatures, since the benefits of this asymmetric procedure are fully exploited here.

Wider use of RSA is handicapped by claims to patent rights in the algorithm being made in several countries, as well as the considerable restrictions on the export and import of devices using it. Smart Cards containing a coprocessor for the RSA algorithm fall under these provisions, which make international distribution very much more difficult.

---

[9] The biggest number which can be represented by 512 bits is:

$(2^{512})-1 =$
13 407 807 929 942 597 099 574 024 998 205 846 127 479 365 820 592 393 377 723 561 443 72
1 764 030 073 546 976 801 874 298 166 903 427 690 031 858 186 486 050 853 753 882 811 946
569 946 433 649 006 084 095.

Realization	512 bits	768 bits	1024 bits
Smart Card with 3.5 MHz	308 ms	910 ms	2000 ms
Smart Card with 3.5 MHz (with Chinese residual class set)	84 ms	259 ms	560 ms
Smart Card with 4.9 MHz	220 ms	650 ms	1400 ms
Smart Card with 4.9 MHz (with Chinese residual class set)	60 ms	185 ms	400 ms
PC (80486, 50 MHz)	180 ms	500 ms	900 ms
RSA-hardware circuitry	8 ms	—	—

Figure 4.17 Example of calculation times in RSA encryption and decryption, depending on key length. The values given here can differ widely, as they depend greatly on the bit structure of the key

The DSS algorithm

In mid-1991, NIST (US National Institute of Standards and Technology) published a draft cryptographic algorithm for message signatures. The algorithm is now covered by a US standard (FIPS Pub 186), is designated DSA (digital signature algorithm) and is covered by standard DSS (digital signature standard). Next to RSA, the DSA algorithm is the second main procedure used in the creation of digital signatures. The background for the standardization of this procedure was the need to have one which could only be used to generate signatures, and which could not be used for data encryption. As a result of this requirement, DSS is more complex than the RSA algorithm. However, a possibility has in the meantime been found to encrypt data with this algorithm too.

In contrast to the RSA algorithm, the security offered by DSS does not rely on the difficulty of factoring large numbers but on the discrete logarithm problem. The fast computation of $y = a^x \bmod p$ is possible even for large numbers. However, in the reverse direction, determining x for given numbers y, a and p requires great effort.

In all signature algorithms it is necessary to reduce the message to a fixed length with a hash algorithm. Therefore, NIST published a suitable one for DSS, known as SHA-1 (secure hash algorithm)[10]. It generates a 160-bit hash value from a message of arbitrary length.

Just like RSA, the DSS algorithm only deals with integers. To calculate a digital signature with DSA, the following global values must first be found:

- p public Prime, 512 to 1024 bits long, the length must be exactly divisible by 64
- q public Prime factor of $p - 1$, 160 bits long
- g public $g = h^{((p-1)/q)}$, h being an integer for which $h < p - 1$ and $g > 1$

---

[10] See also 4.3 Hash functions, and FIPS Pub 180-1.

The secret key x must satisfy the following criterion:

- x secret key $x < q$

The public key y is calculated as follows:

- y public key $y = g^x \bmod p$

After all the required keys and numbers have been calculated, a message can be signed:

1. Generate a random number k	1. $k < q$
2. Calculate a hash value for m	2. $H(m)$
3. Calculate	3. $r = (g^k \bmod p) \bmod q$
4. Calculate	4. $s = k^{-1} (H(m) + x \times r)) \bmod q$

The two values r and s are the message's digital signature: DSA uses two numbers, rather than the one used in the RSA algorithm.

Checking the signature is carried out by the following routine:

1. Calculate	$w = s^{-1} \bmod q$
2. Calculate	$u^1 = (H(m) \times w) \bmod q$
3. Calculate	$u^2 = (r \times w) \bmod q$
4. Calculate	$v = ((g^{u1} \times y^{u2}) \bmod p) \bmod q$

If the condition $v = s$ is satisfied, the message m has not been altered and the digital signature is genuine.

Realization	Checking 512 bit signature	Calculating 512 bit signature
Smart Card with 3.5 MHz	238 ms	126 ms
Smart Card with 4.9 MHz	170 ms	90 ms
PC (80386, 33 MHz)	16 s	35 ms

Figure 4.18 Example of calculation times in DSA, separate encryption and decryption according to frequency. The values shown here can differ widely, as they depend greatly on the bit structure of the key. The speed can be significantly increased through prior calculation

It is impossible to say at the present time which of the two algorithms used to generate digital signatures, RSA or DSS, will be adopted in the long term or is the more secure. The original objective of DSS, that of standardizing a signature algorithm which cannot be used for encryption, has largely failed. The procedure's complexity, too, has not aided its widespread adoption. However, the standard exists, and with it the political pressure to use DSS and SHS to generate digital signatures. This is a very strong argument for many organizations.

4.2.3 Padding

In the Smart Card domain, DEA is extensively deployed to run ECB and CBC, the two block-oriented operating modes. But since not all transmitted data is a multiple of the block length, the blocks must sometimes be extended. This process, in which a data block is filled up with bits until it becomes a multiple of the block length, is known as padding.

The recipient of such an encrypted block, having decrypted it, is now faced with a problem. He does not know where the actual data ends and the padding bytes start. One solution consists in specifying the length at the start of the message, but this would alter message structure which as a rule is undesirable. This would be particularly costly in the case of data which does not always need to be encrypted, since in that eventuality they would not require padding and thus no length specification. Hence, in many cases the message structure must not be changed.

It is therefore necessary to use some other method to identify the padding bytes. A common convention is to specify the following algorithm, based on ISO/IEC 9797. The highest bit in the first padding byte which follows the actual data is set to 1. Thus this byte has the hexadecimal value '80'. If additional padding bytes are needed, they are reset to '00'. The recipient of such a padded message searches backwards from its end for a set bit, in other words for the value '80'. Once found, he knows that this byte and all subsequent ones belong to the padded section and are not part of the message.

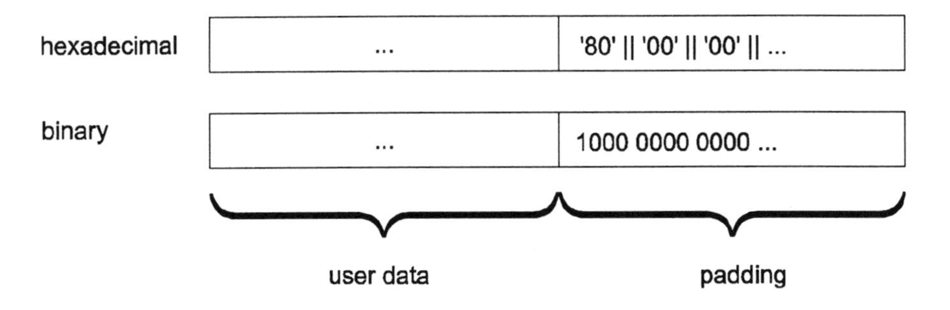

Figure 4.19 data padding

In this context it is important for the recipient to know whether padding always takes place, or only when necessary. If padding only occurs when the length of the encrypted data is only evenly divisible by the block length, then the recipient must also take this into account. That is why one often agrees implicitly that padding always takes place, which of course leads to the disadvantage that in the worst case, one block more has to be encrypted, transmitted and decrypted than is strictly necessary.

For the sake of completeness we must mention here that sometimes padding is performed only with the value '00'. The reason is that with MAC calculations, only padding with '00' is permitted. Using a uniform padding method leads to a saving in program code. However, in this case the application must know the exact structure of the data in order to distinguish between message and padding bytes.

4.2.4 Message authentication code

Often it is more important to ensure message authenticity than its confidentiality. The term authenticity means that a message is unmodified and has not been tampered with. To achieve this, a calculated MAC (message authentication code) is placed after the actual message and both parts sent to the recipient. The latter can also use the message to calculate a MAC, and compare it with the one transmitted. If both agree, then the message was not altered during transmission.

In order to form a MAC, one uses a cryptoalgorithm with a secret key which must be known to both parties. In principle, a MAC is a kind of EDC (error detection code) which, however, can only be checked when the secret key is known. That is why several other terms are used besides MAC, including CCS (cryptographic checksum); but technically, this is completely identical to a MAC. Generally, the difference between the two terms is that MAC is used in the context of data transmissions, while CCS is used in all other areas of application. Often the term "signature" is used instead of MAC; but this is different to "digital signature", which is calculated with an asymmetric cryptoalgorithm.

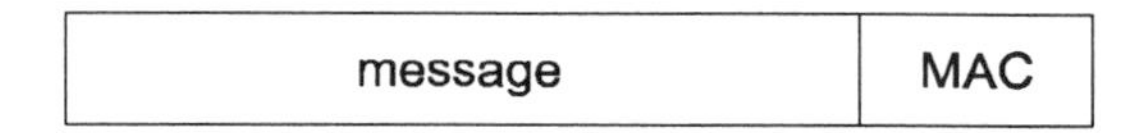

Figure 4.20 Usual arrangement of message and Message Authentication Code (MAC)

In principle, any cryptoalgorithm can be used to compute a MAC. However, in practice almost only DEA is used, and it will serve here to demonstrate the procedure.

DES encryption of a message in CBC mode links all blocks with their respective predecessor. Thus, the last block depends on all the previous ones. This block, or part thereof, represents the MAC of a message. But the message itself remains in plaintext, that is, it is not transmitted in encrypted form.

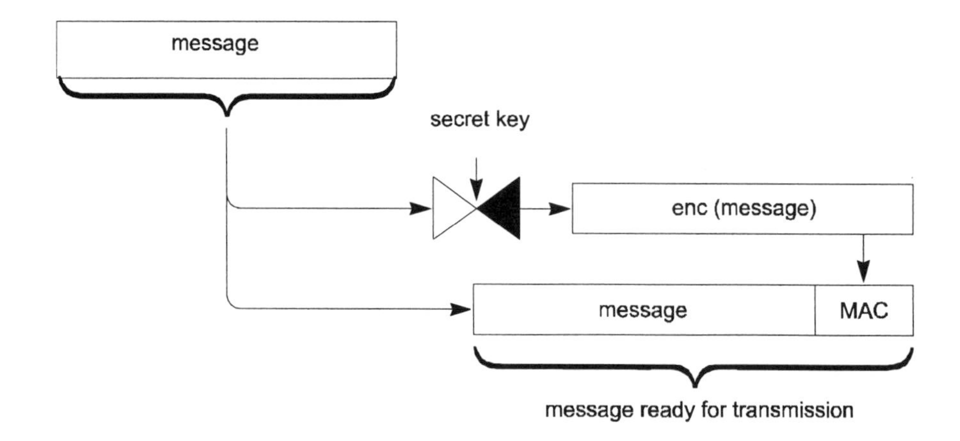

Figure 4.21 Example of MAC calculation procedure

Several other important conditions are necessary for MAC generation with the DES algorithm. If the message length is not a multiple of 8 bytes, it must be extended to such a value. This is usually termed padding; but in this case, it usually only involves filling up with '00' bytes (ANSI X9.9 – Message Authentication). This method is permitted here, since the MAC's length and position within the message must be agreed beforehand. The MAC itself consists of the four most significant bytes of the last block encrypted in CBC mode. However, the padding bytes are not transmitted as part of the message, only the protected data bytes and the following MAC. This minimizes the amount of transmitted bytes.

4.3 HASH FUNCTIONS

Even powerful computers need a great deal of time for the calculation of a digital signature. Many signatures would be required for larger documents, since the length of a document which needs a signature cannot be arbitrary. Here one can make use of a trick. First the document is compressed to a much shorter, fixed length, and then the signature is calculated for the compressed data. It is of no importance whether the compression can be reversed, since it can always be reconstructed from the original document. The functions used in this calculation are called one-way hash functions.

Expressed in general terms, then, a one-way hash function is one which derives a fixed-length value from a variable-length document, thus reproducing the document's contents in a compressed and irreversible form. In the Smart Card domain, these functions are used exclusively to calculate the input values for digital signatures. If the document's length is not a multiple of the block length used by the hash function, it must be padded appropriately.

For a hash function to be efficient, it should have a number of properties. The output length should be fixed, for its subsequent use by the signature algorithms to be optimized. Since large quantities of data normally need to be processed, the hash function must have a high throughput, that is, calculation of the hash value must be simple. On the other hand it must be difficult, or ideally impossible, to derive the original document from a known hash value. Finally, the hash function must be collision-resistant, which means that for a given document it must be difficult to find a second, modified document with the same hash value.

Nevertheless, without any doubt there must exist other documents whose hash value is the same. After all, every possible message with any length between nil and infinity is imaged onto a finite hash value. It is a necessary consequence of this fact that collisions do take place. That is why the term used here is collision-resistant, and not collision-free.

What would be the effects of a collision? There would be two different documents with the same hash value, and hence the same digital signature. The result would be fatal, making the signature worthless, since one could make changes to the document without them being discovered. This is precisely what happens in one of the two typical attacks possible with hash functions. One searches systematically for a second document whose hash value is the same as the original one. If the document's contents also make sense, then one has managed to undermine the digital signature based on the hash value. Since in this case the two documents are interchangeable, the signature becomes worthless. And it does really make a huge difference whether a conveyancing contract relates to a £95 000 house or a £750 000 one.

The second attack on hash values is somewhat more subtle. One prepares in advance two documents whose hash values are equal but whose contents differ. This is not particularly difficult, when one considers the special symbols available in the character set. The result is that there is a valid digital signature for both documents, and it is impossible to prove which document was signed to begin with.

It is not as difficult as may appear at first sight, to find two documents whose hash values are equal. It is possible to exploit the birthday paradox, which is well known in statistical theory. It revolves around this question: how many people must there be in a room so that the probability of one of those present sharing a birthday with the quiz-master exceeds 50%? The answer is simple: one only needs to compare the quiz-master's birthday with those of everyone present in the room. There must be at least 183 (= 365/2) people.

The next question reveals the paradox, or rather the surprising result in this comparison. How many people must there be in the room so that the probability of two people sharing a birthday exceeds 50%. The solution is only 24 people. The reason is that although only 24 people are present, one can form a total of 552 pairs in order to compare their birthdays. The probability of a shared birthday results from the number of these pairs.

It is precisely this paradox which is exploited in the attack on the hash function. It is very much simpler to create two documents sharing a common hash value than to alter a document until it corresponds to a given value. The consequence is that the outputs of hash functions must be sufficiently large if they are to be successful in preventing both these types of attack we have described. Thus, most hash functions generate an output value whose length is at least 128 bits, which at present is generally regarded as sufficient protection.

Many hash functions have already been published, and some of them are governed by standards; but existing functions are repeatedly subject to modifications once successful attacks come to light. The following is a short summary of hash functions currently in common use. Their internal structures are, unfortunately, beyond the scope of this book.

Standard ISO/IEC 10118-2 deals with a hash function based on an n-bit block encryption algorithm (e.g. DEA). The hash value may be n or $2n$ bits long. The function MD4 (message digest) and its later development MD5 were published by Ronald L. Rivest in 1990/91. They are based on a stand-alone algorithm, and both generate a 128-bit hash value. NIST published a hash function for DSS in 1992, known as SHA. After the discovery of certain weaknesses it was improved, and since mid-1995 it is known as SHA-1 and covered by a standard (FIPS Pub 180-1).

Since data transmission to the Smart Card is generally slow, the hash function is executed in the terminal or in a computer connected to it. This drawback is balanced by the fact that this makes the function interchangeable. Besides, in most cases it would not be possible to store a hash function in a card, since the 4 kbytes of Assembler code needed for the program would almost always exceed the space available. The required execution speed is very high, usually at least 300 kbyte/s, which is met by an 80386 computer running at 33 MHz.

4.4 RANDOM NUMBERS

Random numbers are routinely needed as part of cryptographic procedures. Typical applications in the field of Smart Cards are ensuring the uniqueness of a session during authentication, as padding in the course of encryption or as a starting value in transmission counters. The length of these random numbers is usually between 2 and 8 bytes. The maximum 8-byte length is the one made necessary by the DES algorithm's block size.

Name	Block length	Length of hash value
ISO/IEC 10118-2	*n* bits (64, 128 bits)	*n* / 2*n* bits (64, 128 bits)
MD4	512 bits	128 bits
MD5	512 bits	128 bits
MDC-4	64 bits	128 bits
RIPEMD	512 bits	128 bits
SHA-1	512 bits	160 bits

Figure 4.22 Summary of common hash functions

The security of these procedures relies on random numbers which cannot be predicted or influenced from outside. The ideal solution would be a hardware-based random number generator in the card's microprocessor. However, it would have to be completely independent of external effects such as temperature, supply voltage, radiation, etc., otherwise it could be interfered with. This would make it possible to compromise some of the procedures whose security relies on true randomness.

Since current technology makes it almost impossible to construct in the processor a good, semiconductor-based random number generator which is immune to external influences, operating system designers have to fall back on software implementations. The result is pseudo-random number generators, which usually produce very good randomized outputs. Nevertheless, these are not genuinely random numbers, as they are calculated by strictly deterministic algorithms and thus are predictable if the algorithm and its input values are known. That is why they are termed pseudo-random.

It is also very important to ensure that all the cards in one production batch generate different sequences of random numbers, so that numbers from one card do not assist in predicting those from another of the same batch. This is achieved during the manufacture of the card's operating system, by entering a random number as the seed (starting value) for the random number generator.

4.4.1 Random number generation

There are many different methods of random number generation using software. However, since the available storage capacity in Smart Cards is extremely limited, and the time allowed for the calculation should be as short as possible, the number of options is severely restricted. In practice, a routine will be implemented only if it can be used by the operating system's other functions, and thus requires only a very small amount of additional program code.

Naturally, interruption of a session through resetting or removal of the card from the terminal must not affect the quality of the random numbers. In addition, the generator must be so constructed that the same sequence of random numbers is not generated during every session. Although this sounds trivial, it does require at least one write access to the EEPROM, in order to store a new seed for the generator's next session. The RAM, of course, is not suitable for this purpose, since it can only retain its contents under power. An attack could be based on generating random numbers for long periods, until the EEPROM cells which store the seed break down. Theoretically, this would result in the same sequence of random numbers occurring in every session, thus making them predictable and helping the attacker. If this EEPROM store is built as a ring buffer and all further actions are prohibited once a write error occurs, this type of attack can easily be averted.

Another very important criterion for a software-implemented random number generator is to ensure that it never runs in an endless loop. The consequence of such an event would be a significantly reduced repetition period for the random numbers. Then it would be easy to predict the numbers with some degree of certainty, and the system would be cracked.

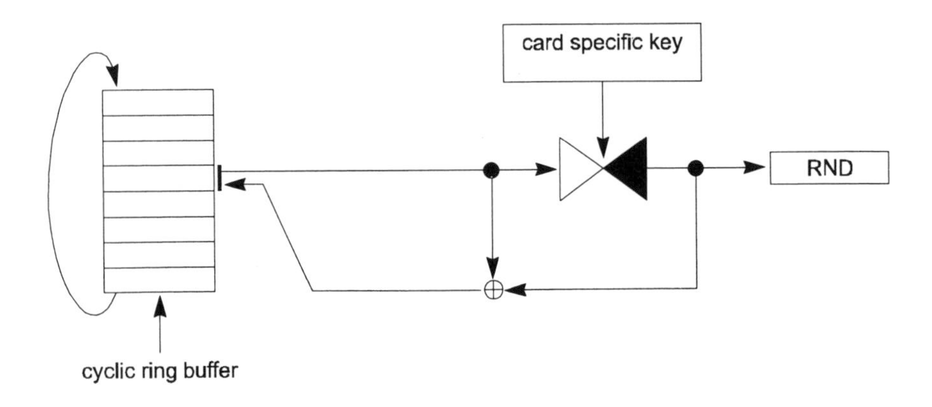

Figure 4.23 Example of DEA random number generator

Almost all Smart Card operating systems contain an encryption algorithm for authentication. It seems obvious to use it as the starting point for a random number generator. One needs to be aware that a good encryption algorithm mixes the plaintext as thoroughly as possible, so that it cannot be derived from the ciphertext without knowing the key. This principle, also known as the avalanche criterion, states that, on average, altering one input bit should alter half of the output bits. This property can be usefully exploited for a random number generator. The exact details of the generator's construction may vary from one implementation to the next.

Figure 4.23 illustrates one option. This generator uses the DES algorithm with a block length of 8 bytes, and the output is fed back to the input. Of course, any other encryption algorithm could be used.

The generator works as follows. The value of a ring buffer element is encrypted by DES, using a key unique to the card. The ciphertext so produced is the 8-byte random number. This number, XOR-ed with the previous plaintext, provides the new input into the EEPROM's ring buffer. Now the next value in the cyclic ring buffer is selected. Mathematically, the relationship can be represented as:

- $RND_n = f(key, RND_{(n-1)})$

During production, each Smart Card is provided with its unique DEA key, and at the same time random numbers entered into, for example, a 12 x 8 byte long ring buffer as seeds. This ensures that each card produces its own sequence of random numbers. The ring buffer increases the generator's life-span by a factor of 12. If one assumes that the guaranteed number of EEPROM write cycles is 10 000, then this generator allows at least 120 000 8-byte random numbers to be produced.

Erasing and writing eight bytes into the EEPROM takes about 2x2x3.5 ms, and DES execution at 3.5 MHz takes about 17 ms. The remaining processing time is negligible. Hence, the card needs about 31 ms for the entire generation of a random number.

4.4.2 Testing random numbers

The numbers produced by a random number generator must be thoroughly tested for quality. In particular, this cannot be achieved simply by printing a few out, inspecting and comparing them. The testing of random numbers can be carried out with conventional statistical, i.e. mathematical, methods. It goes without saying that a large number of 8-byte random numbers is needed for this. Between 10,000 and 100,000 numbers should be generated and evaluated for a reasonably reliable test. Thus, the only way to check these numbers is with the help of computerized test programs.

In connection with the quality of random numbers, it is necessary to mention the distribution of the values produced. If it is very non-uniform, and if some values are significantly more frequent as a result, then it is precisely these numbers which can be exploited for predictability. In other words, Bernoulli's theorem should be satisfied as well as possible: it states that the occurrence of a number should be unaffected by its history, only by the probability itself of the number occurring. So, for example, the probability of a 4 in a six-sided die is always 1/6, regardless of which number was thrown previously. This is also termed the independence of events.

Another important criterion relates to the random numbers' period and how often the sequence of generated numbers repeats itself. Naturally, the period must be as long as possible, and at any rate greater than the generator's life-span. This prevents, in very straightforward and reliable fashion, an attack made by recording all the random numbers which constitute one period.

The science of statistics offers an almost unlimited number of tests which can be used to investigate the randomness of events, but in practice one can limit oneself to a few simple and easily interpreted methods[11].

One which is particularly easy to set up and interpret is to count the single-byte values which occur in a large sample of random numbers. When represented graphically, this provides an overview of the numbers' distribution. The abscissa is marked with the possible values, and the ordinate with the frequency of occurrence (Figure 4.24).

---

[11] See also D. E. Knuth: The Art of Computer Programming, Addison-Wesley, Reading, MA, 1973.

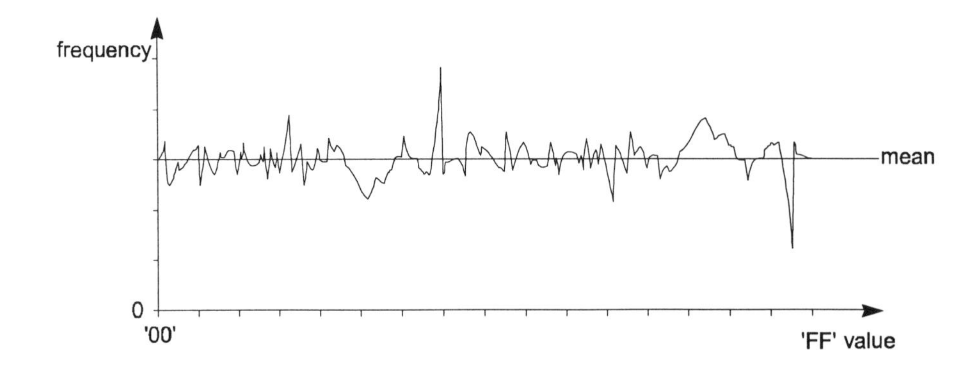

Figure 4.24 Statistical distribution of values produced by typical Smart Card pseudo-random number generator

Where 8-byte numbers are to be tested with the aid of this diagram, the abscissa should nevertheless still only record one- or at most two-byte values, otherwise the sample needed for a statistical analysis would be extremely large. As a guideline, each value should occur about five to ten times to make a reasonably reliable decision possible. This provides a quick summary which shows whether the generated random numbers span the entire available bandwidth of one byte. If this is not the case, and a few values are very definitely preferred, this could at least be exploited by someone as a first line of attack.

Unfortunately, this test says nothing about the random numbers' sequence of occurrence, only something about their distribution. Thus it is conceivable that the generator simply counts cyclically from 0 to 255. This would achieve a superbly uniform distribution, but the numbers would be totally predictable. Other tests must be used to test this quality criterion as well.

A much more powerful approach uses the well-known χ^2 test. Although it tests the same property as the graphic uniformity check described above, it is significantly more precise since it is performed by calculation.

Starting from the assumption that the random numbers are uniformly distributed, the mean and standard deviation can be derived. Using a χ^2 distribution, the deviation from a normal distribution can be computed. This allows a numerical evaluation of the random numbers' distribution.

However, even this test cannot be used to evaluate the sequence. Further statistical tests can be used to assess this property, such as evaluating gaps between patterns which occur in random numbers (serial test). Alternatively, the gaps between non-occurrence of patterns can be obtained (gap test), the χ^2 distribution of occurring patterns (pattern test) or the χ^2 distribution of non-occurring patterns (coupon collector test).

Execution and evaluation of the above-mentioned tests are the minimum requirement for ensuring a reliable assessment of a random number generator. Additional calculations and tests can be used to underpin the results. Only then can a sufficiently meaningful judgement be made of the random numbers' quality.

On the other hand, a huge investment in a Smart Card random number generator is usually unjustified when one considers how the numbers are used. For example, the effect on security of predictable random numbers during authentication would be rather minor. An attack would not be possible without the secret key used to encrypt the random number. However, a much greater problem would arise if the generator could be manipulated, so as always to produce the same numbers, for example. Attacks through reinsertion of recorded sessions would then not only be possible, but actually successful. This would also be the case if the period of repetition is very small. Thus in each individual case it is necessary to make a judgement about the random numbers' main required characteristics, since this would of course be reflected in the generator chosen. Although high investment in this area may lead to a very high quality of randomness, normally this also leads to an increased use of storage space, which is precisely one of the limiting factors in Smart Cards.

4.5 DATA STRUCTURING

Data storage and transmission necessarily require an exact definition of the data and its structure. Only thus is it possible to identify and interpret the data elements following the above-mentioned operations. Fixed-length data structures with a non-modifiable sequence regularly cause systems to "crash". The best example of this phenomenon is the conversion of German postcodes from four to five digits. All systems and data structures whose length was not susceptible to modification had to be extended to five digits at great expense. The same problem applies to many Smart Card applications. Fixed data structures which need to be extended or shortened eventually cause considerable expense.

However, the data-structuring problem has been around for many years, and there are plenty of solutions. One of these, very popular in the world of Smart Cards and also used more and more in other fields, derives from data transmission. It is known as abstract syntax notation one, ASN.1 for short. This is a coding-independent description of data objects, originally developed for the transmission of data between different computer systems.

In principle, ASN.1 is a form of artificial language which is suitable not for describing programs, but rather data and its structure. The syntax is standardized in ISO/IEC 8824, and the coding rules are covered by ISO/IEC 8825. Both standards are a development of Recommendation X.409, defined by CCITT.

ASN.1's fundamental concept consists in providing data objects with an unambiguous designation and length specification, placed in front of the object. The descriptive language's quite complex syntax also makes it possible to define one's own data types, as well as to nest data objects. The original idea, that of creating a generally valid syntax on the basis of which data can be exchanged between fundamentally different computer systems, is hardly exploited in Smart Cards. Currently, only a very small part of the available syntax is used, mainly as a result of the very limited storage space in Smart Cards.

One area of application which will attain a great deal of significance in future is that of secure messaging. The various methods of secure data transmission can be dealt with much more simply and flexibly with ASN.1 objects than has been the case up to now with very rigid structures.

The coding of ASN.1 objects takes place in the classic TLV structure. T, standing for 'tag', denotes the object's designation; L is its length, and V, or value, is the actual data.

The first field in a TLV structure is the tagging of the data object contained in the subsequent V field. So as to obviate the need for each user to define his or her own designations and thus open the door to incompatibility, several standards exist which lay down agreed tags for various commonly used data structures. For example, ISO/IEC 7816-6 defines tags for objects used in general industrial applications. On the other hand, those for secure messaging are specified in ISO/IEC 7816-4. It is by no means the case that a given tag is always used for the same data element, but a process of standardization is essentially in progress.

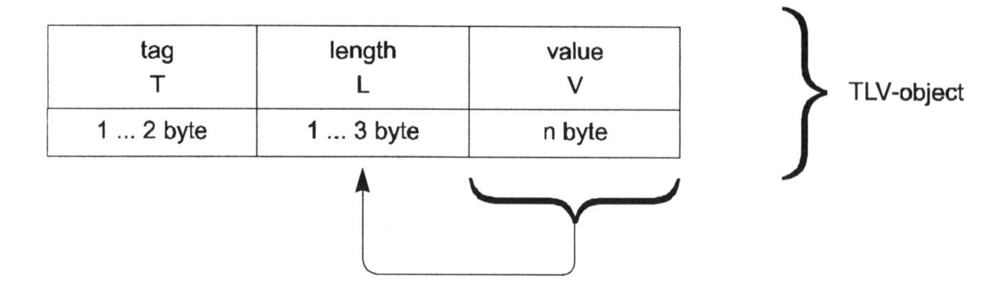

Figure 4.25 Principle behind TLV tags

ISO/IEC 8825 summarizes the basic encoding rules (BER) for ASN.1. The data objects derived from them are described as BER-TLV encoded objects. Figure 4.26 shows the tag bits as specified by these encoding rules.

The two highest bits encode the class to which the following data object belongs. The class indicates whether the data object belongs to a universal class, such as integers or character strings. The application class indicates whether the object belongs to a particular application or standard (e.g. ISO/IEC 7816-6). The other two classes, context specific class and private class, fall under the heading of non-standardized applications. The bit after the two class bits indicates whether the tagged object is constructed of further data objects. The five lowest bits are the actual designation. Since as a result of the restricted address space only the values 0 to 30 can be encoded, it is possible to point to the following byte by setting all five bits. All values from 31 to 127 are then permissible in that byte. Bit 8 points to future applications and therefore must never be set.

b8	b7	b6	b5	b4	b3	b2	b1	Meaning
0	0	...	...	...	...	...	...	universal class
0	1	...	...	...	...	...	...	application class
1	0	...	...	...	...	...	...	context specific class
1	1	...	...	...	...	...	...	private class
...	...	0	...	...	...	...	...	primitive data object
...	...	1	...	...	...	...	...	constructed data object
...	....	...	X	X	X	X	X	code number (0 to 30)
...	...	...	1	1	1	1	1	the byte following (byte 2) determines the length

Figure 4.26 Tag bits specified by ASN.1 (byte 1)

b8	b7 to b1	Meaning
0	31 to 127	Tag number

Figure 4.27 Tag bits specified by ASN.1 (byte 2)

The number of necessary length bytes is illustrated in Figure 4.28. The standard also defines the term 'template'. This is a data object which serves as a container for further objects. ISO/IEC 7816-6 defines the tags for possible data objects in the domain of industry-wide applications. ISO 9992-2 covers the domain of Smart Card banking transactions.

Byte 1	Byte 2	Byte 3	Meaning
0 to 127	—	—	Necessary length for these values 1 byte
'81'	128 to 255	—	Necessary length for these values 2 bytes
'82'	256 to 65535		Necessary length for these values 2 bytes

Figure 4.28 Number of necessary length bytes in ASN.1

This method of data encoding has several characteristics which are of particular benefit in the field of Smart Cards. Since storage space is too small as a rule, the use of data objects as per ASN.1 leads to considerable space savings: TLV encoding makes it possible to transmit and store variable-length data without too many complications. This leads to great storage economies and can be illustrated by a further example, the TLV encoding of a name (Figure 4.29).

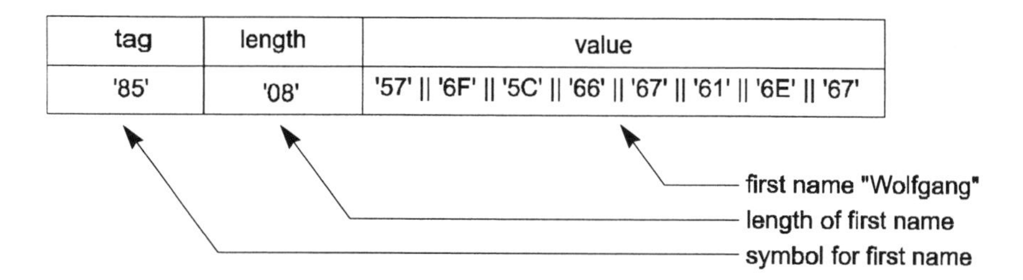

Figure 4.29 TLV encoding of a name

Later extensions of data objects can be implemented very simply with ASN.1, since only an additional TLV encoded data object needs to be inserted into the existing data structure. Full compatibility with the older versions is retained, so long as the earlier TLV objects are not removed. This is just as true of new versions of data structures, in which changes have been made as against the previous encoding. This is straightforward, and is done by modifying the tags. Equally, the same data can be represented very easily with different encodings. All in all, these advantages are the reason why the ASN.1 syntax, based on TLV encoding, is so popular, particularly in the Smart Card industry.

The main weakness of ASN.1 data objects lies in the considerable space used up in administrative data, even for a small amount of user data. For instance, if the latter only amount to 1 byte, two additional bytes (tag and length) are needed to manage it. However, the longer the user data, the more favourable the relationship. The structured ASN.1 data in the German health insurance card illustrates this situation well. There are between 70 and 212 bytes of user data. The administrative data needed amounts to 36 bytes, that is, between 17% and 51% of the former.

Let us recap all the above by way of a further example. Let us assume we wish to store surnames, first names and titles in a file with a transparent data structure. Independently of a correct ASN.1 description, the TLV encoded data will have the structure shown in Figure 4.30. The tags used in the example were chosen arbitrarily, and thus do not follow any particular standard.

	T	L	V	T	L	V	T	L	V
variant 1	'85'	'06'	"George"	'87'	'05'	"Smith"	'84'	'04'	"BSc."

	T	L	V	T	L	V	T	L	V
variant 2	'84'	'04'	"BSc."	'85'	'06'	"George"	'87'	'05'	"Smith"

	T	L	V	T	L	V	T	L	V
variant 3	'87'	'05'	"Smith"	'85'	'06'	"George"	'84'	'04'	"BSc."

Figure 4.30 Example of independence of a sequence within a TLV structure

When evaluating this data structure, the computer compares the first tag with all the ones known to it. If it finds a match, then it recognizes the first object as a first name. Its length is read off from the next byte. The subsequent bytes are then the actual object, i.e. the first name. This is followed by the next TLV object, whose first byte is the tag for surname. In recognizing it, the computer follows exactly the same procedure as in the first data object.

If it becomes necessary to extend the data structure, e.g. with an additional name, then a new element can simply be inserted into the existing structure. The point of insertion is unimportant. The extended structure remains fully compatible with the older version, since the new element receives its own tag and thus is unambiguously designated. Programs which only know the old tags are not hampered by the new one, since they fail to identify it and so by definition it can be skipped. Other programs which do know the new tag as well, can thus evaluate it but neither does the old structure present them with any problems.

4.6 STATE AUTOMATA

An automaton is commonly thought of as a machine into which one inserts a coin and then presses a button. Afterwards a drawer can be opened and some object removed.

In slightly more abstract terms, an automaton defines a chain of events involving various transitions between states. In the initial state, the automaton waits for money to be inserted, and all other actions such as pressing a key do not result in anything happening. Only acceptance of a coin causes transition from the ground state to the “Money inserted” state. The next transition occurs as a result of pressing the button, when the automaton permits the opening of the drawer.

In information theory, state automata can be effectively visualized by using graphs or Petri networks. Besides modelling state automata, the systems thus described may be investigated for certain properties. The objective is to identify any deadlocks which may occur during operation, and to ensure that the correct sequence of instructions is given.

4.6.1 Fundamentals of automata theory

This section provides a summary and an introduction to the use of graphs for the description of states in Smart Card applications.

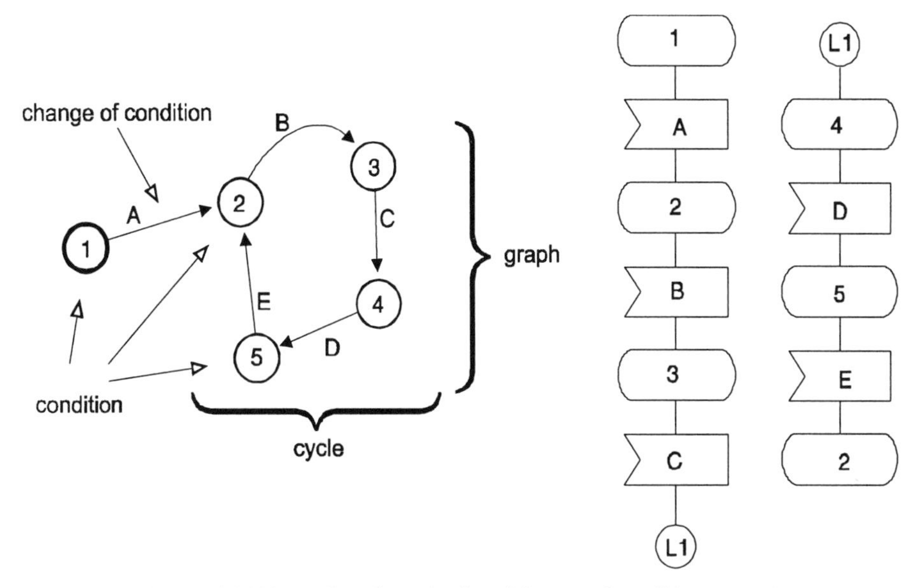

Figure 4.31 Examples of graphs describing set of conditions, part 1
(left: directed graph, right: SDL diagram)

	Condition				
Change of condition	1	2	3	4	5
1	—	—	—	—	—
2	A	—	—	—	E
3	—	B	—	—	—
4	—	—	C	—	—
5	—	—	—	D	—

Figure 4.32 Examples of graphs describing set of conditions, part 2
(in the form of a table)

A graph represents both a set of states, and the relationships between these states. The states are shown as nodes, their relationships as edges. If the edges are characterized by a direction, that is, they are drawn as arrows, then we speak of directed edges and hence a directed graph. The arrow specifies the direction in which a state transition is permitted. The precise location of nodes and edges in the graphical representation plays no part in its interpretation. A sequence of nodes connected by edges is also termed a path. If the first and last nodes are identical and there is more than one node, the path is termed a cycle.

This is only a very small part of graph theory, but it is essentially all we need to know to describe the states and the relevant automata in Smart Card applications.

4.6.2 Practical application

A further advantage of Smart Cards compared to memory cards consists in the fact that the sequence of instructions can be determined in advance. In other words, it is possible to specify precisely all the instructions, their parameters and their sequence. This also constitutes an additional access protection, in parallel to the object-oriented access authorization to files. Nevertheless, the facilities offered by Smart Cards in this respect vary greatly. In the simplest case, we cannot define a state automaton. In contrast, in very modern operating systems it is possible to define application-specific state automata involving instruction parameters.

Smart Card state automata may be divided into so-called micro- and macro-automata. Micro state automata merely define a short instruction sequence and only become active once the first instruction in a sequence has been sent to the card.

A typical example are the two instructions necessary to authenticate a terminal. The first instruction asks the card for a random number. This activates the micro-automaton, which permits an authentication instruction as the only possible next instruction. If the card receives it, the sequence of instructions is complete and any other instruction is permitted once again. If this is not the case, that is, the card receives an instruction other than the expected authentication instruction, the micro-automaton generates an error message and the sequence is aborted. It has to be restarted from the beginning, and cannot be continued from its intermediate position.

Micro state automata are only a subset of all possible state automata, but have several great advantages as far as Smart Cards are concerned. The very small number of permitted instructions in a rigidly defined sequence, means that storage and program overheads are minimal. In many applications it is sufficient to protect the file contents with the object-oriented access mechanisms, and otherwise to permit free sequencing of all instructions. Only the sequence of a few procedures, such as authentication, must be prescribed. This can be achieved with high storage economy through a micro state automaton.

Macro-automata can be regarded as an extension and generalization of micro-automata. They make it possible to check all instructions, and all their parameters, within a defined graph before execution. Depending on the automaton's details, it may even be possible in certain circumstances not to have object-oriented file access protection, since the automaton which precedes the execution of every instruction could take over all the necessary checks. One problem is that an error in the state graph's definition could have fatal consequences for the system's security. As the absolute absence of errors in the state definitions of complex automata can only be demonstrated at great expense, in practice the additional file access protection remains in place. The true description of all sequences and all instructions in a Smart Card is very costly, and often must be obtained, at least in part, empirically.

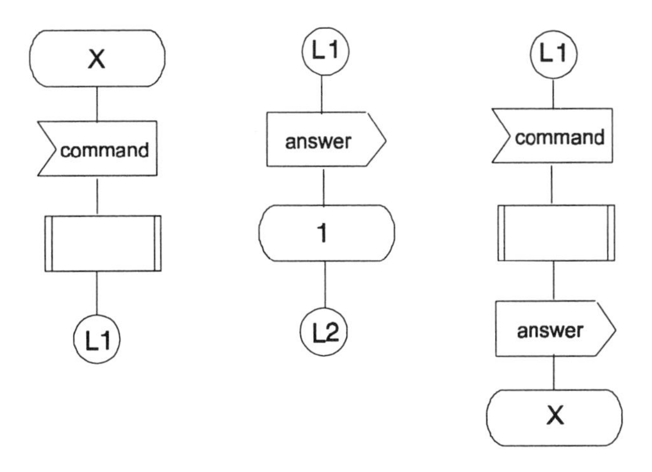

Figure 4.33 Example of micro condition automaton (conditions: X, 1)

After describing the advantages of macro-automata, we must now turn to their shortcomings. The implementation of a macro-automaton with the required power is very costly, both in design and in the subsequent programming. The demand for program storage for the automaton alone, controlled as it is by the stored representation of a graph, is huge. In addition to this program code, the appropriate graph must also be stored. The amount of storage space naturally depends on the complexity of the graph that needs to be executed. The amount of information contained in a graph with many states, and corresponding transitions, can become very large in relation to the size of Smart Cards.

The graph for a small application, reproduced as Figure 4.34, summarizes the possibilities offered by a macro-automaton. Its function can be described as follows.

After resetting, the Smart Card is in the ground state, denoted by 1. In this state, each file in the directory may be selected (SELECT FILE); this does not alter the state. All other instructions, except VERIFY PIN, are prohibited, and the card responds to them with an error message. After successful verification of the PIN, the automaton switches into state 2.

Two instructions are permitted in this state. The first path leads via SELECT FILE to state 3, where the chosen file may be read. The second path, branching off from state 2, leads to state 4 after a request by the terminal from the card for a random number (ASK RANDOM). Any instruction other than EXTERNAL AUTHENTICATE leads back to ground state 1. When the terminal is successfully authenticated, the card reaches state 5. The graph defines this state as one in which files may be selected and written to (SELECT FILE, UPDATE BINARY).

Neither of states 3 and 5 can be exited from in this session, as defined by the graph: they represent the two final states. A transition to state 1 is only possible by resetting the card. However, this is not shown in the graph, since all state automata are only "conscious" during the actual session. No information is carried forward from one session to the next within the automaton.

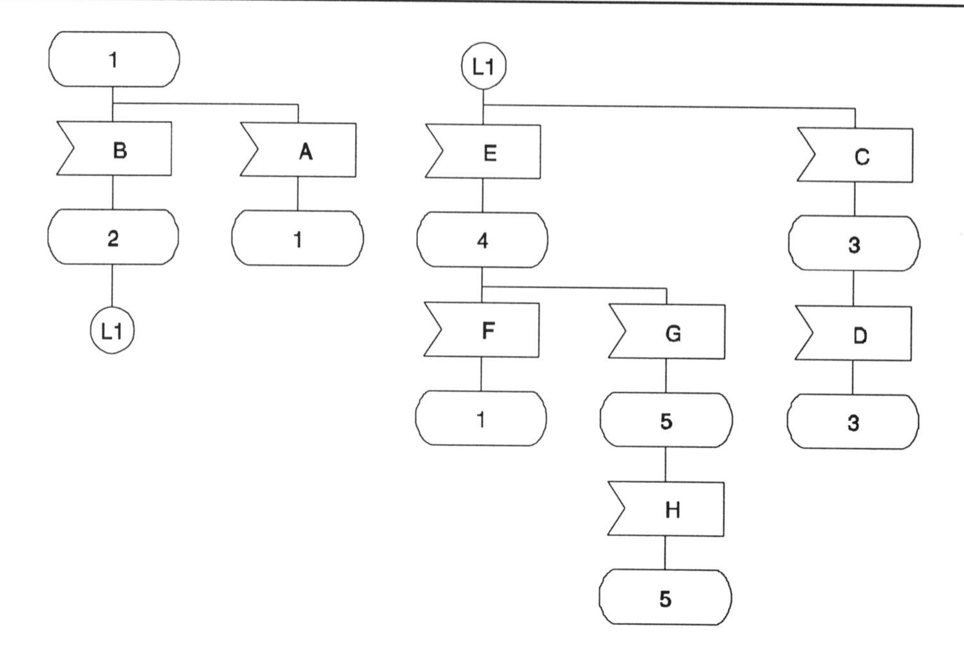

Figure 4.34 Example of macro condition automaton: 1 – ground state; 2,4 – transition states; 3, 5 – final states. A – SELECT FILE, B – VERIFY PIN, C – SELECT FILE, D – READ BINARY, E – ASK RANDOM, G – EXTERNAL AUTHENTICATE, H – SELECT FILE/UPDATE BINARY, F – all commands except G

4.7 SDL SYMBOLISM

This book uses SDL symbolism to describe states and state transitions. For some years this approach has been adopted ever more frequently in the Smart Card domain to describe state-oriented mechanisms, such as in communication protocols. SDL stands for Specification and Description Language, and is described in detail in CCITT Recommendation Z.100.

SDL symbolism is similar to that used in flowcharts. However, it describes not program flow but states and transitions. SDL diagrams are made up of individual standard symbols, interconnected by lines. The diagrams are always read from top left to bottom right, hence the lines connecting individual symbols do not need to be identified by arrows denoting start and end.[12]

To put it more simply, the notation can be thought of as describing a system consisting of a certain number of processes. Each process in turn is a state automaton. If the automaton is in a stable state, it can receive an external signal. A certain new state is then reached, depending on the input data. Additional intermediate actions may be present, such as the reception and sending of data or the computation of a value.

---

[12] For a detailed example of an SDL diagram, see 6.2.2 Transmission protocol T=0.

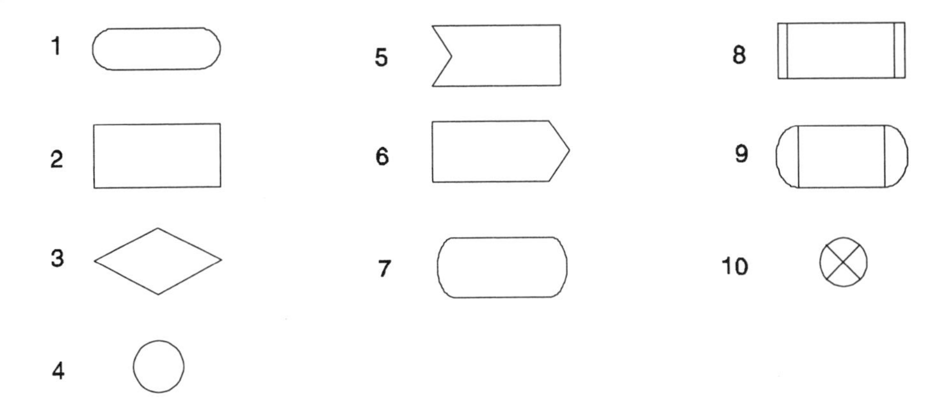

Figure 4.35 Symbols in SDL notation in accordance with CCITT Z.100: 1 - Start, 2 - Task, 3 - Decision, 4 - Connector, 5 - Input, 6 - Output, 7 - Condition, 8 - Subprogram, 9 - Start subprogram, 10 - End subprogram

Figure 4.35 shows the ten symbols used in this book. They are merely a selection from the much larger set defined in Z.100, but they suffice as a basic set for use in Smart Cards.

The Start symbol (no. 1) denotes the beginning of a process. In most cases it initiates an SDL diagram. The symbol for a task (2) is used, together with the explanatory text contained in it, to specify a certain activity. In the case of this symbol, there is no further detailed description in the form of a subprogram. The next symbol, Decision (3), permits a query during a state transition, to which the answer may be "YES" or "NO". The connector (4) connects to another SDL diagram, and is used mainly to divide large diagrams into several small ones.

The next two symbols, Input (5) and Output (6), represent interfaces to the outside world. The exact input/output parameters are described inside the symbol. The Condition or State symbol (7) is used to describe a state, and specifies the one just reached.

The next three symbols describe subprograms. Number 8 – Subprogram –indicates that the contents of this box are described in more detail elsewhere. The two symbols Start (9) and End (10) of a subprogram delimit this more detailed description.

Operating system architecture

It may seem presumptuous to refer to the few thousand bytes of program code in a Smart Card microprocessor as an operating system, but this program does fully justify the name. According to German standard DIN 44300, an operating system is neither more nor less than "The programs in a digital computer system, which together with the computer installation's features constitute the basis for the operating modes the system is capable of, and in particular, control and supervise program execution".

Thus the term "operating system" is not necessarily restricted to large programs and data, rather it is quite independent of size since it serves exclusively to define functionality. It is important not to associate the concept of an operating system with the multi-megabyte programs for DOS or Unix computers. These are just as specifically designed for a particular man-machine interface using a colour monitor, keyboard and mouse, as are Smart Card operating systems for their bidirectional, serial terminal interface.

At the end of the day, an operating system depends only on the functionality resulting from the cooperation of library routines, which are designed to fit together and call on each other. In this context it is also important to note that an operating system is an interface between the computer's hardware and the actual applications software. This also offers the great advantage for applications programs that they do not need to access the hardware directly, and thus retain at least some portability, though often in practice a very limited one.

The operating system's structure also depends very critically on the manufacturer's philosophy. A very pragmatic style is often preferred in certain countries and in certain firms, where the only important factor at the end of the R&D effort is that the application works. On the other hand, some firms maintain a regime of strict scientific methodology. In the future, Smart Cards will increasingly require a certificated operating system, which is the only way to take the security aspect fully into account.

So as to keep our discussion firmly anchored in current practice, this chapter describes not a fictional operating system but the one developed by Giesecke und Devrient and the Society for Mathematics and Data Processing (Gesellschaft für Mathematik und Datenverarbeitung), and known as STARCOS. This system, undergoing development since 1990, makes it possible to load, operate and manage several applications on one Smart Card. STARCOS is highly suitable for demonstrating the general mechanisms of modern

Smart Card operating systems. This chapter also covers fundamentals and procedures derived from a variety of specifications, standards and descriptions of Smart Card software, so that it serves as a general cross-section of current operating systems.

The following sections deal with the fundamentals of practical and functional operating systems, and the mechanisms built upon them.

5.1 HISTORY

The evolution of Smart Card operating systems reflects the same developmental phases as all other computer systems. The initial, special programs for a single application have been continuously generalized and extended, the end-result being an operating system which can be used simply and in a structured fashion, and implemented everywhere.

The Smart Card microprocessor programs which were around during the early days of this process, around 1980, can strictly speaking not yet be regarded from today's perspective as operating systems. They were really user software stored in ROM. But since the manufacture of mask-programmable microprocessors is expensive and time-consuming, a need was soon perceived for generally implementable core routines, on which special user software stored in EEPROM could build as required. This, naturally, led to an increase in storage requirements, which resulted in several firms moving in the opposite direction, that of special software for a single application.

However, since market demand for tailor-made solutions is still increasing apace, operating system manufacturers are more or less obliged to offer programs which meet this demand. Only in the case of applications sold in very large numbers is the specially developed ROM software approach used. The conventional operating system is the one based on standardized instructions, which is so designed that it can be generally implemented for any application. Where this is impossible for some special reason, then at least it is so structured that it can be modified at little expense and in a short time to satisfy the requirements of any given application.

The "historical" development of Smart Card operating systems from 1980 onward can be clearly illustrated through those used by the German mobile-phone network. The card used in the C-network since 1987 employs a system optimized for this particular application. It also involves its own transmission protocol, special instructions and a file structure tailored for the application. All in all, the card certainly has a full operating system, but it is entirely tailor-made for the C-network domain.

The next step was a transition from the specialized solution to a somewhat more open system architecture. This is exemplified by the first GSM cards, whose structure is multifunctional and significantly more open. At the same point in time when specifications for GSM cards were being written, draft standards for Smart Card instruction sets and data structures were already in existence, thus laying the foundation for compatibility between individual operating systems. Further developments were built on this foundation. Modern GSM systems display functions such as memory management, a range of file structures and state automata which bring them very close to the possibilities offered by multi-application operating systems. They can manage several applications independently of each other, without any interactions between them. Usually they also have very extensive state automata, a large instruction set and sometimes, also, several transmission protocols.

Foreseeable developments in the field of Smart Card operating systems are likely to lead, through many intermediate steps, to an international quasi-standard such as is today the case in many other systems. A so-called industry standard always gets established after a number of years, which all commercial competitors must support if they are to continue operating successfully in the marketplace. Such a standard does not yet exist in the Smart Card industry, but the first indications are already in sight. However, in contrast to the DOS industry, here it is based on international standards or specifications such as ISO/IEC 7816-4 and GSM 11.11.

5.2 FUNDAMENTAL PRINCIPLES

In contrast to the generally known operating systems, Smart Card systems do not feature user interfaces or the possibility of accessing external storage media, since they are optimized for quite different functions. The first priorities are secure program execution and protected access to data.

Due to memory restrictions, the amount of code that can be stored is very small, typically between 3 and 24 kbytes. The lower limit is that used by special applications, while the upper one is for multi-application operating systems. The mean memory requirement is around 10 kbytes.

Program modules are written as ROM code, which rather restricts the programming methods used since many procedures common in RAM code are impossible to exploit (e.g. self-modifying code). The ROM code is also the reason why no more processor modifications can be made after programming and ROM manufacture. Error correction, therefore, becomes very expensive, and involves a turnaround of 10 to 12 weeks. Where the card has already reached the end-user, such errors can only be rectified by large-scale recalls and exchanges, which are capable of destroying the reputation of a Smart Card-based system. Thus, "quick and dirty" programming is ruled out by definition. This normally makes the time invested in testing and quality assurance substantially higher than in the programming itself.

In addition to this requirement for an extremely low error rate, the operating system must also be very reliable and robust. An external instruction should not be capable of interfering with its operation and above all, with its security. System crashes or uncontrolled reactions due to a faulty instruction, or as a result of failed EEPROM sections, must not occur under any circumstances.

Unfortunately, the execution of certain system mechanisms is influenced by the hardware used. Above all, it is the physically cleared status of the EEPROM which affects system design in a small but nevertheless noticeable way. For instance, all error counters must be so designed that their highest value coincides with the EEPROM's cleared state. Should this not be the case, then when writing to the error counter it would be possible to reset it to its original value by deliberately switching off the power supply. This is possible since the EEPROM must be cleared before certain write actions take place. If the power supply is switched off at precisely the right moment between clearing and writing, then the EEPROM section containing the error counter is in the cleared state, and if the operating system is incorrectly designed the error counter would be back at its original value.

The concept of "secure operating system" relates to yet another aspect. Trapdoors and other backdoors used by system programmers, which occur frequently and in fact are quite conventional in large systems, are completely ruled out in the case of Smart Card systems.

In other words, it must be impossible to read unauthorized data through bypassing the operating system in some way.

Nor can the required performance be ignored. The cryptographic functions forming part of the operating system must run very fast. Thus it is usual, during development, to spend weeks of detailed programming on optimizing the algorithms in Assembler. It should then be apparent that multitasking cannot be employed, due to the hardware platforms used and the necessary reliability. A particular downside of this is the fact that because of the restriction to a single executed task, it is impossible to use additional protective routines to supervise aspects of system operation and boundary conditions.

To sum up, a Smart Card operating system has the following main tasks:

- data transmission to and from the card
- control of instruction execution
- data management
- management and execution of cryptographic algorithms.

Instruction processing

Typical instruction processing within the Smart Card operating system proceeds as follows. All instructions sent to the card arrive via the serial I/O interface. If required, the I/O manager carries out error recognition and correction mechanisms totally independently of the other, higher layers. Once a complete and error-free instruction has been received, the secure messaging manager must decrypt it as necessary or check it for integrity. Where a secured data transmission is not taking place, this manager is completely transparent for both instruction and response.

After this process, the overlying layer, the instruction interpreter, attempts to decode it. If this is not possible, the return code manager is called up, generates the appropriate return code and sends it to the terminal via the I/O manager. However, if the instruction could be decoded, the logical channel manager selects a channel, switches over to its parameters and, if successful, calls up the state automaton.

The latter now tests whether the instruction which was sent to the card, with the parameters set to their current state, is permitted. If so, then the actual program code in the application instruction which takes over execution, is carried out. If the instruction is prohibited in the current state, or its parameters are not allowed, the terminal receives an appropriate message via the return code and I/O managers.

If it is necessary to access a file during instruction execution, this only happens via the file manager which converts all logical addresses into the card's physical ones. It also supervises the address range, and checks the access conditions in the relevant file.

The file manager itself uses a further memory manager, which takes over the whole administration of the physically addressed EEPROM. This ensures that true physical addresses are only used in this program module, which considerably increases the whole operating system's portability and security.

Generation of the return code is the job of the central return code manager, which produces the complete response for each program section calling it up. This layer is responsible for the administration and generation of all return codes for all other parts of the system.

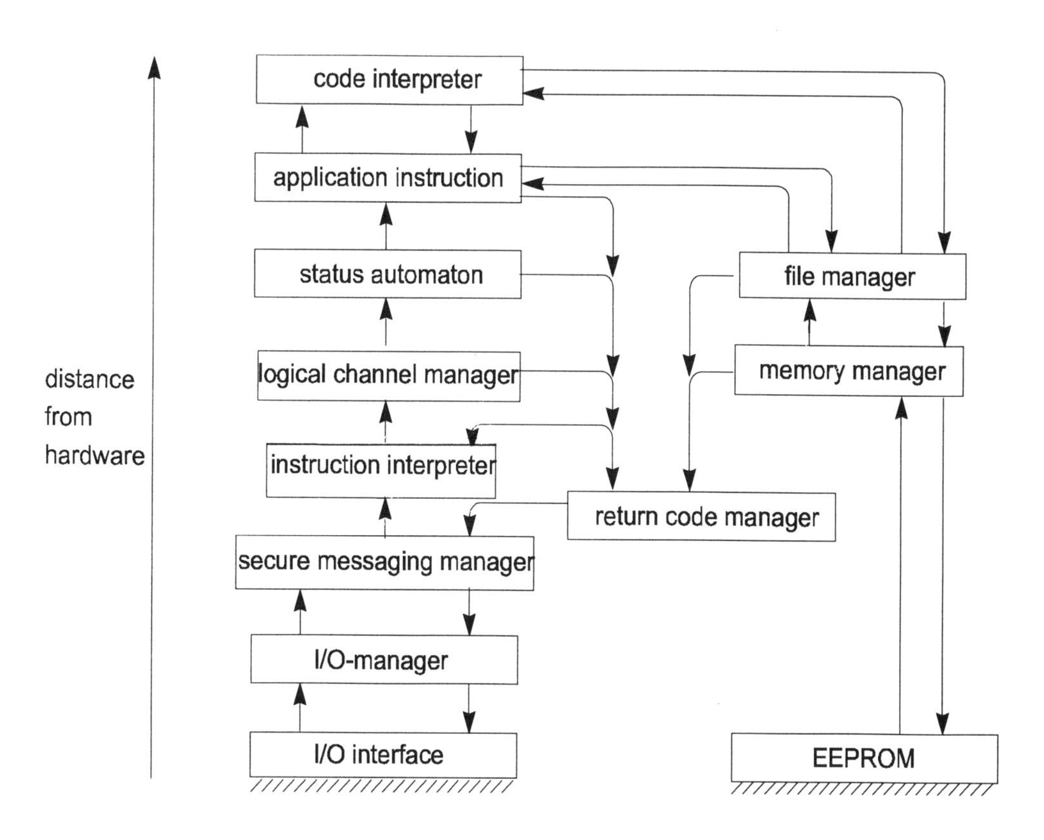

Figure 5.1 Instruction processing in Smart Card's operating system

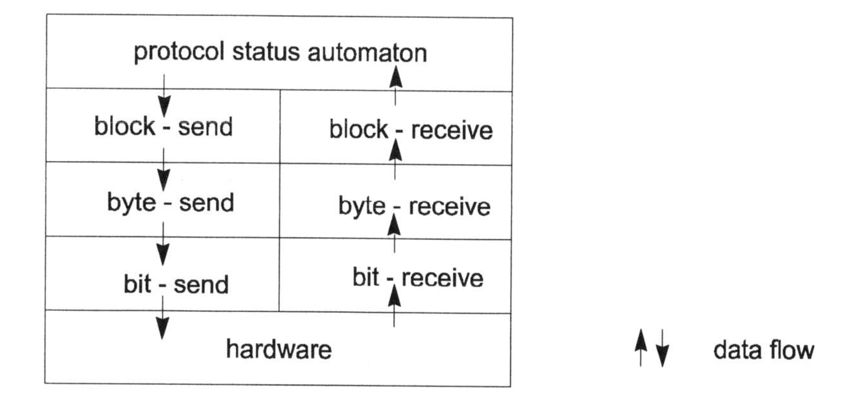

Figure 5.2 Partial representation of instruction processing by I/O managers in Smart Card operating system

In addition to these layers there may be, above the application instructions, an interpreter or a test program for executable files. It supervises the programs contained in these files and executes them or interprets them. The precise structure and its implementation depend on whether files with executable code have been provided, and whether they store machine code for the processor or code which needs interpreting.

Smart Card profiles

In contrast to PC operating systems, memory space is so severely restricted in Smart Cards that, often, not all standardized instructions and file structures can be implemented. For this reason, so-called Smart Card profiles have been introduced in the two relevant standards for generally implementable operating systems (EN 726-3 and ISO/IEC 7816-4). Each of these defines a subset of instructions and file structures in the relevant standard.

The subset is the minimum contained in a Smart Card of a particular profile. However, the description of card profiles is designated in the appendix as informative and not normative, hence it only represents a recommendation to the system designer. Figure 5.3 summarizes the five profiles covered by ISO/IEC 7816-4.

5.3 DESIGN AND IMPLEMENTATION PRINCIPLES

It is well known that design errors only manifest themselves during implementation, and then multiply the cost which would have been borne by a better and more error-free design. But this is simply a given fact in all software projects. In order to prevent such errors, it is worth observing a few principles during the design and implementation stages of Smart Card operating systems.

By definition, a Smart Card operating system needs to be a secure system, one which manages and in particular protects the confidentiality of data. Equally, no software modifications or updates can be possible during operation. This defines the primary principle. A Smart Card system must be extremely reliable, and hence also extremely error-free, though in reality it is impossible to achieve absolute freedom from error since even the small systems used here are too large to permit the testing of all possible permutations.

Nevertheless, a strictly modular design contributes crucially to the detection and correction of errors during the implementation phase. This modularity, which increases reliability by a large factor, need not necessarily involve huge additional overheads in terms of program code. Another advantage of modularity is the fact that any system crashes generally do not affect security to the same extent as in the case of highly optimized and memory-saving code. This in turn means that the repercussions of any errors are contained locally, and the system as a whole becomes more robust and stable.

Error susceptibility is increased by the fact that implementation must be completely carried out in Assembler. The construction, based on single completely testable modules, contributes greatly through the defined interfaces to the early recognition and containment of programming errors. The end-result is the operating system shown in Figure 5.1, built up of several layers. The greater investment in planning and coding is always offset by the considerably simpler tests and checks. The consequence is that almost all current systems possess the illustrated internal structure, or at least a very similar one.

Profile	Description	
Profile M	Data structures:	• transparent • linear fixed
	Commands:	• READ BINARY, UPDATE BINARY no implicit selection and maximum length up to 256 bytes • READ RECORD, UPDATE RECORD no implicit selection • SELECT FILE using FID • VERIFY • INTERNAL AUTHENTICATE
Profile N	like Profile M plus use of an AID with SELECT FILE	
Profile O	Data structures:	• transparent • linear fixed • linear variable • cyclic
	Commands:	• READ BINARY, UPDATE BINARY no implicit selection and maximum length up to 256 bytes • READ RECORD, UPDATE RECORD without automatic selection • APPEND RECORD • SELECT FILE • VERIFY • INTERNAL AUTHENTICATE • EXTERNAL AUTHENTICATE • GET CHALLENGE
Profile P	Data structures:	• transparent
	Commands:	• READ BINARY, UPDATE BINARY no implicit selection and maximum length up to 65 bytes • SELECT FILE using AID • VERIFY • INTERNAL AUTHENTICATE
Profile Q	Data transfer:	• secure messaging
	Data structures:	• —
	Commands:	• GET DATA • PUT DATA • SELECT FILE using AID • VERIFY • INTERNAL AUTHENTICATE • EXTERNAL AUTHENTICATE • GET CHALLENGE

Figure 5.3 Summary of Smart Card profiles covered by ISO/IEC 7816-4. The data structures and commands shown represent the minimum requirements

The usual procedure followed during system design is the module-interface concept. The system's and application's tasks are broken up into their functions as far as possible, and the functions encapsulated in modules. Provided the modules' interfaces are precisely described, each one can now be programmed by a different person. Since the volume of program code in Smart Card systems is very small, this very pragmatic method can be exploited without too much difficulty. It makes full use of the benefits of a minimum in planning requirements, distribution of programming tasks between several people, and straightforward reuse of code. These are offset by the difficulty in demonstrating system correctness, and where modifications are involved, sometimes by extensive repercussions on numerous modules.

In future, the development of Smart Card systems will move away from pure programming in Assembler towards a hardware-related high-level language such as C. The system's actual core will continue to be based on machine-dependent Assembler routines, whilst all superordinate modules such as file manager, state automaton and instruction interpreter will be written in C. This would reduce development time, the programs would become more portable, and above all, using a high-level language would make the code much easier to test, which as a rule leads to a reduced error rate. Unfortunately, program code generated even by an efficient self-optimizing C compiler requires between 20% and 40% more memory than that written by a good Assembler programmer, given the same functionality. However, the worst problem is not the additional memory needed in ROM, but the space required in RAM. Since this is only available on Smart Cards in extremely limited quantities, and to make things worse, requires the most area on the chip in proportion to cell size, we can see why no high-level languages have been used thus far.

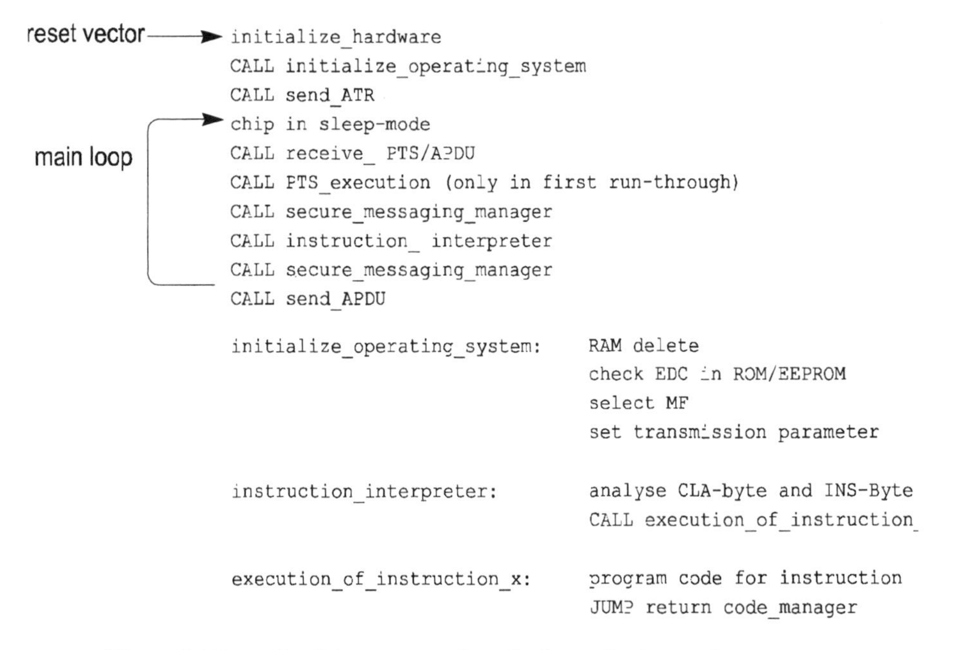

Figure 5.4 Example of the programming of a Smart Card operating system

Since Smart Cards find application in very security-conscious fields, the issuer or application provider must have a great deal of confidence in the integrity of the operating system's manufacturer, since the latter has every opportunity of compromising the entire system through a deliberate security failure. For example, consider an electronic purse on a Smart Card, whose crediting instruction was so manipulated that in certain circumstances it may be credited without authorization.

Such scenarios are the reason why only a few system manufacturers have managed to carve themselves a niche on an international scale. The risk of purchasing a supposedly secure operating system from a small and unknown manufacturer, only to discover it contains a Trojan horse, is considerably greater than in the case of a firm well known in the industry.

5.4 PROGRAM CODE SECTIONS

The life-cycle of a Smart Card operating system is divided into two parts - before and after completion. Before completion, when the microprocessor leaves the semiconductor production line with an empty EEPROM, all program sections run in ROM. No data are read from the EEPROM, nor are programs executed there. If it turns out at this point that an error is present in the ROM code which makes completion of the system impossible, the entire microprocessor batch must be destroyed as there is no further use for the chips.

In order to reduce the likelihood of such an error, one could implement in the ROM only a small loading routine for the EEPROM, and then load the actual operating system in the EEPROM. But since the area needed per bit in the EEPROM is four times larger than for a bit in ROM, and this is reflected more than proportionally in the chip's price, as much code as possible must be stored in ROM for purely economic reasons. Hence, all system core routines as well as the rest of the system's essential parts are held in ROM in their entirety. Only a few jumps to the EEPROM are still provided. Some systems run completely in ROM even after completion, and only the data are stored in the EEPROM in order to minimize the size of this expensive type of memory. Naturally, this saving in storage area is offset by severe limitations on system flexibility.

During completion, the ROM's sections are tailored to the application proper. The ROM is, as it were, a large library, which through the EEPROM is expanded and linked to a functional application. In addition, almost all operating systems allow, during completion, program code for further instructions or special cryptographic algorithms to be loaded into the EEPROM. This is independent of any executable files which may be present, since their contents can be loaded at a later point in time. The programs loaded in EEPROM during completion are entirely associated with the system and are used directly by it.

5.5 MEMORY ORGANIZATION

The three different types of memory used in a Smart Card microprocessor have totally different properties. The ROM can only be programmed as a whole in the form of a mask during microprocessor manufacture, and remains static during the card's entire lifetime. Due to the manner of construction, the likelihood of an inadvertent alteration of ROM contents is practically nil.

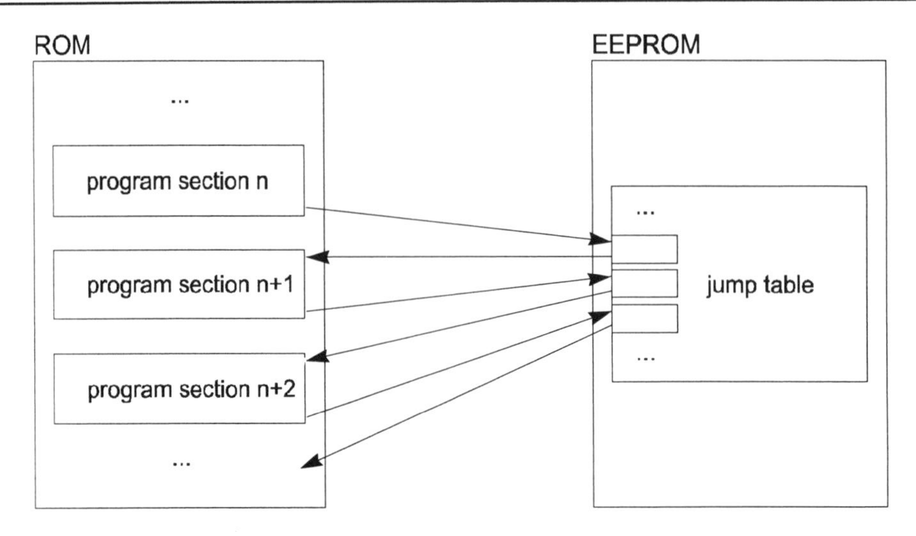

Figure 5.5 ROM program code is loaded into the EEPROM during the initialization of the operating system via a jump table

In contrast, RAM contents are only preserved while voltage is applied to the card. Voltage failure results in complete loss of all RAM data. On the other hand, RAM data can be written at the processor's full working speed, and also erased an unlimited number of times.

An EEPROM can store data even without an external voltage supply. It suffers from three shortcomings, though: it has a limited life-span, writing and erasing take very long (*ca.* 1 ms/byte) and it is divided into sections.

Program code stored in ROM for the operating system must not be controlled by any higher structure, other than interrupt vectors. Individual program sections can be connected to each other in arbitrary sequence, where the aim is always to minimize the distance of software jumps and so save storage space.

Figure 5.6 summarizes the conventional division of a 256-byte RAM. It is divided into areas for the register, stack, general variables, working space for cryptographic algorithms and I/O buffer. If, for instance, a 256-byte I/O buffer is required, or additional variables need to be stored in RAM, the available storage space is quickly used up. The problem is solved through a working space in EEPROM, which is used as RAM. The disadvantage is the writing time, longer by a factor of *ca.* 10***000 than when accessing a true RAM. Another drawback is the restricted life-span of an EEPROM cell, which in contrast to RAM cells cannot be written to an unlimited number of times. However, often the transfer of RAM contents into the EEPROM is the only solution, when I/O buffers are needed which are larger than the whole available RAM.

EEPROM structure is incomparably more complicated and expensive than the other two types of memory. In a modern operating system, it is basically divided up as follows: various production data, e.g. a number which is only used once and thus is unique, can be written to the section at the start of the EEPROM, which in many microprocessors is specially hardware-protected. Often these sections are designed only for WORM access (write once, read many), i.e. they can be written to once and subsequently only read.

register	10 bytes
stack	26 bytes
general variables	50 bytes
working space for cryptographic algorithms	70 bytes
I/O-buffer	100 bytes

Figure 5.6 Example of division of a 256-byte RAM

After this section, which frequently is 32 bytes long, come the tables and system pointers. These are loaded into the EEPROM during completion, and together with the ROM programs form the final system. In order to ensure that the system is always operated in a secure and stable condition, this section is protected by a checksum (EDC), which is recalculated before each access. Discovery of a memory error during this check always results in the relevant EEPROM sections not being used again in subsequent operation, since correct system functionality can no longer be ensured.

The system's secure section is followed by one containing additional application program code. In certain circumstances, this section is also protected by a checksum against alteration. This section can contain application-specific instructions or algorithms which do not need to stay in ROM, or were not loaded in it for reasons of lack of space.

The adjacent file section contains all file structures, that is the whole of the file tree which is visible from outside. This section is not protected as a whole by a checksum, but usually has file-oriented protection. The internal structure is shown again in detail in Figure 5.7.

There may be a spare memory section at the end of the EEPROM, administered by its own manager. Often, however, this memory is allocated to individual applications contained in the file section, and can be used there within the applications for new files which need to be set up. Otherwise it forms part of the general file section, where it is only available for new applications loaded in their entirety.

It is a fundamental principle of memory organization within the file section, that the operating system can supervise memory use by an application during run time, and ensure it does not exceed its boundaries. In addition to a pure software solution, this can be efficiently achieved by an MMU module (memory management unit) in the processor[1]. Therefore, all parts of a DF must be contained in a physically continuous area of memory. Since modern operating systems make it possible to install new files after they are personalized, each data file must possess its own spare memory. This can be used as a file memory when necessary, i.e. when a new file is installed.

---

[1] See also 3.4.3 Supplementary hardware.

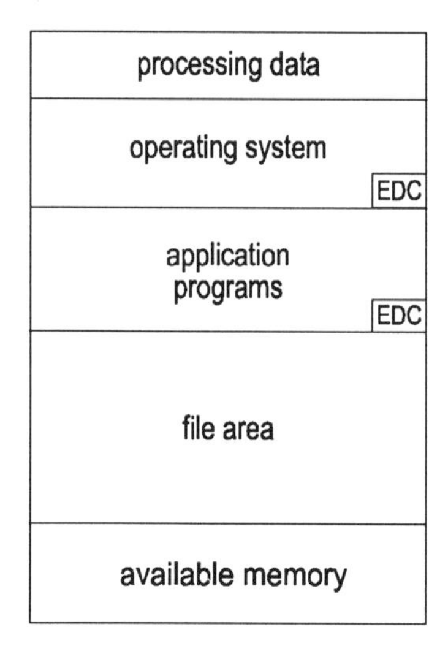

Figure 5.7 Example of division of EEPROM

If a file is deleted, the liberated space is reallocated to the spare memory. With this two-part arrangement, file memory and spare memory, only the memory used by the last file can be reallocated to the spare memory on deletion. This rather restricts memory management, but due to the limited space available for program code on a card, cannot be implemented in any other way.

The ideal situation would be true memory management with double linked lists, one each for the used memory and the spare memory. This would be a straightforward method, similar to DOS or UNIX, of creating new files and deleting them as necessary. The overheads in terms of file management programs and file descriptors, however, would no longer be a sensible design for Smart Card operating systems. The limitations of the chosen solution are not too painful, since it is very rare for files to have to be installed or deleted within applications.

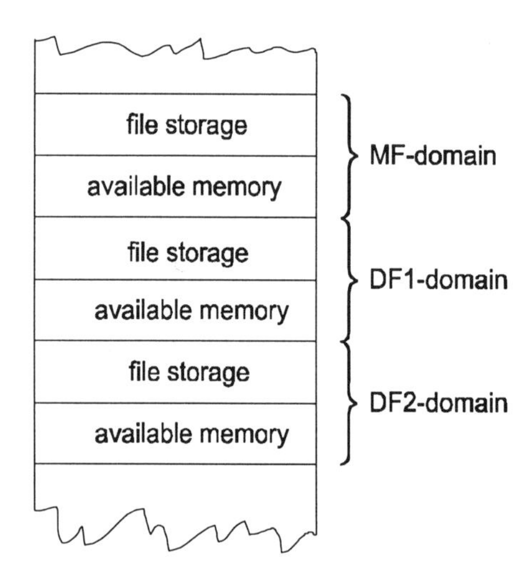

Figure 5.8 Example of data structures

5.6 DATA STRUCTURES IN THE SMART CARD

In addition to containing mechanisms for identification and authentication, Smart Cards also store data. Their advantage compared with discs is that file access can be linked to conditions.

The first cards had only more or less directly addressable memories, in which data could be written or read. Access required the use of physical memory addresses. Current cards have complete and hierarchically organized file management systems, with symbolic and hardware-independent addressing.

These file managers have certain properties specific to Smart Cards: the most evident is the absence of a man-machine interface. All files are addressed by hexadecimal code, and the other instructions are also based on communication strictly between two computers. Another characteristic of these file managers is the fact that they are designed to minimize memory use. Whenever possible, the last redundant byte is dispensed with. Since the "user" connected to the terminal is a computer, this does not really constitute a disadvantage.

Memory management programs are normally kept to a minimum in order to reduce memory use as far as possible. When a file is deleted, and even this is only possible in very few operating systems, it by no means follows that the liberated space can be taken up by another file. All files are usually created during initialization or when the card is personalized. Later modifications are limited to file contents.

Naturally, the properties of the memory types used affect the nature of file management. EEPROM sections can only be written to and erased a fixed number of times. As a consequence, there are special file attributes in order to store data redundantly, and if relevant, in such a way that they can be corrected.

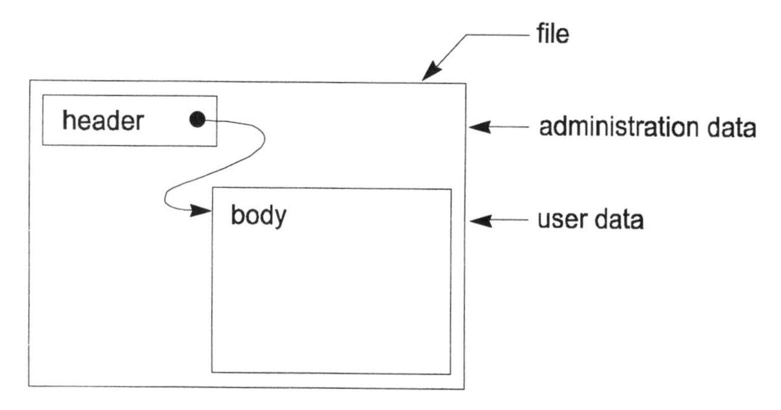

Figure 5.9 Internal structure of Smart Card file systems

The more recent Smart Card operating systems are usually object-oriented. This means that all data concerning a file are stored in this file itself. Another effect of this principle is that a file must always first be selected before any action takes place. Therefore, in object-oriented systems, files are always split into two parts. The one known as the header contains data about file structure and access conditions; and the body, which is associated with a pointer, contains the variable user data.

In addition to improved data structuring, this approach has the further benefit of greater physical data security. The EEPROM which stores all the files is divided into sections, and can only support a finite number of write/erase cycles. Header and body are always located in separate sections of the memory. Since as a rule, the header which stores all the access conditions is never modified, a write or erase error in the body cannot affect it. Had the header and body been located in the same section of the EEPROM, then it would have been possible to alter the access conditions through a deliberate write error, so that eventually confidential data could be read from the file body.

5.6.1 Types of files

The structure of Smart Card file systems is laid down in ISO/IEC 7816-4, and is similar to DOS and UNIX. Various directories exist which serve as a kind of folder, and can contain several related files.

MF

The root directory, which is implicitly selected after a card reset, is designated "master file", or MF for short. It contains all other directories and all files. The master file represents all the memory available in the card for file storage.

DFs

If necessary, dedicated files can exist at the next level. These are directories which contain further files. They may even contain further DFs. The nesting depth is unlimited in principle. However, the very restricted storage space in a Smart Card means that it is rare for more than one level of DFs to be present.

EFs

The user data needed by an application are contained in elementary files. These may be ordered directly below the MF level, or within a DF. The operating system also supports various internal EF structures.

Internal system files

In addition to the EFs, there are internal system files which may store data for the operating system itself, the execution of an application or secret keys. Access to these files is specially protected by the operating system. There are two variations on the way in which these internal system files are integrated into the file manager. In the ISO procedure, these files are concealed in the relevant application DF and cannot be selected. The Smart Card's operating system manages them in completely transparent fashion, in a manner similar to that of resource files in System 7. In the ETSI model, these system files are given a regular file name (FID), and can be selected with it. Essentially this corresponds to DOS file management. Both systems have similar advantages and drawbacks, but fulfil the same functions in somewhat different ways.

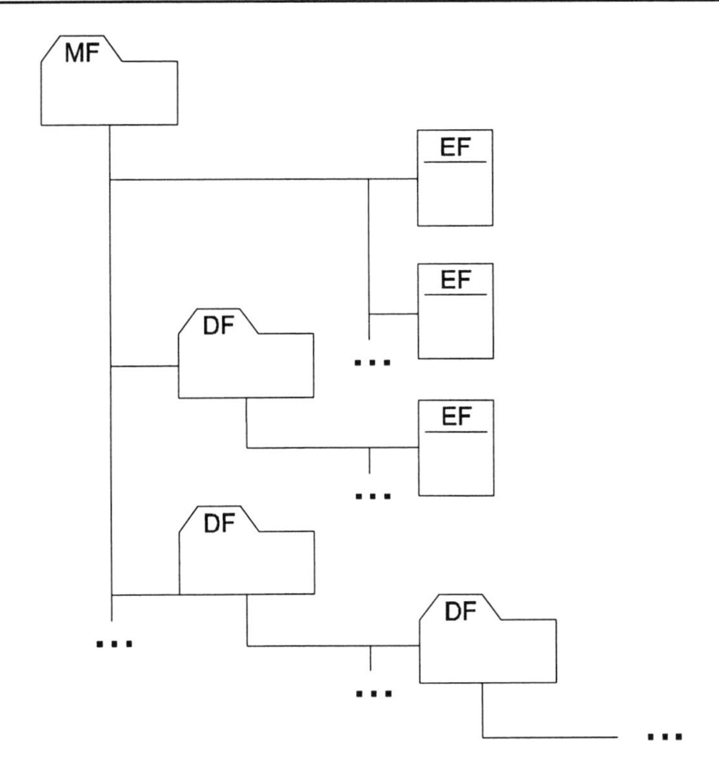

Figure 5.10 Various Smart Card file types

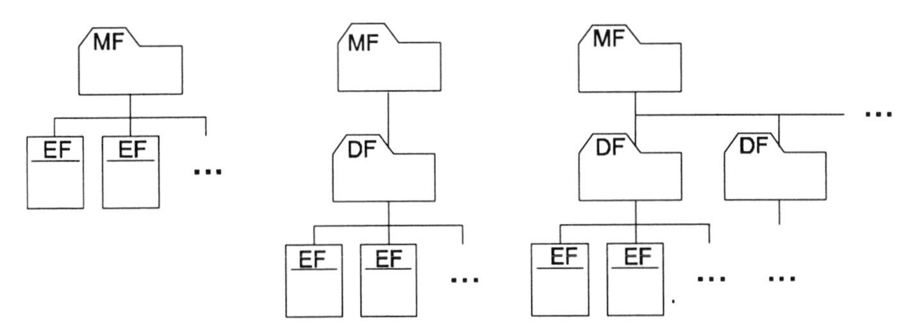

Figure 5.11 Differences in data organization between single and multi-application Smart Cards (left, middle: single application card;right: multi-application card)

It is conventional to collect all files containing user data, i.e. EFs, which belong to a given application, in one DF. This provides a clear and organized structure, and a new application can be inserted into the card in simple fashion by creating the appropriate DF. This approach offers another security advantage: during each access to memory, the system can check through the application whether the boundaries of that application's DF have been exceeded, and if so, prohibit access.

In a typical Smart Card running a single application, all EFs are placed either directly under the MF or within a single DF. Multi-application cards possess the corresponding number of DFs, in which the relevant EFs are located.

5.6.2 File hierarchies

The root directory which is implicitly selected after a Smart Card reset is always the MF. All other files are part of the MF, that is, they are contained within it.

The next level below the MF may be made up of EFs or DFs. All EFs which form part of one particular application are located in one specific DF. Further DFs may be present at a lower level below such an application DF. For example, a DF dedicated to the application "Traffic management system" may be present immediately below the MF. Further nesting within the application's DF may contain the possible languages, such as "English" or "German", in their own DFs.

5.6.3 File names

Without exception, the files in modern Smart Card operating systems are addressed logically and not called up via direct physical addresses as was quite usual in the world of Smart Cards in earlier times, and is still to be found occasionally. In the case of simple applications which specify precise addresses, this type of access can be a very space-saving one. Since access to all files is carried out by a computer in the terminal, this does not involve a loss of user-friendliness. However, direct addressing in no way conforms to modern software design criteria, and also causes severe problems with software extensions and Smart Card microprocessors with different address spaces. The concept of logical file addressing is substantially more efficient, and above all allows much easier extension.

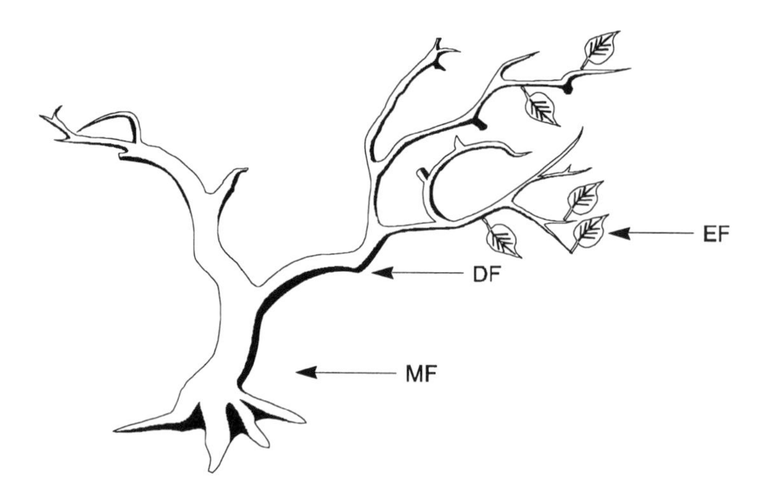

Figure 5.12 Smart Card file hierarchies: MF = trunk, DF = branches, EF = leaves

FID

The system described here is based on ISO/IEC 7816-4, and in principle is also a component of all other international Smart Card standards. All files, including the directory, possess a 2-byte long file identifier (FID), which is used to select them.

For historical reasons, the MF has the FID '3F00', which is exclusively reserved for the MF throughout the logical address space. The logical file name 'FFFF' is reserved for future applications, and may not be used. The lowest 5 bits in the FID (= 1 to 31) are also designated short-FID, and used for implicit selection of files within access instructions.

MSB								LSB								Meaning
b8	b7	b6	b5	b4	b3	b2	b1	b8	b7	b6	b5	b4	b3	b2	b1	
X	X	X	X	X	X	X	X	X	X	X	X	X	X	X	X	FID
0	0	0	0	0	0	0	0	0	0	0	X	X	X	X	X	Short FID

Figure 5.13 Structure of file identifiers (FID)

The FIDs in the file tree must be so chosen that the files can be uniquely selected. That is, two EFs within the same DF may not have the same FID; nor may a DF have the same FID as an EF located directly below it, since otherwise the operating system would have to decide whether the directory or the file is to be selected first.

The following three rules are suitable for ensuring that FIDs are unique:

- Rule 1: Files within a directory may not have the same FID.
- Rule 2: Nested directories may not have the same FID.
- Rule 3: Files within a directory may not have the same FID as the subordinate or superordinate directory.

AID structure and coding

The DFs serve to organize files for individual applications. They act in the manner of directories or folders, and can contain further directories or EFs.

In future, the available address space with the 2-byte long FID may become too small for this purpose. Hence, in addition to their FID, DFs have a so-called application identifier (AID). The AID can be 5 to 16 bytes long, and is constructed of two data elements defined by ISO.

The first element is the registered identifier (RID), with a fixed length of 5 bytes. It is allocated by a national or international registration office, and contains a country code, an application category and a number identifying the application provider. This numerical sequence results in a uniquely allocated RID, which can be used worldwide to identify a particular application.

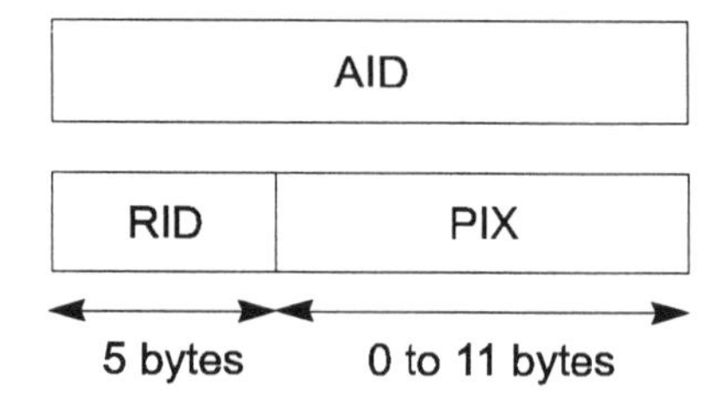

Figure 5.14 Structure of application identifiers (AID)

RID			Meaning
D1	D2 to D4	D5 to D10	
X	...	...	Registration category: 'A'– international registration 'D' – national registration
...	X	...	Country code, in accordance with ISO 3166
...	...	X	Application provider number allocated by national or international registration office

Figure 5.15 Code of 5 byte (= 10 digit) long registered identifier (RID)

If necessary, the application provider can add a proprietary application identifier extension (PIX) after the RID, and this is the optional second part of the AID. This can be up to 11 bytes long, and may be a serial or version number, and thus used for administrative purposes.

FID coding

The GSM application is a typical illustration of the fact that various parts of the FID cannot be freely selected. In the GSM 11.11 specification, the most significant byte is determined by the file's position in the file tree. The codes themselves have been chosen historically, and are based on the earliest French Smart Cards. In GSM, the first (i.e. most significant) byte in DFs is '7F'. EFs directly below the MF have the value '2F' as the first byte in their FID, and EFs below a DF have the value '6F' in this position. The least significant byte is numbered consecutively.

This feature is specific to GSM. It derives from the original Smart Card operating systems. In modern file managers, the full FID address space of 2 bytes can be exploited, and is not subject to any restrictions.

5.6.4 Addressing

Object-oriented file management systems require the file to be selected before being accessed. This informs the operating system which file is to be addressed from now on. The successful selection of a new file results in the previous selection becoming void. This means that only one file can be selected at any one time. Due to the freely selectable FID, certain restrictions must be placed on file addressability, otherwise it could easily happen that several files with the same FID are available for selection in the file tree, and the operating system would have to decide which file is meant. Hence file selection is restricted, which avoids this ambiguity and leads to independence from the search algorithm used by the card's file manager.

The situation would be different if all FIDs used in the file tree were unique. Then it would be easy to select the required file across several directory boundaries. But it is precisely this which cannot always be guaranteed.

MF and DF selection

The MF can be selected from anywhere within the file tree, using its unique '3F00' FID. This selection reproduces the state which applies after a card reset, when the MF is selected implicitly by the operating system.

The DFs can be selected either via their FID, or through the registered and hence unique AID.

Explicit and implicit EF selection

Two methods are available in principle for the selection of EFs. During explicit selection and before actual access, a special instruction is sent to the card with the 2-byte FID as a parameter for identifying the required file. This allows access to the file.

If, during file access, one of the instruction's parameters simultaneously selects the file by using the short identifier, then this is referred to as an implicit selection. However, implicit selection suffers from several limitations. It only works with EFs within the currently selected DF or MF. It is impossible to select a file across directory boundaries. Furthermore, it is only possible with certain access instructions which accept the short identifier as a parameter. Finally, the file FID must be so constructed that the file can be unambiguously selected with the short identifier. This means that the FID may only use the range '0001' to '001F'.

The major advantage of implicit selection lies in the fact that one instruction can select the file and simultaneously access it. This leads to a simplification in instruction execution, and due to the reduced communication requirements, to an increase in data transmission speed.

Selection through path specification

In addition to direct file selection, two further methods exist for explicit selection through path specification. In the first variant, the path from the currently selected file to the target file must be sent to the operating system. The second method provides for a path from the MF to the target file.

Selection options

As a result of the absence of unique file names within the file structure, selection is always possible only within certain limits, otherwise the non-ambiguity of the file being selected can no longer be maintained. The MF can be selected from anywhere within the file tree, since its FID is unique. Selection of first-level DFs (that is, the level immediately below the MF) is only possible from a DF of this level or from the MF. Figure 5.16 illustrates possible and prohibited selections.

5.6.5 File structures

In contrast to DOS files, Smart Card EFs have an internal structure. This structure can be chosen individually for each EF, depending on its purpose. This is of great benefit for the external user, since these internal structures make it possible so to construct data that they can be accessed very quickly and efficiently.

The management of these data structures requires considerable overheads in terms of program code on the card. This is why the data structures are not all symmetrical with each other, but only exist in variants which are frequently needed in practice.

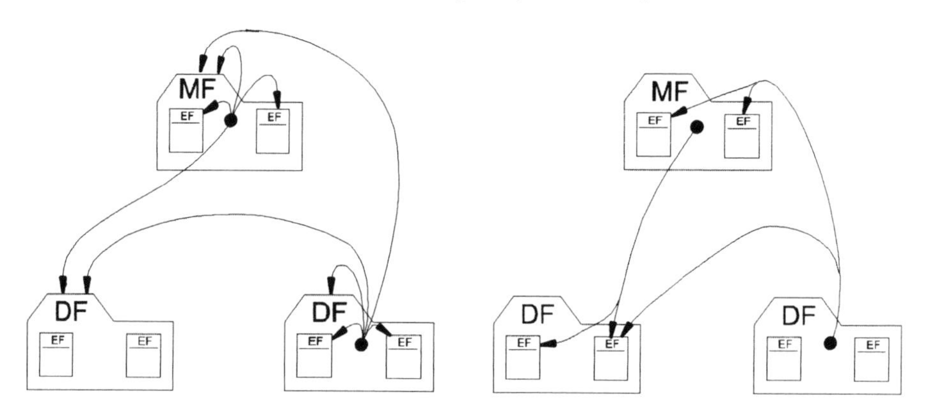

Figure 5.16 Examples of authorized (left) and prohibited selections (right) (selection through path specification is not shown)

Transparent

The "transparent" file structure is often also known as binary or amorphous structure. In other words, this means that a transparent file does not possess internal structure. The data contained in the file can be read- or write-accessed by byte or block with an offset. The instructions used are READ BINARY, WRITE BINARY and UPDATE BINARY.

The smallest size of an EF with a transparent structure is 1 byte. The maximum size is not specified explicitly in any standard. However, a size of up to 65792 bytes is possible through the maximum number of bytes to be read or written (256) and the largest offset (65536). Naturally, in currently-available cards this figure is illusory due to the size of the available memory. In practice, transparent files are rarely larger than a few hundred bytes.

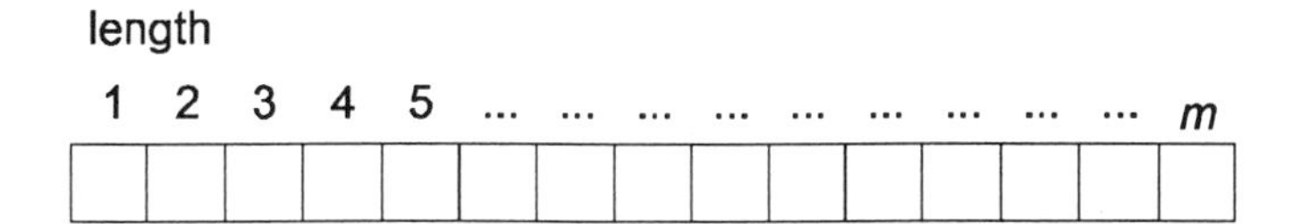

Figure 5.17 Transparent file structures

For instance, if 5 bytes are to be read from a 10-byte file with a 3-byte offset, access takes place as in Figure 5.18.

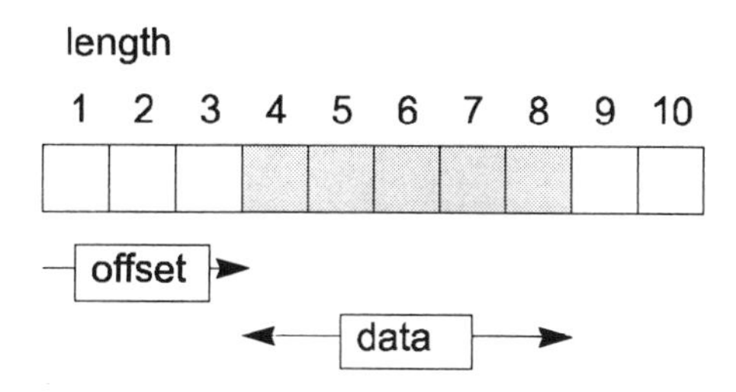

Figure 5.18 Reading 5 bytes with a 3-byte offset from transparent file

The application of this structure is mainly with unstructured or very short data. A typical use is a file containing a digitized passport photograph that can be read from the card.

If necessary, this linear and one-dimensional data structure can be used to simulate all other structures. However, designing the terminal's access strategy then becomes somewhat more complex, since the file also needs to store the structure parameters.

Linear fixed

The linear fixed data structure is based on linking records of equal length, where a record is a sequence of individual bytes. Individual records in the structure can be accessed at will. The smallest accessible unit is a single record, i.e. it is not possible to access parts of a record. The instructions used to read and write to this structure are READ RECORD, WRITE RECORD and UPDATE RECORD.

The first record is always identified as number 1. The highest record is numbered 'FE', i.e. 254, since 'FF' is reserved for later extensions. The size of individual records can be in the range 1 to 254 bytes, depending on the accessing instructions used, but all records must be of equal length.

A typical application for this file structure would be a telephone directory, in which a name is always followed, from a specific point onward, by the corresponding telephone number.

Linear variable

In the linear fixed file structure, all records are of equal length. Often this leads to wasted memory space, since many record-oriented data are of variable length – for example, the names in a telephone directory. The linear variable structure is designed to solve this need to minimize memory use. Here, any individual record can take on an arbitrary but specified length. The result is that each record has an additional data field specifying its size. Other than that, the format is similar to the linear fixed structure.

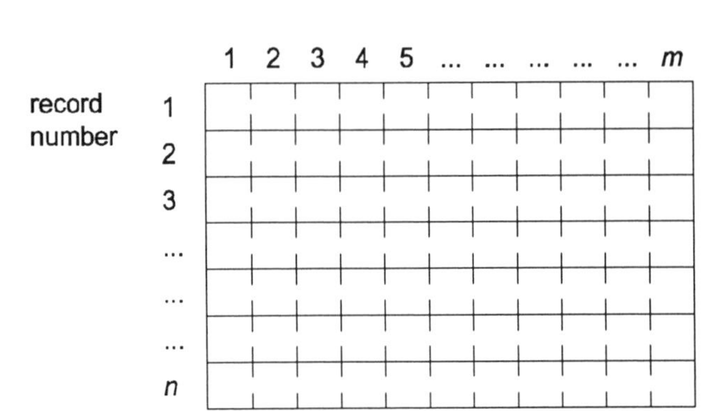

Figure 5.19 Linear fixed file structure

The first set of data is number 1, and a maximum of 254 sets can be held in a file. The length of individual records is determined by the accessing instructions, but can be anywhere in the range 1 to 254 bytes. The instructions are as for the linear fixed structure, i.e. READ RECORD, WRITE RECORD and UPDATE RECORD.

This approach is used when it is necessary to store data sets of very variable lengths, and at the same time save space on the card. For example, the telephone directory described above can be optimized by having records identical in length to the actual entries, rather than all being of the same length.

Cyclic

This is based on the linear fixed data structure. It therefore consists of a given number of records of equal length. In addition, the EF contains a pointer which always points to the last data set written to, which in principle is identified as number 1. Once the pointer reaches the last data set in the EF, then at the next write access it is automatically reset by the operating system to the first data set. In other words, it behaves like an analogue clock with an hour hand.

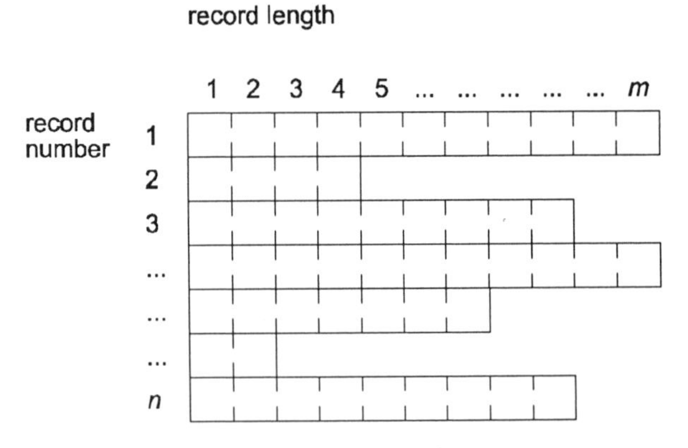

Figure 5.20 Linear variable file structure

Let the cyclic file contain n records. The last one written to is number 1. The following record is number 2, and the previous record is number n. This data structure can be accessed by addressing the first, last, previous or next record. This is true of all three record-oriented structures[2].

The number and length of data sets is completely analogous to those in the linear fixed structure; that is, because of the restrictions on read and write instructions, up to 254 data sets with a maximum length of 254 bytes each may be created.

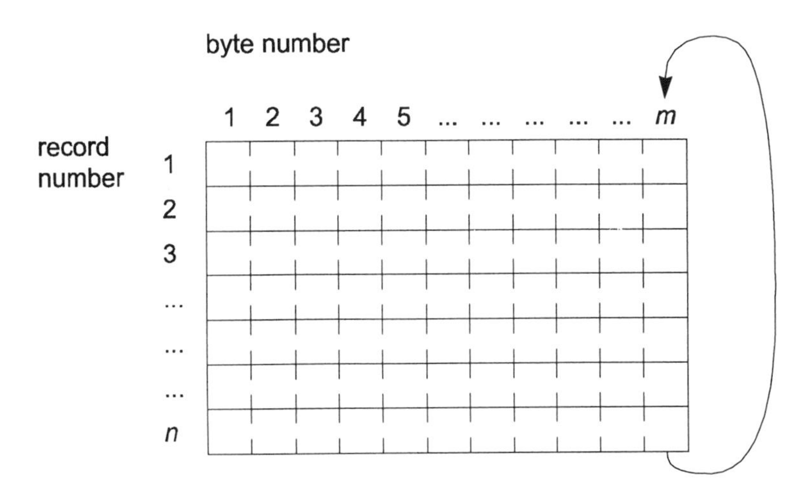

Figure 5.21 Cyclic file structure

The typical application of this structure is protocol files within the Smart Card. The oldest entry is always overwritten by the most recent one. This is ideally suited to the recording of protocols.

Execute

Strictly speaking, the following file structure is not a separate one, since it is based on the transparent data structure. It is described in European standard EN 726-3, and offers great extension possibilities within the operating system. The execute structure is not designed for data storage, but for holding executable program code[3]. Execute files can be accessed with the instructions appropriate to transparent file structures. It may be noted that this creates a trapdoor, since anyone who can write to this file is thereby able to load his own program code into the card, including Trojan horses.

---

[2] See also 7.2 Read and Write instructions.

[3] See also 5.10 Code programmed in circuit.

5.6.6 Access

As part of their object-oriented structure, all files contain data which regulate access within the framework of file management. Physically, the relevant coding is always contained in the file header. The entire Smart Card file management security relies on file access authorization management, since this controls access to the contents of the file.

Access authorization is determined when a file is created, and as a rule may not be modified thereafter. The range of possible file access conditions varies enormously between operating systems, depending on the instructions available in them. For instance, there is no point in defining access conditions for READ RECORD if this instruction does not exist in a particular operating system.

In contrast to EFs, no access data (read or write authorization) are stored for the MF and DFs, but rather – *inter alia* – the conditions for creating new files. That is, various access conditions are stored depending on the file type: access to file contents in the case of EFs, and the conditions which obtain within those organizational structures in the case of MFs and DFs.

One needs to distinguish here between state- and instruction-oriented operating systems. In the former, the necessary state for access is laid down. However, this may not necessarily be a particular state, but also a relative one; e.g., a read access may be permitted as from state 5, and prohibited for all lower states.

In contrast, an instruction-oriented operating system defines the instructions to be executed before access is authorized. Above all this covers authentication and identification instructions. For example, writing to a file may only take place after a PIN is successfully tested with the VERIFY instruction.

All possible types of access to an EF must be precisely regulated by access rights. The number of different instructions varies with the system. Hence we can only summarize the most common ones:

- APPEND — Enlarge a file
- DELETE FILE — File deletion
- INCREASE / DECREASE — Calculations within a file
- INVALIDATE — Block a file
- LOCK — Final locking of a file
- READ / SEEK — Read / search in a file
- REHABILITATE — Unblock a file
- WRITE / UPDATE — Write to a file

DF access conditions differ fundamentally from those relating to an EF. They specify the conditions which must be met to allow certain instructions to be executed within the relevant directory. The following summarizes the three most important access conditions:

- CREATE — Generate a new file
- DELETE FILE — File deletion
- REGISTER — Register a new file

5.6.7 Attribute

Within the object-oriented definition, all EFs possess special attributes in order to define additional file properties. However, this depends on the operating system and also on the card's field of application. The attributes define EF properties which mostly relate to the EEPROM storage medium. This is due to the potential insecurity of data and possible write errors in EEPROM operations. Attributes are defined during the creation of a file, and usually cannot be modified thereafter.

WORM attribute

One attribute based on EEPROM properties is known as WORM (write once, read many). Where a file has this attribute, data may only be written to the file once, but can be read an unlimited number of times. This feature may be supported by the EEPROM's hardware, or be implemented as a software function. The WORM attribute finds its application, for instance, in the one-off and irreversible writing of a serial number in a file. Other uses of this file property can be found in the area of personalizing a card, where fixed data (e.g. name, expiry date) are written to the card irrevocably.

The purpose of this attribute is the protection of sensitive data from overwriting. Naturally, this protection is only optimized when WORM access exists at the hardware level: i.e. when the EEPROM is hardware-protected thus only permitting a single writing episode. However, even software implementation offers much better protection than other comparable mechanisms.

Multiple write attribute

One attribute which is mainly defined and used in the GSM domain is the one for "high update activity". It only exists because of the restricted number of write/delete cycles on an EEPROM. Files with this attribute can be written to very frequently, without write errors affecting file contents. This is achieved by multiple writing of the data, and a majority decision when reading them. The usual approach is triple parallel data storage during writing, and a 2-out-of-3 majority decision during reading. An alternative mechanism involves a switching of the multiple data fields in a way that is externally transparent.

Attribute for EDC use

This attribute is used for particularly sensitive data, and relies on special protection of user data through an EDC (error detection code). This allows at least the detection of switched bits in EEPROM. EDC protection, combined with multiple writing of data, also allows error correction of switched data bits. This ECC (error correction code) property is used in the main in the area of electronic purses. Here the switching of bits means a real financial loss, since the actual balance is stored in the card. EDC or ECC attributed files are used to minimize the serious repercussions of switched bits.

Error recovery attribute

A mechanism which is often integrated into modern Smart Card operating systems, this ensures that a file is either completely written to or not at all[4]. Since this mechanism more than doubles write access time, it should not in principle be used for all files. A special attribute selectively determines the write mechanism for each file.

5.7 FILE MANAGEMENT

All Smart Card files are stored in EEPROM. This is the only storage medium on a card which retains data without an external power supply, and also allows them to be altered if needed and when a power supply is available. It is the only way to store data between sessions, since the RAM loses its contents after the card is deactivated, and ROM content cannot be modified after manufacture.

In early cards, files were accessed directly using physical addresses. Actually, these were not files in the strictest sense, rather the whole memory was linearly addressable from outside and could be accessed with read and write instructions. However, for security and application-related reasons this is no longer provided in modern operating systems. The currently standard method is to use object-oriented file managers, with data about access conditions directly related to the file. The administration and organization of these files is the task of the "file manager" part of the operating system.

Thus, an object-oriented structure requires each file to have a file descriptor, which contains all relevant data about the file itself. In Smart Card technology, the file descriptor is also designated a file "header". The contents, i.e. the user data, are held in the so-called file "body".

The data contained in the file descriptor strongly depend on the possibilities offered by the file manager. However, the following must be present in the file descriptor in all cases:

- File name (e.g. FID = '0001')
- File type (e.g. EF)
- File structure (e.g. linear fixed)
- File size (e.g. 3 records @ 5 bytes)
- Access conditions (e.g. READ = after PIN input)
- Attribute (e.g. WORM)
- Connection to file tree (e.g. directly below the MF)

When the file is an EF or the MF, the file name is the 2-byte FID (file identifier). If the file is a DF, the file descriptor also contains an AID (application identifier). The file type, that is EF, DF or MF, must also be indicated.

---

[4] See also 5.9 Atomic routines.

The header contains a data element which depends on the file type, and describes the internal structure (transparent, linear fixed, linear variable, cyclic, executable). All data concerning the length of the transparent data part or the number and size of the records, are in their turn affected by the structure.

Once all fundamental file properties have thus been described, the operating system still needs detailed information about access conditions, i.e. which instructions may access how in which state. These must be separately specified for each possible instruction. If the file manager supports them, special attributes such as high update activity, WORM or EDC protection can also be specified.

All the above relate to the file itself as an object. In order to determine its position in the file tree, a few pointers are also needed to specify its exact position within the MF or DF.

DF separation

A general problem facing the file manager is the EEPROM's limited number of write/delete cycles, and its division into separate sections. This considerably affects the entire file manager's structure and also the internal file structures. That is why data concerning file administration must be strictly separate in memory from the actual data contents. Were this not the case, undesirable interaction might occur between administrative data in the header and the user data in the body, possibly destroying the entire security structure within the card's operating system. This can be illustrated with the following example.

Assume that a file's access conditions and secret and unreadable keys are held in the same section of memory as the publicly accessible (read and write) user data of another file. If the writing routine into this file is interrupted, e.g. by pulling the card out of the card reader, this would affect the access conditions stored in the same section of memory. In the worst case, no access conditions would be present after this process, and the file and its secret keys would be readable by anyone. Hence it is of fundamental importance to divide the internal file structures which deal with administrative and user data between separate memory sections.

Free memory management

The small memory also forces on operating systems considerable restrictions in EEPROM free memory management. It is only relatively recently that systems have existed which can create or delete files after personalizing the card. True free memory management with file relocation, allocation and releasing of memory areas cannot be implemented, due to the limited memory available for program storage.

Data integrity

Another important point concerns data integrity. The file manager should be able, at any point in time, to test whether the data in memory have been changed inadvertently. Not all data need to be protected by checksums: one would want to minimize the administrative overheads for these functions. But data redundancy and any supervisory functions must increase hand in hand with the importance of the data involved.

5.8 EXECUTION CONTROL

If a state automaton needs to be implemented in an operating system, this can be done in a variety of ways. However, several fundamental principles must always be satisfied, regardless of the system chosen and its manufacturer.

The state automaton, whether a micro- or macro-automaton, must be located after the instruction interpreter and before the instruction's actual execution, within the layer model of the operating system described above. Its task is to determine, by referring to a table, whether the received instruction may or may not be executed in the current state. The basic principle, as elsewhere in the Smart Card industry, is to use the minimum of memory for the processing of state data. In addition, the information must be so structured that the automaton itself can be made to use as little memory as possible.

The state automaton needs a certain amount of information to analyse the instruction held in the I/O buffer. Figure 5.22 shows a possible Smart Card state table.

The first data element (initial state) contains the state for which the next part of the data structure should be processed. It may contain a number which directly defines the state in which all further data must be considered. A sub-table then lists the instructions permitted in the initial state. Here it should be possible to allow single instructions, groups of instructions, none or all of them, as appropriate.

In the table, instruction definitions are followed by the corresponding permitted parameters. These data elements should be able to define individual values as well as ranges. For example, the instruction field may hold the code for READ BINARY, and parameter fields 1 and 2 the minimum and maximum offset for a read access to the transparent file. Parameter 3 would then store the limits for the minimum as well as maximum length. Since several entries may be present in this sub-table for one state, it is possible that further instructions are defined after READ BINARY together with all their parameters.

A table entry ends with the new state reached when successful execution of the instruction was possible. With this type of data structure, some other state could be defined for the event that execution fails. In order to maximize flexibility within the state automaton, both absolute and relative outcome states should be permitted, for both success and failure. "Relative" in this context means a new state reached by adding or subtracting a value to or from the initial state. With absolute specification, the new value is set directly without reference to the initial one.

...	...	...	...
Initial state	Instruction definitions	New state (successful execution)	New state(failed execution)
...	...	...	...

Instruction definitions			
Instruction	Parameter P1 (min, max)	Parameter P2 (min, max)	Parameter P3 (min, max)
...	...	...	...
...	...	...	...

Figure 5.22 Possible Smart Card state table

In principle, the options available to a state automaton are unlimited. The data structure presented here is quite suitable for implementation in a somewhat advanced Smart Card system.

With the data structure described, and an appropriate state automaton, any possible state graph can in principle be represented in a Smart Card. Of course, the individual files are additionally protected through the instructions' access conditions against unauthorized reading and writing. However, and in order to complement the object-oriented file protection, the instructions' execution control offers another, and higher, system security mechanism. This is where Smart Card state automata really score.

5.9 ATOMIC ROUTINES

It is often the case that the card's microprocessor is required to run certain parts of its software either completely or not at all. Routines which are indivisible and fulfil this requirement are therefore called atomic. They always occur in connection with EEPROM write routines.

Atomic routines are based on the idea that when writing to EEPROM, one wants to ensure that the relevant data are never written only in part. For example, this would be the case where the user removes the card from the reader at the wrong moment, or a sudden power failure occurs. Since the card does not possess any electrical power buffers, its software immediately grinds to a halt.

It is in the case of Smart Card electronic purses, above all, that file entries must be complete and correct under all circumstances. For example, it would be fatal if removal of the card from the reader caused the balance to be incompletely updated to its new value. The corresponding entries in the protocol files must also be complete at all times.

Since Smart Card hardware does not support atomic routines, they must be implemented in software. The methods utilized here are anything but new in principle, as they have been in use for databases and hard disk-drives for a long time. One variant used in Smart Card operating systems is described below in general terms. This error recovery routine is transparent as far as the outside world is concerned, and thus does not result in modifications to already existing applications.

To demonstrate the method, assume that data are sent to the card across the interface, and that the destination is a file. This, for example, would be the typical procedure in an UPDATE BINARY instruction. Using Figure 5.23 and the following explanation, the concrete steps taking place in the card can be followed.

A buffer is installed in the system's EEPROM section, which is sufficiently large to hold all necessary data. There is a state flag, also in EEPROM. It may be set either to "Buffer data valid" or to "Buffer data invalid". In addition to the buffer, there needs to be a corresponding storage space for the destination address and the actual length of the buffer data.

The concrete sequence of events is as follows: in the first step, the data from the destination address onward, e.g. in a file, are copied into the buffer with their physical address and length. Now the buffer contents flag is set to "Buffer data valid". In the next step, the operating system copies the new data into the required address, and resets the flag to "Buffer data invalid". When the operating system is booted before ATR, the flag is interrogated. If it is set to "Buffer data valid", an automatic process takes place in which the buffer data are written into the (also stored) address.

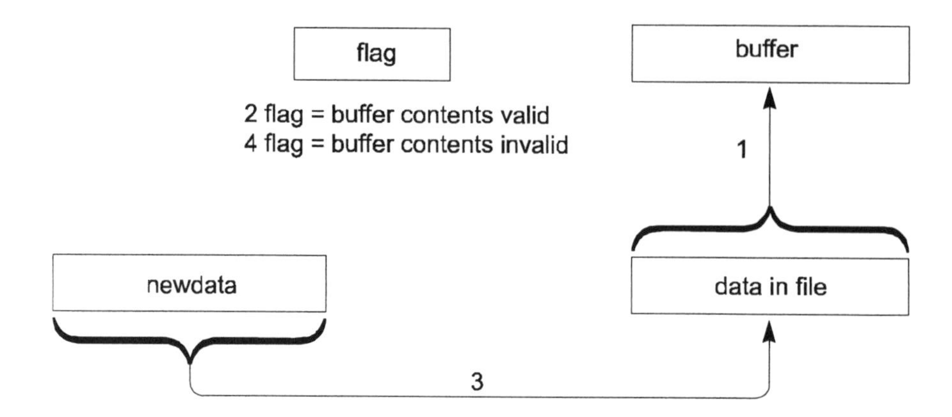

Figure 5.23 Example of atomic routines in Smart Card operating systems

This mechanism ensures, in any case, that the file contains valid data. If the routine is interrupted anywhere during program execution, the data held in EEPROM can be restored to the card. If, for instance, the user removes the card during step 3, when the new data are written to EEPROM, these new data are only partly held in the file. When the card is switched on at the next session, the operating system notes that the buffer contains valid data, and copies them to the appropriate place. This reproduces the original state, and all entries in the EEPROM files are consistent. A very suitable time for these corrections is the initial waiting time between individual bytes in ATR[5].

However, the procedure described above suffers from two serious disadvantages. Of all EEPROM areas, the buffer is the one most affected by writing and erasing. Since the number of EEPROM write/delete cycles is limited, it is very likely that the important buffer will be the first EEPROM section to exhibit a write error. This would mean that the card can no longer be used, since its data consistency cannot be guaranteed. This problem can be partially overcome by making the buffer cyclic, so that writing need not always take place at the same location. Unfortunately, the buffer would then require far too much EEPROM.

Another shortcoming of this implementation of atomic routines is the extension of program execution time, due to the obligatory write access to the buffer. In the worst case, this mechanism makes access three times longer than with direct writing to the EEPROM memory. Therefore it is usual not to buffer all EEPROM access, but only the writing to certain files or data elements. This can be specified through an attribute in the relevant file's header.

5.10 CODE PROGRAMMED IN CIRCUIT

In contrast to all other computer operating systems, it is not normal procedure in the Smart Card industry to store programs as files in the file tree, and execute them there as necessary. Nevertheless, in addition to data storage this is one of the main functions of all

---

[5] See also 6.3 Answer to reset.

operating systems. Naturally there are good reasons why in particular this function is largely absent in the case of Smart Cards.

Technically and functionally, executable EFs do not represent any problem. More recent operating systems, therefore, also offer the option of managing files containing executable code, and even loading them into the card at some point in time after it is personalized. Thus it is possible, for example, for an application provider to execute program code in the card that the operating system's manufacturer does not know about. The former can store on the card an encryption algorithm only known to him, and execute it there. The end result is that knowledge about the system's security functions is distributed between various parties, which is a fundamental criterion of security systems.

An important reason for using the mechanism of code programmed in circuit is the possibility it offers for removing programming errors (i.e., bug fixing) in completely personalized cards. Errors discovered in the operating system can thus be corrected or at least improved in cards already issued.

This function can be implemented in two different ways. In the first variant, program code is stored in an executable EF. Having been selected, the EF is sent the instruction EXECUTE. Depending on the application, this may first require an authentication stage. Run-time parameters are contained in the EXECUTE instruction sent to the card. The response generated by the program in the EF is returned to the terminal as part of the response to the instruction.

The other variant is based on a somewhat different principle. It uses a strictly object-oriented approach. This is also described in EN 726-3, *inter alia*, as "application specific instructions" (ASC). According to this standard, a DF contains the complete application together with all its files and application-specific instructions. The DF contains an area internally managed by the operating system, into which code can be programmed in circuit. This is achieved with a special instruction, which sends all the necessary data to the card. When the relevant DF is now selected and an instruction sent to the card, the operating system checks whether it belongs to the code, and if so calls up the program code held in the DF straight away. On the other hand, if another DF is selected then the instruction is simply treated as non-existent.

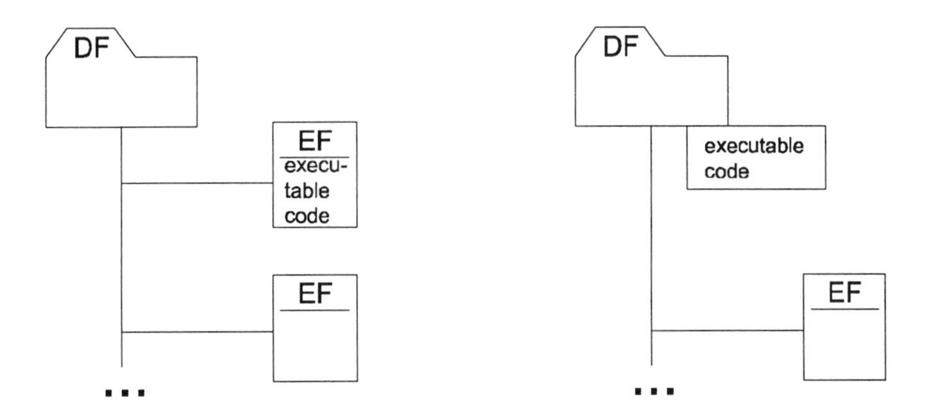

Figure 5.24 Two variants of executable program codes in Smart Card (left: executable Efs; right: application specific commands –ASCs)

Several fundamental conditions must be satisfied to allow the implementation of either variant to proceed. The code must either be so programmed that it is relocatable, or the card itself must relocate it during loading. The relocation requirement (i.e. moving the program code in memory) is made because only the card's operating system, and not the outside world, knows the precise destination address in memory. However, several other points must be taken into consideration in this connection.

Without exception, Smart Card microprocessors do not have memory-protection mechanisms or supervisory capabilities of any kind. As soon as the program counter is located within machine code which is foreign to the processor, the entire control over all memory sections and functions is held by this executable code. There is then no possibility of restricting the functions of this executable program. All memory managers or handlers can be circumvented, and every addressable memory location can be read and those in the EEPROM also written to. Of course, these memory contents can also be sent over the terminal interface.

This is the very weak point as far as executable programs and code programmed in circuit are concerned. If anyone were permitted to load programs, or if it were possible to do so by circumventing the protection mechanisms, no more security would exist for secret keys or data anywhere within the memory. This would be an ideal way to break into a Smart Card. To the outside world, the card would behave like a non-tampered one, and a special instruction would make it possible to read the whole memory or write to parts of it.

If loading is only permitted to a few application providers – which is easily manageable through reciprocal authentication before program code is loaded – the problem is still not eradicated. The application provider could access confidential data relating to other existing applications, without restriction and beyond the limits of the DF allocated to him. The system would be breached once again.

However, yet another sound argument can be made against executable files programmed in circuit. In order to be able to use the operating system, which is absolutely necessary, the manufacturer of the files programmed in circuit must know all entry addresses and call-up parameters. But operating system manufacturers consider it necessary, for security reasons, that as little as possible is known about internal routines or program code addresses. Furthermore, it is necessary to ensure that the imported code performs exactly how it is supposed to, and does not contain a Trojan horse, for example. This can only be checked by an independent organization.

One solution to this problem is represented by an interpreter, which checks during execution which memory areas are being addressed. This process must be very fast, since slowly executed program code offers no benefits. Equally, the interpreter's implementation should require as little memory as possible, since we know that the latter is strictly limited.

A different solution consists in examining the executable code by a special program during importation, to discover its calls to memory. But just like an interpreter, this requires relatively large programming overheads in the operating system. Nevertheless, this approach appears to be more realistic than an interpreter, since the program can then run at full speed during execution, and slow interpreting of the instructions contained in the program is no longer necessary.

The most elegant variation, and also the one which holds most promise for the future, is one which exploits an MMU (memory management unit) in addition to the card's processor. It checks the program code during run-time, using hardware circuitry, to discover if it remains within its allocated limits. This is the only way to allow each application provider to load program code into the card without previous checks by the card issuer and without loss of security. This user is allocated the physically contiguous memory area of a DF. The MMU checks the permitted memory boundaries during the call-up of a program loaded into the DF. If they are exceeded, program run can be stopped immediately via an interrupt, and the application blocked until further notice[6].

---

[6] See also 3.4.3 Supplementary hardware.

Data transmission to the Smart Card

The possibility of two-way communications is a precondition for all interactions between the Smart Card and the terminal. However, only a single channel is available. Digital data is exchanged between card and terminal across this electrical connection. Since only one channel exists, the card and terminal can only transmit in turn, and the other party must always be in reception mode. This intermittent sending and receiving is known as a 'half-duplex' procedure.

The full-duplex procedure, in which both parties can transmit and receive simultaneously, is not implemented in the Smart Card industry at this point in time. However, as most Smart Card processors have two I/O ports, and two of the eight contact fields are reserved for future applications (e.g. a second I/O connection), full-duplex could become technically feasible. This will certainly become the case for hardware operating systems in the medium term, since it is the only possible method of decrypting the card's data in real-time. Preliminary proposals for standardization already exist.

Communication with the card is always initiated by the terminal: the card only responds to the terminal's instructions. That is, the card never transmits data without an external request. This results in a client—server relationship, with the terminal as server and the card as client.

When a card is inserted in the terminal, its contacts are first connected to those of the terminal. Then the five contacts involved are electrically activated in the correct sequence[1]. Following this, the card automatically executes a power-on-reset, and sends an answer to reset (ATR) to the terminal. The latter evaluates the ATR, which pinpoints various card parameters, and then sends the first instruction. The card processes the instruction and generates a reply, which it transmits back to the terminal. This back-and-forth interplay of instruction and response continues until the card is deactivated.

---

[1] See also 3.3.6 Booting/shutdown sequence.

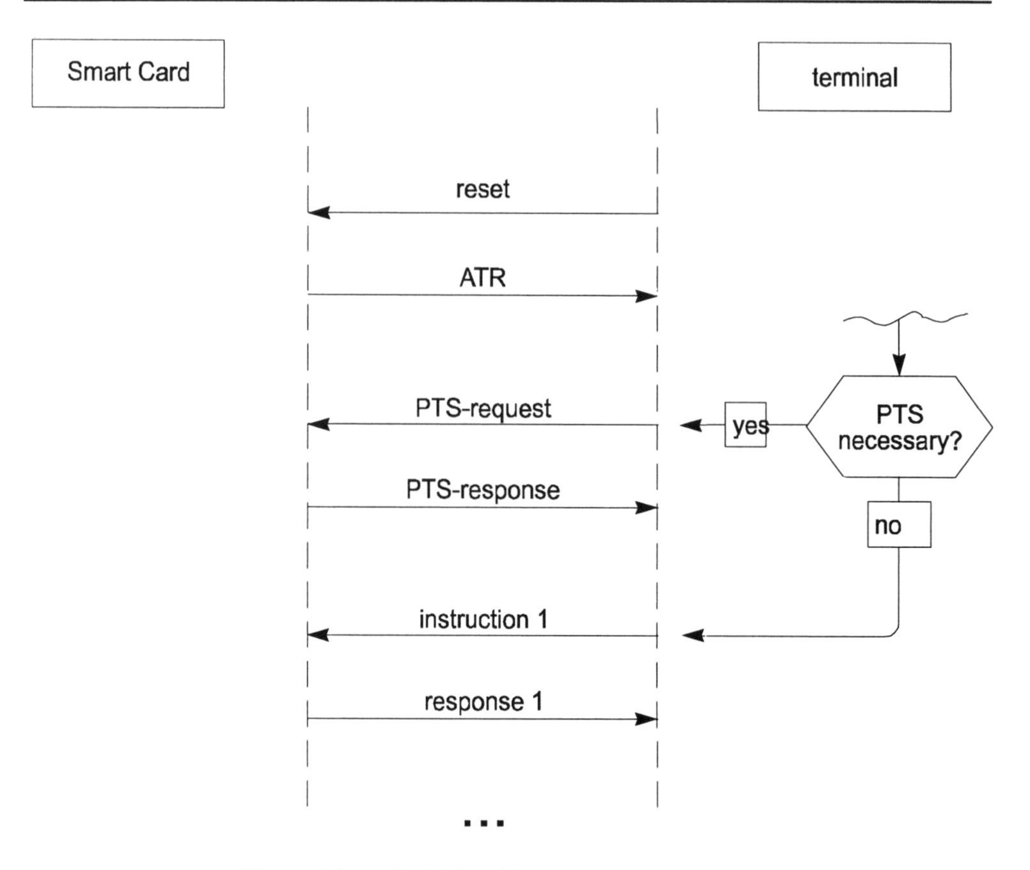

Figure 6.1 ATR, PTS and first instruction sent to card

Between the ATR and the first instruction sent to the card, the terminal can also transmit a protocol type select instruction (PTS). This instruction, which like the ATR is independent of the transmission protocol, can be used by the terminal to set various transmission parameters relating to the card's protocol.

The whole data transmission process to and from the Smart Card can be represented as part of the OSI layer model. Here one distinguishes between the electrical events on the I/O channel, the logical routines in the actual transmission protocol and the behaviour of the applications based on these. The behaviour and interactions between those layers are set down in several international standards. The relationships are illustrated in Figure 6.2.

The functions of the asynchronous transmission protocols described below, follow the relevant standard respectively. All parameters and settings possible within the protocol's framework are specified. In practice, however, a card might not support all the various protocols, since the available memory is insufficient.

Seen functionally, one may regard the various possibilities simply as a range of options, from which one chooses the optimal solution for a particular application or card. It is important that the parameters chosen are not too unusual, so that cards can communicate with all terminals, if possible.

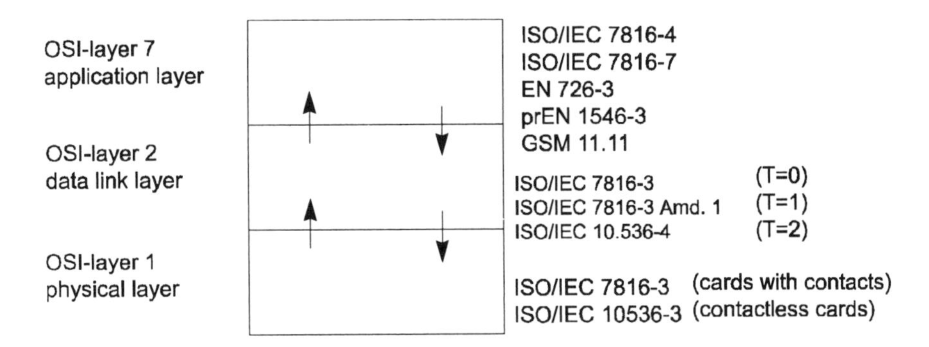

Figure 6.2 OSI communication model between terminal and Smart Card

The situation is somewhat different in the terminal. Here it is common for the functions described in the relevant standard to be fully implemented, since sufficient memory is available.

6.1 PHYSICAL TRANSMISSION LAYER

The physical transmission layer is specified in the international Smart Card standard ISO/IEC 7816-3, with all its universal parameters. This is the fundamental standard for all communications options at the physical level.

The entire data exchange with the Smart Card takes place digitally, using only the logical values ‘0’ and ‘1’. The voltage levels used are those conventional in digital technology, 0 and +5 volts. The new microprocessors for the 3-volt range do, of course, also support it in the area of data transmission. Which of the two levels, 0 volts or 3V/5V, represents logic ‘1’ can be decided arbitrarily, and is indicated by the card in the first ATR byte. In this context, direct convention refers in this context to logic ‘1’ at the +3/5-volt level; ‘inverse convention’ means that the +3/5-volt level is used to represent logic ‘0’. Either way, in the ground state, when no data is being transmitted, the I/O channel is always at the high level.

Communication between a Smart Card and the outside world takes place serially. This means that data processed byte-wise must be converted to a bit-serial datastream. A byte is separated into its eight individual bits, which are sent down the line one after the other. The bit order is also governed by a convention. In the direct convention, the first data bit after the start bit is the lowest in the byte. In the case of inverse convention, the highest bit in the byte is sent immediately after the start bit.

Data transmission between card and terminal proceeds asynchronously, which means that each transmitted byte must be provided with additional synchronization bits. The beginning of each serially transmitted byte is supplemented with a start bit, which signals to the recipient the start of the sending sequence. At the end of each byte, the sender also adds a parity bit for error detection, as well as one or two stop bits. The period relating to the stop bits is designated in protocol T=0 as guard time, which in principle is also a kind of stop bit. During this interval, both the recipient and the sender have time to prepare for the next transmitted byte. The parity of each byte must always be even. The parity bit, therefore, has the logical value ‘1’ when the number of 1’s in the byte is odd, and ‘0’ when even.

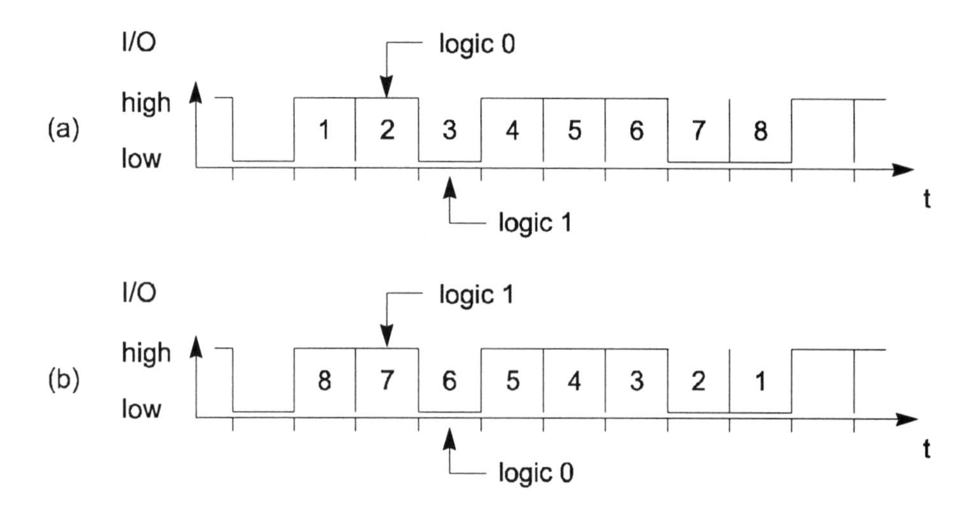

Figure 6.3 Data transfer conventions: a) data transfer using direct convention; b) data transfer using inverse convention

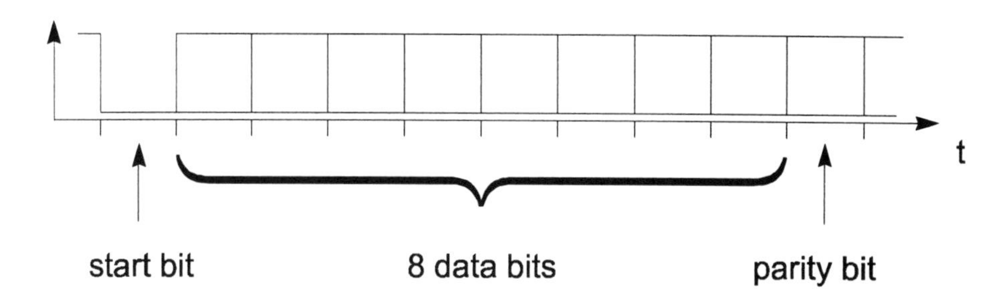

Figure 6.4 Character formation in data exchange

Since Smart Card microprocessors do not have a clock independent of the applied clock-pulse, the duration of a data bit cannot be given in absolute terms. It is specified as a function of the applied clock. A divider is defined, which gives the number of pulses per bit. The duration of one bit is designated as an etu (elementary time unit).

It is, therefore, nonsensical to specify a fixed data transmission rate for Smart Cards (e.g. 9600 bit/s), since this is directly proportional to the applied clock. However, there are essentially only two dividers in use worldwide: one has the value 372, and the other 512. In order to maximize transmission speeds, progressively smaller dividers have been coming into use for some time, but so far they are still the exception. The situation will change in the future, since the data transmission speed is one of the bottlenecks in instruction execution. With small dividers, it becomes increasingly difficult for the card's operating system to receive and to transmit data, since the processor has very little time left for this task. Thus, when receiving data with a 256 divider, the processor has only 256 pulses during which to recognize a bit and place it in the card's I/O buffer.

To calculate the transmission rates which can be achieved with standard dividers, we only need to divide the clock rate by the divider's value:

$$\frac{3.5712\ \text{MHz}}{372} = 9\,600\ \text{bit/s}$$

$$\frac{4.9152\ \text{MHz}}{512} = 9\,600\ \text{bit/s}$$

So for example, with both common clock frequencies, 3.5712 MHz and 4.9152 MHz, one obtains a data transmission speed of exactly 9600 bit/s. This 9600 bit/s is also the reason for the awkward divider values. In the early days of Smart Card technology there were only very few frequencies for which economical quartz oscillators were available. Therefore cheap standard oscillators originally produced for television were used, and defined the dividers employed in cards so as to achieve the transmission speed of 9600 bit/s, this already being a common value at the time.

If we take 10 MHz as the largest practical value for the clock, and 32 as the smallest divider, we obtain the current upper limit for transmission speed:

$$\frac{10\ \text{MHz}}{32} = 312\,500\ \text{bit/s}$$

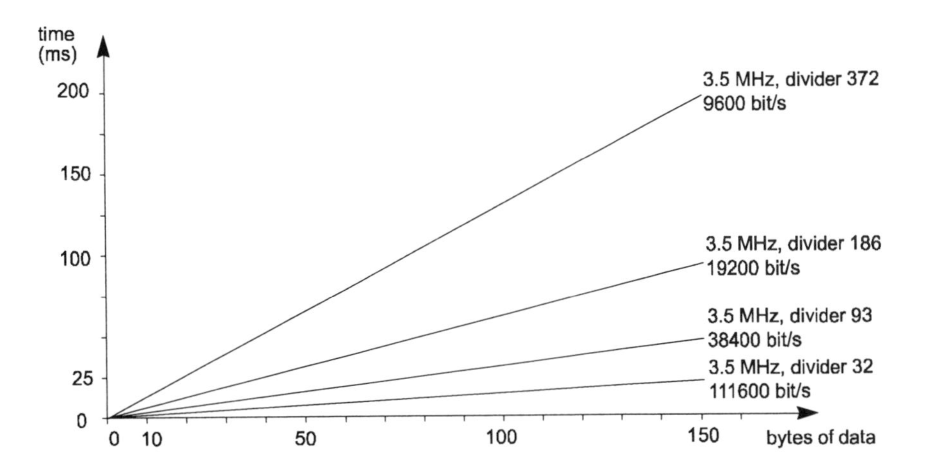

Figure 6.5 Diagram to calculate data transmission rate at typical transmission speeds. Assumption: 1 start bit, 8 data bits, 1 parity bit and 2 stop bits per byte

Of course it is possible to reduce the divider even further, to increase transmission speed. However, this requires considerable additional overheads in terms of program code in the card, which is normally not implemented due to the limited amount of memory available. In many new Smart Card microprocessors, data exchange is carried out across the serial interface through a hardware module (UART) built into the card. The cost of transmission software in this type of card is minimal, and the hardware solution allows transmission speed to increase sharply. The standardized transmission speed of 111.6 kbit/s is easily achieved by an interface module on the chip[2]. This high transmission speed is the upper limit for a pure software data transmission solution, even in the case of very high execution speeds of machine instructions in the microprocessor.

The time needed for one bit can be calculated from our figures for clock-frequency and dividers. With a 3.5712 MHz clock frequency and a 372 divider, we obtain a time of 104 µs, which by definition is one etu (elementary time unit) for this divider. This allows us to draw up the time diagram shown in Figure 6.6.

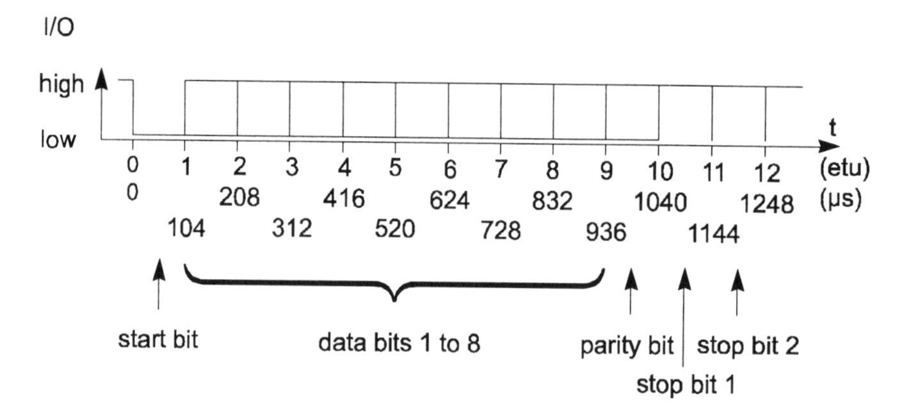

Figure 6.6 Time diagram of a character at 9600 bit/s (e.g.: clock frequency: 3.5712 MHz, divider: 372)

The tolerance in this serial data transmission need not be zero, and, for technical reasons, may be subject to certain deviations. Since many current Smart Card microprocessors do not have interface modules, in practice the permitted tolerance needs sometimes to be exploited within the software solutions. The timing error between the start bit's falling slope and the final slope of the nth bit may not exceed ±0.2 etu. As far as the sender is concerned, this means that an individual bit may well be permitted an error of ±0.2 etu, but a larger error is not permitted even across several bits. Bit timing errors may not, therefore, sum up to the point that the permitted tolerance is exceeded.

---

[2] See also 15.4.3 Calculation table for transmission speed.

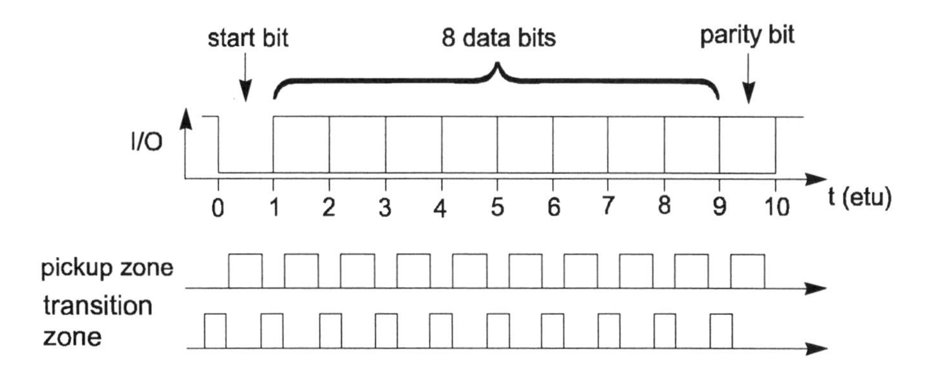

Figure 6.7 Single capture of incoming byte

Signal fading or spikes often occur, especially in the case of data transmission over a physical conductor. Therefore, often one relies not solely on simple capture of the incoming signal, but also on repeated testing. A typical procedure involves a triple capture with a subsequent 2-out-of-3 decision. This permits inexpensive compensation for small distortions in signal levels. Increasing the number of repetitions to five or seven, however, would make little sense in view of the generally fairly good data transmission quality in Smart Cards and of the required overheads.

The three signal captures should, as far as possible, be distributed across the incoming bit, in order to optimize the smoothing of transient signal fading. The optimal timing is based on the bit's centre point and on the permitted tolerance applying to byte transmission, i.e. the upper and lower limits determine when the three signal captures should take place. However, this is not covered by any standards[3].

6.2 TRANSMISSION PROTOCOLS

Once the Smart Card has sent out an ATR and eventually a PTS has taken place, it waits for the terminal's first instruction. The following routine always corresponds to the client—server principle, with the terminal as server and the card as client. In concrete terms, the terminal sends an instruction to the card, the latter executes it and subsequently returns a response. This back-and-forth interplay of instruction and response is never interrupted.

There are, nevertheless, various options for structuring Smart Card communications. Equally, if a transmission failure occurs, a number of different procedures exist for re-establishing synchronized communications. The precise implementation of these instructions, the corresponding responses and the procedure followed in the event of transmission errors are described in the so-called transmission protocols.

A total of 15 transmission protocols are provided, and their basic functions defined. They are designated "T=" plus a serial number, and are summarized in Figure 6.9.

---

[3] See also 15.4.4 Table for sensing points.

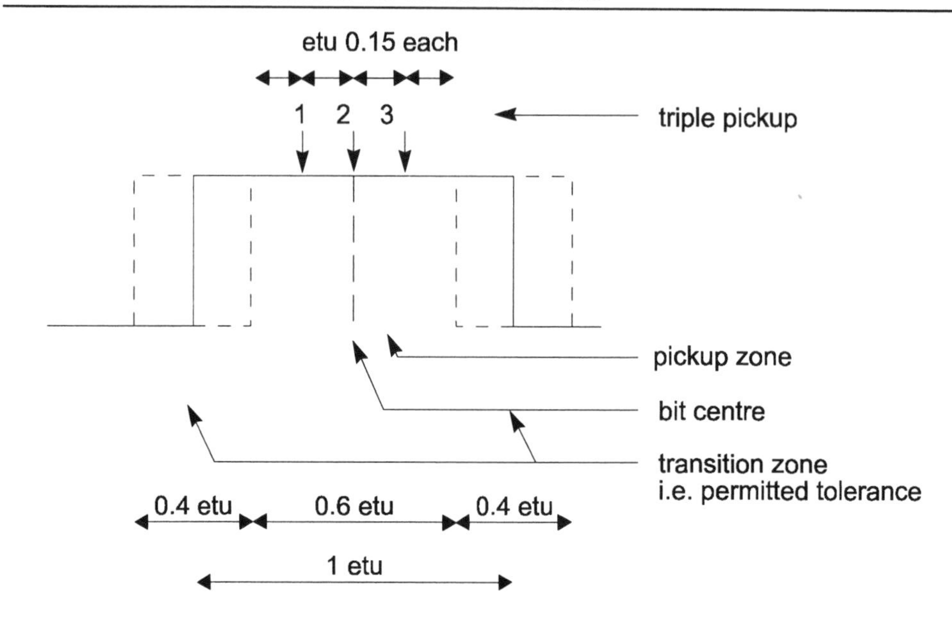

Figure 6.8 Example of three-signal capture across incoming bit

Transmission protocol	Meaning
T=0	Asynchronous, half-duplex, byte oriented, covered by ISO/IEC 7816-3
T=1	Asynchronous, half-duplex, block oriented, covered by ISO/IEC 7816-3 Adm. 1
T=2	Asynchronous, full-duplex, block oriented, covered by ISO/IEC 10536-4
T=3	Full-duplex, not yet covered
T=4	Asynchronous, half-duplex, byte oriented, expansion of T=0, not yet covered
T=5 to T=13	Reserved for future functions, not yet covered
T=14	For national functions, no ISO norm
T=15	Reserved for future functions, not yet covered

Figure 6.9 Overview of transmission protocols covered by ISO/IEC 7816-3

Two of these transmission protocols are currently in common use worldwide. The first is the T=0 protocol, internationally standardized in 1989 (ISO/IEC 7816-3), and the other is T=1, introduced in 1992 as a schedule to an international standard (ISO/IEC 7816-3 Amd. 1). The full-duplex transmission protocol T=2, which relies heavily on T=1, is currently in preparation and will become available in a few years' time as an international standard.

A third protocol, designated T=14, is used in Germany in the widely available cardphones, and is specified by DBP Telekom as a company standard.

The data units carried by transmission protocols are denoted by TPDU (transmission protocol data unit). They are, so to speak, protocol-dependent containers, which transport data to and from the card. The actual user data is embedded in them.

In addition to the technically complex Smart Card transmission protocols, a further series exists of very simple synchronous protocols for memory cards. They find typical application in telephone cards, health insurance cards and similar. However, they do not have error-correction mechanisms, and are based on hard-wired logic in the chip.

6.2.1 Synchronous data transmission

Synchronous data transmission finds no application in microprocessor-based Smart Cards, since they only communicate with the terminal asynchronously. However, it is standard practice in memory cards, which are produced in very large numbers in the form of debit cards (value cards) for cardphones or as electronic purses. This widespread use is the reason for specifying synchronous data transmission in this case.

Synchronous transmission in memory cards is very closely linked with the chip's hardware, and is designed for the greatest possible simplicity. The transmission protocol exhibits no layer separation nor is there logical addressing, so that the application in the terminal must access the chip's memory addresses directly. The data held there can be physically addressed by the protocol, and then be read or written to. Thus actual data transmission is also linked to the functions of memory addressing and management.

Equally, there is no procedure for detecting or correcting errors during data transmission; though incidentally, such transmission errors between card and terminal occur very rarely. If, nevertheless, the terminal application discovers such a transmission error, it must re-read the relevant area in the card's memory. All these limitations, though, allow the transmission of data from card to terminal and vice versa with a small number of logic gates and at high speed.

Since synchronous transmission is only used so as to allow data transfer in as simple a manner as possible, that is, with a minimum of logic circuitry in the chip, it almost necessarily means a strong dependence on the hardware used. Synchronous transmission protocols are anything but uniform, and sometimes vary greatly from chip to chip. Only the ATR is standardized. Therefore a terminal which needs to communicate with a range of memory cards requires several different implementations of synchronous transmission protocols.

The exact definition of memory card data transmission reads "clock synchronous serial data transmission". This describes clearly which basic conditions apply to this type of communication. As in the asynchronous case, the data is transmitted between card and terminal serially, that is, bit by bit. However, they are synchronous with an additional transmitted clock pulse. This makes the transmission of start and stop data superfluous.

In the case of a simple memory card, this also involves the absence of error recognition data, thus neither parity bit nor an additional checksum are transmitted. The low probability of transmission errors is related to the very slow applied clock rate. It ranges from 10 to 100 kHz. Since one bit is transmitted per cycle, a clock frequency of (e.g.) 20 kHz results in a transmission rate of 20 kbit/s. The effective rate is lower, though, since additional address information needs also to be transmitted.

In order to describe memory card synchronous data transmission in detail, we first need to understand a few fundamentals of their construction. In their simplest form, they possess a two-part memory consisting of a fixed ROM and a write/erase supporting EEPROM. Both areas are bit-addressable and can be read at will, and in the case of the EEPROM also written to and erased.

The client—server behaviour is even more pronounced in memory cards than in microprocessor Smart Cards. For example, the terminal completely takes over the physical addressing of memory. The card itself can only block certain areas globally against erasing. This is controlled by hard-wired logic in front of memory, this logic also being in charge of the very simple data transmission.

6.2.1.1 Telephone chip protocol

Data transmission is illustrated here with the help of a phonecard containing a Siemens SLE4403 chip. The memory in this module is bit-oriented, which means that all operations are carried out on individual bits. Other chip types may have protocols with different specifications. However, the basic principles of data transmission are the same in all synchronous cards.

Transmission occurs with the aid of three channels. The bidirectional data channel can be used by both card and terminal to exchange bit-length data. The clock channel transmits the pulse generated by the terminal to the card, forming the basic step in the synchronous data transmission. The third connection necessary for transmission is the control channel which determines the chip's actual operation, depending on the state prevailing on the other two channels.

In principle, complete control of a memory card requires four different functions which need to be decoded by the chip's logic circuitry: these are read, write, clear memory and increment the address pointer. Memory cards possess a global memory pointer, with which all areas of memory can be addressed bit-wise. When the pointer reaches the last address it resets to zero. With a bit-oriented chip structure, it would then point to the first bit in memory. One of the functions of synchronous data transmission is to set this pointer to an initial value, which is normally zero.

The next function is to read data from memory. Writing and erasing EEPROM bits complete these functions. Erasing, and hence the possibility of rewriting, are of course blocked in phonecards, otherwise they could always be re-credited.

Resetting the address pointer

The address pointer is reset by the card's booting circuitry to its initial value of zero, when the clock channel and the control input are simultaneously at a high level. However, the control pulse must be applied for a somewhat longer period than the clock pulse, in order to prevent the address from immediately being incremented. The address pointer should be reset to its original value after each booting sequence, otherwise it would be pointing to an undefined address.

Address pointer incrementation and the read function

When the control channel is at low level and the clock pulse has a rising slope, the card's internal logic increments the address pointer by one. With a falling pulse slope, the contents of the address specified by the pointer are sent to the data channel. When the pointer reaches its maximum value, which depends on memory size, it resets to zero.

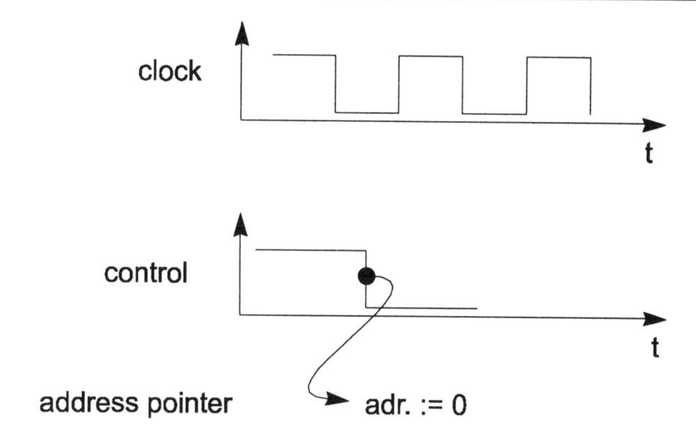

Figure 6.10 Resetting the address pointer to zero

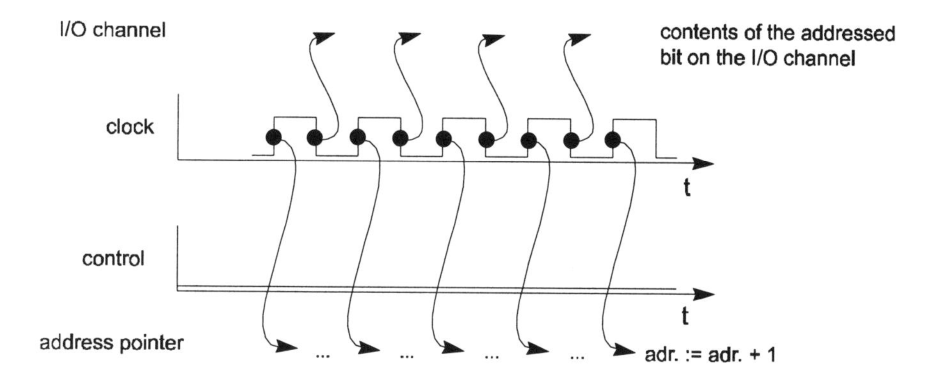

Figure 6.11 Address pointer incrementation and the read function

Writing to an address

When the address pointer specifies an EEPROM area where writing is permitted, the value held in the data channel can be written to EEPROM with a high level on the control channel and a low one on the clock channel. The length of the write cycle is determined by the duration of the clock pulse immediately following. If the writing procedure took place correctly, the contents of the written memory cell appear at the data output.

Erasing bytes

Part of EEPROM memory in a typical phonecard is always organized as a multi-digit octal counter. If a byte needs to be erased in this counter because of a carry to the next digit, this is carried out by the logic circuitry. That is why erasing a byte in memory is somewhat more complex. The procedure is as follows: if a bit is written twice consecutively, the chip's hardware logic automatically erases the corresponding less significant byte. This ensures that a carry has occurred to the higher digit, and the lower one is thus erased without the possibility of a fraud taking place.

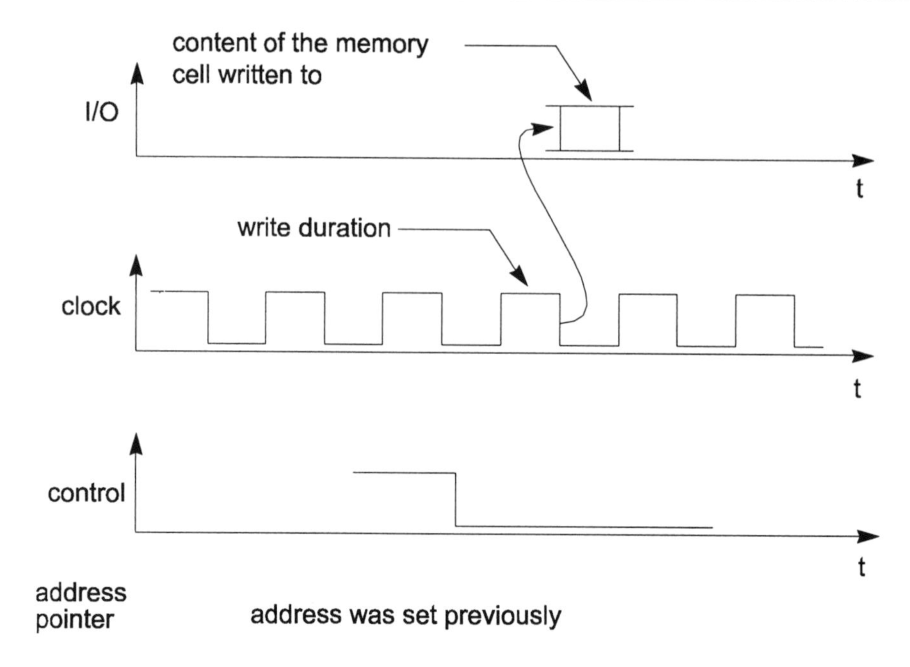

Figure 6.12 Writing a bit to an EEPROM address

The four types of access described above may vary from chip to chip, and also between manufacturers. Another type of data transmission, which however is standardized, can be found in the I^2C bus. Many of the newer memory cards use this bus for communicating with a terminal. Naturally, this has the advantage that different modules, made by different manufacturers, can be used in parallel with each other in the same system. Any problems which may be due to the need to cope with several transmission protocols to the card, are thus obviated since all chips are compatible with each other at the transmission interface.

6.2.1.2 I^2C bus

Since serial and clock-synchronized data transmission protocols are uncomplicated and can be utilized in manifold ways, they are used relatively frequently. Modules containing the I^2C bus, as developed by Philips, have been available since 1990. This bus, whose abbreviation stands for inter-integrated circuit, is based on a serial and bi-directional data channel and a serial clock channel. Both the hardware, that is the two channels, and the software, in the sense of the data transmission format, are defined in this bus. Each module in the bus can take over control and interrogate further modules connected to it.

Since memory cards are also controlled in a clock-synchronous way, the I^2C bus has very quickly established itself in the Smart Card industry. A wide range of memory modules has become available for incorporation in the card. The following example uses the SGS-Thomson ST24C04 memory chip. It has a 512-byte EEPROM with freely selected writing and reading capability. Timing of the EEPROM's programming is performed internally by the chip, and does not need to be controlled externally.

I^2C bus operation assumes the existence of hardware, viz. a double wire connection from terminal to card. The SCL (serial clock) channel transmits the pulse, whose frequency can extend to 100 kHz. This results in a data transmission speed of up to 100 kbit/s, which is relatively high for Smart Cards. The other channel, SDA (serial data), is used bidirectionally for data exchange between card and terminal. The SDA channel is connected in the terminal to the supply voltage V_{cc}, via a pull-up resistance. Both communicating parties can merely earth this conductor. Sending a high level therefore follows passively, by the sender switching his output to a high resistive state (tri-state), and the pull-up resistance pulls the SDA channel to the supply voltage level.

In the Smart Card context, the terminal is always the server of the I^2C bus and the card always the client. Data transmission is based on one-byte packets. The highest bit (bit 8) in the byte is sent first. Each transmission down the SDA channel is initiated by a start signal and completed with a stop signal. The start signal is obtained from a falling slope during a high level on the SCL channel. Conversely, a rising slope during a high level indicates a stop signal. Each reception of a byte must be confirmed by an acknowledgement sent by the recipient. This is done by earthing the SDA channel for one clock cycle.

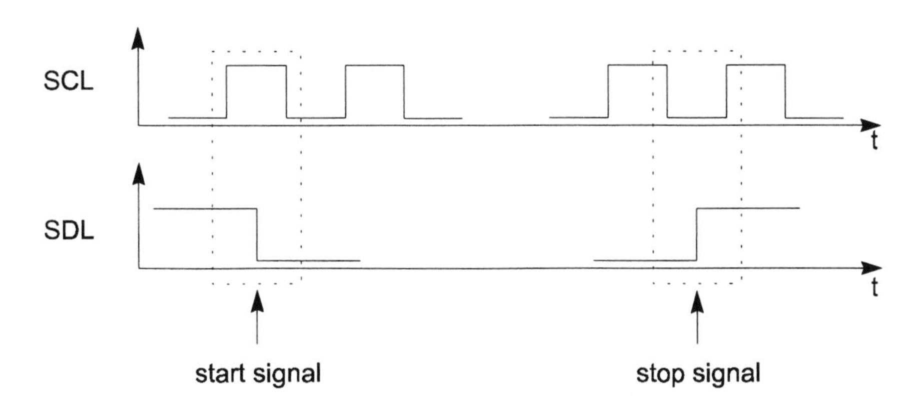

Figure 6.13 Start and stop signals in I2C bus

After the start of communications, the first seven bits of the first transmitted byte are the recipient's address. In our example, we assume for simplicity that the address has the binary value 1010000x. Of course, this may vary between different chip types, and in some memory modules can be chosen arbitrarily within certain limits. The last bit in the address indicates to the recipient whether data is to be read or written. The value 1 is understood as a reading operation, 0 is writing.

The following examples illustrate the general functions of the I^2C bus in Smart Cards.

Reading an address

There are several ways to access the card's EEPROM in order to read data from memory. In the one described here, one byte is read at a time. However, it is also possible to read several bytes one after the other.

The read sequence is initiated by the start signal. This is followed by the card's address, which also sets the signal for the writing option. This indicates to the card that the data about to follow should be temporarily stored in an internal buffer. This buffer is nothing but a byte-oriented address pointer in EEPROM memory. After the card receives this first byte, it transmits an acknowledgement. To this purpose, it earths the SDA channel for the duration of one clock cycle. This is followed by the terminal sending the address in EEPROM that should be read. Once again, the card acknowledges reception. Then the terminal transmits a start signal and the card address, with the reading signal set. On receipt, the card sends to the terminal the data addressed by the pointer. The terminal does not need to acknowledge receipt of the data: it only sends to the card a stop signal. This completes the reading sequence for one byte.

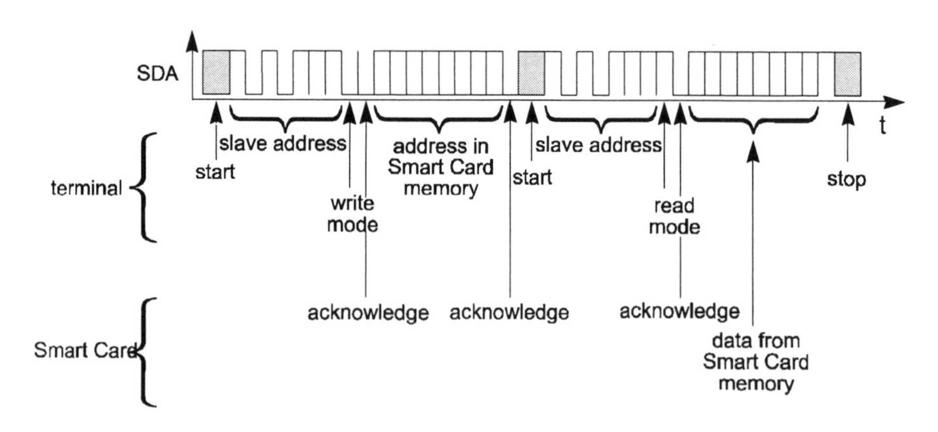

Figure 6.14 Reading a byte from memory with the I^2C bus

Writing to an address

In analogy to the reading of data from the card's EEPROM, there are also various writing modes. The simplest of these is described below, where a single byte can be written anywhere in memory.

Again, the sequence starts with a signal from the terminal. This is followed by the card's address, with the write option set. The card acknowledges receipt and then receives from the terminal the address in the EEPROM which should be written to.

This too is acknowledged by the card, which then receives the data. After the third acknowledgement, that of receipt of data by the card, the terminal sends a stop signal. After this the card starts to write the received data in the EEPROM, which does not require external timing signals. This completes the writing sequence, and the byte is now written in the EEPROM.

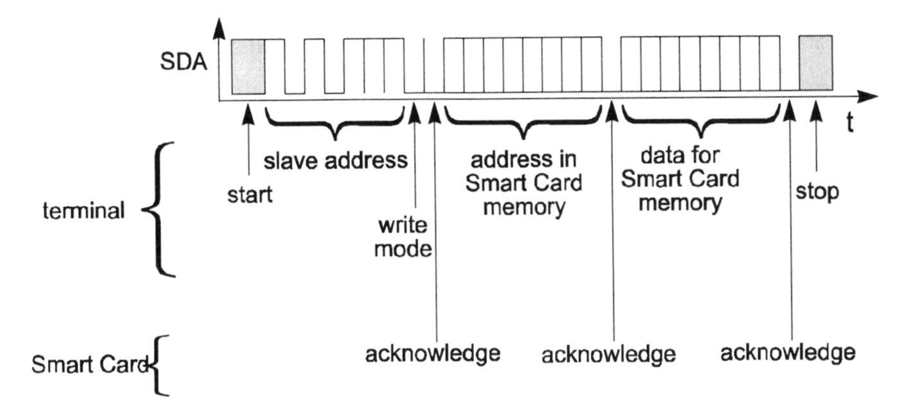

Figure 6.15 Writing a byte in memory with the I^2C bus

6.2.2 Transmission protocol T=0

This transmission protocol was used in France during the initial phase of Smart Card development, and was also the first internationally standardized Smart Card protocol. It was created in the early years of Smart Card technology, and hence designed for a minimum of memory use and a maximum of simplicity. This protocol is used worldwide in the GSM card, and as a result enjoys the greatest popularity of all current Smart Card protocols.

The T=0 protocol is byte-oriented, which means that the smallest unit processed by the protocol is a single byte. The transmitted data units consist of a header, which incorporates a class byte, an instruction byte and three parameter bytes. This is followed optionally by a data section. In contrast to the APDUs covered by ISO/IEC 7816-4, only the parameter P_3 exists as a length datum. It specifies either the length of the instruction data or that of the response, and is also covered by the ISO/IEC 7816-3 standard.

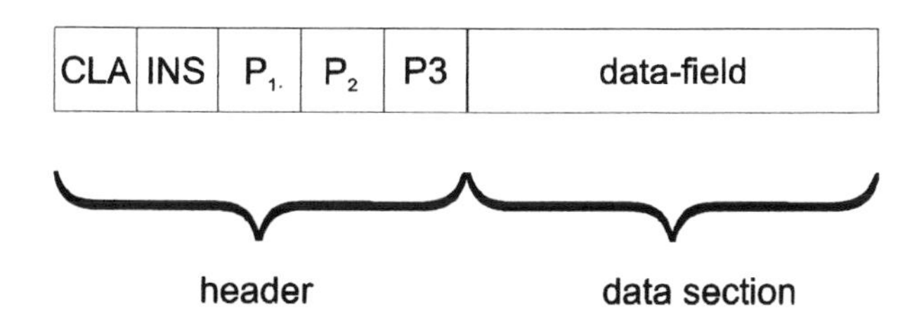

Figure 6.16 Creating an instruction with T=0

Due to the protocol's byte orientation, if a transmission error is detected then the incorrect byte must be requested again. In the case of an error occurring in block protocols, in contrast, the entire block (i.e. a sequence of bytes) must be sent again. Detection of transmission errors in T=0 takes place exclusively with the help of a parity bit, which is appended to each transmitted byte.

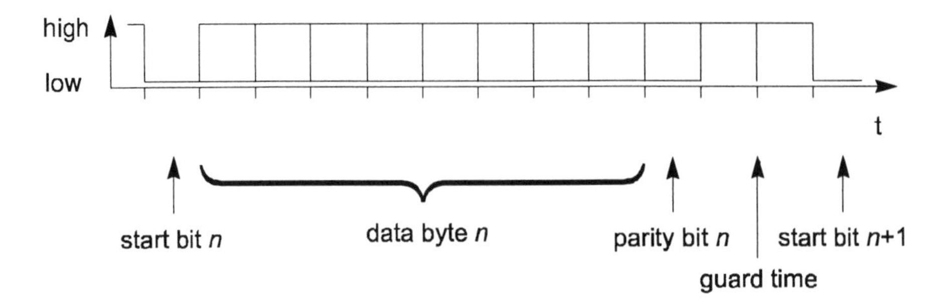

Figure 6.17 Illustration of error-free transmission of bytes with T=0 on the I/O channel

If the recipient detects a transmission error, his I/O channel must be set to a low level for the duration of one etu; this starts halfway into the first bit of the guard time corresponding to the faulty byte. This signals to the other party that the most recent byte must be retransmitted. This "byte repetition" mechanism is very simple, and has the advantage that it is selective, and only the incorrectly received byte needs to be repeated.

Unfortunately, this procedure suffers from a severe drawback. Most interface modules treat the etu time interval as the smallest detectable unit, so that they cannot recognize the low level on the I/O channel if set after half a stop bit. This is why standard interface modules are unsuitable for the T=0 protocol. However, where each bit is received separately by software, the problem does not arise.

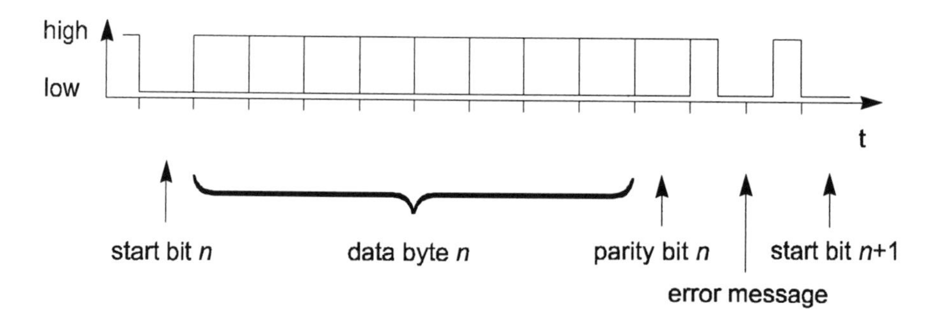

Figure 6.18 Illustration of transmission error in T=0 due to a low level in the I/O interface during the Guard time

The T=0 protocol also provides for the option of switching an external programming voltage for the EEPROM, or EPROM as relevant, on and off. To this end, the value one is added to the received instruction byte and it is sent back to the terminal as an acknowledge byte. This is also the reason why only even instruction bytes are permitted, since otherwise the above mechanism would not work. However, the provision of an external switched programming voltage has become technically outdated in recent times, since all modern Smart Card microprocessors generate their programming voltage on the chip itself. Hence we will not discuss this topic any further.

The guard time's main function is to separate the individual bytes during transmission. This allows both the sender and the recipient more time to carry out the transmission protocol's functions.

To illustrate the T=0 instruction—response sequence, let us assume that the terminal transmits to the card an instruction containing a data section, and the card returns data and a response code.

The terminal first transmits to the card a 5-byte instruction header consisting of a class byte, instruction byte and P1, P2 and P3 bytes. If this is received correctly, the card returns an acknowledge (ACK) as a procedure byte (PB). This acknowledgement is coded as the received instruction byte. On receipt of the procedure byte, the terminal sends exactly the number of data bytes indicated by the P3 byte. Now the card has received the complete instruction, and it can process it and generate a response.

If, in addition to the 2-byte return code, the response also contains data, the card informs the terminal via a special return code and the amount of data in SW2. After receiving this message, the terminal sends to the card a so-called GET RESPONSE instruction, which consists only of an instruction header with the quantity of data to be transmitted. The card now sends to the terminal the data generated after the first instruction, at the required length and with the corresponding return code. This completes one instruction sequence.

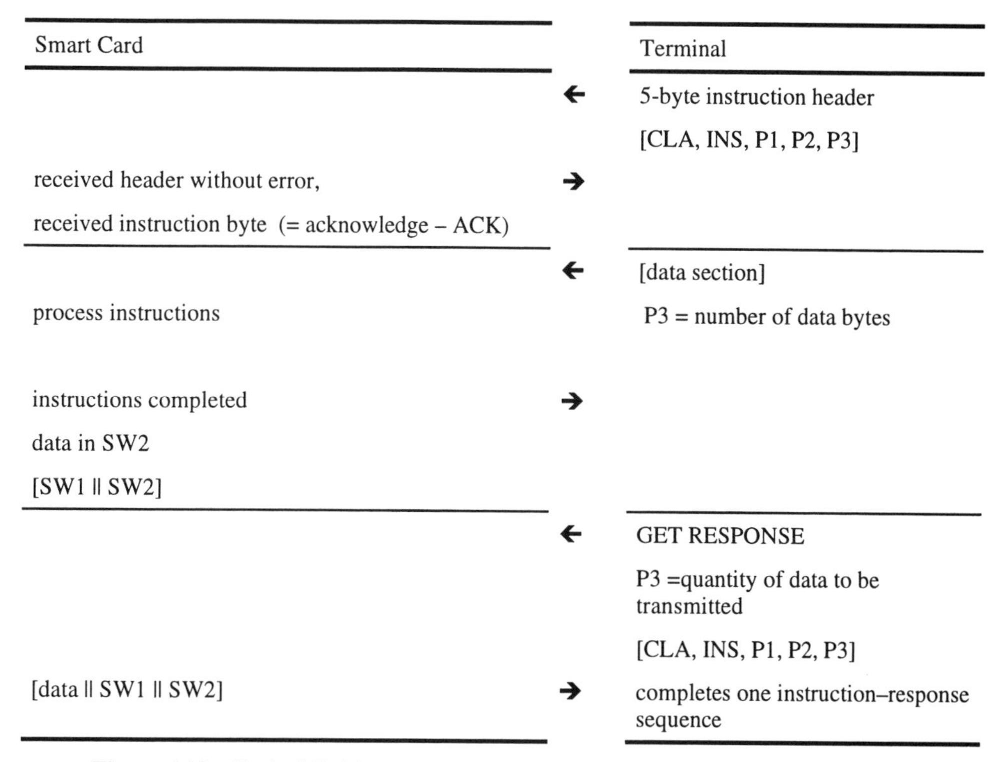

Figure 6.19 Typical T=0 instruction—response sequence (e.g. instruction MUTUAL AUTHENTICATE)

If an instruction is sent to the card and the latter only generates a return code without a data section, then the GET RESPONSE part does not apply. Since in order for this action to be carried out, i.e. the fetching of data relating to a previous instruction, a further instruction from the application layer is needed, naturally there is no longer a strict separation between the protocol layers. It is necessary to use application layer instructions (GET RESPONSE) to support the channel layer, which to some extent affects the relevant application.

The above-mentioned routines may appear complicated at first sight, therefore they are shown again graphically in Figure 6.20.

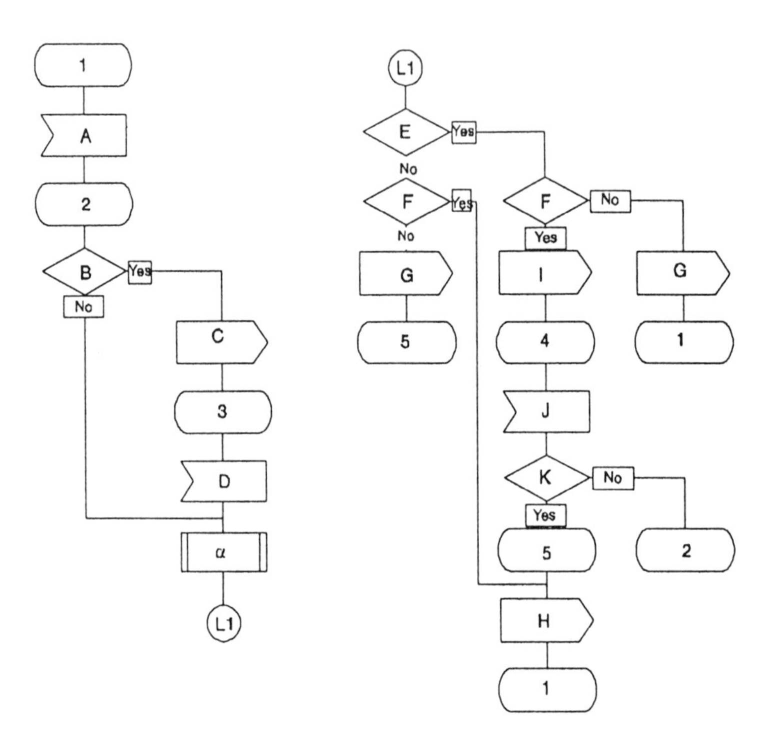

Figure 6.20 Smart Card communication process at T=0 without error correction.
α – Processing instruction

1 – Basic state
2 – Header with CLA, INS, P1, P2, P3 received
3 – Wait for data sections (P3 = number of bytes)
4 – Wait for instructions (header with CLA, INS, P1, P2, P3) (P3 = quantity of data)
5 – SW1, SW2 sent / GET RESPONSE received

A – Receive header (5 byte)
B – Data section (P3 != 0)?
C – Available data section, send procedure bytes to terminal
D – Receive data section (P3 = number of bytes)
E – Did the instruction contain a data section? (i.e. run through C and D)?
F – Is there a return code (i.e. no errors)?

G – Send SW1 and SW2
H – Send return codes and SW1, SW2
I – Send SW1 and SW2 (SW2 = number of return codes
J – Receiving instructions (header = 5-byte)
K – Is the received instruction GET RESPONSE?

The T=0 protocol allows the card, after receiving the header, to receive the bytes in the data section individually. To this end, it only needs to transmit the inverted instruction byte to the terminal as a procedure byte, whereupon the latter sends a single data byte. The next data byte follows the next procedure byte. This byte-wise transmission can continue until the card has received all the bytes in the data section, or until it sends the non-inverted instruction byte to the terminal as a procedure byte. Thereupon the terminal transmits all the remaining data bytes to the card, and the latter has now received the complete instruction.

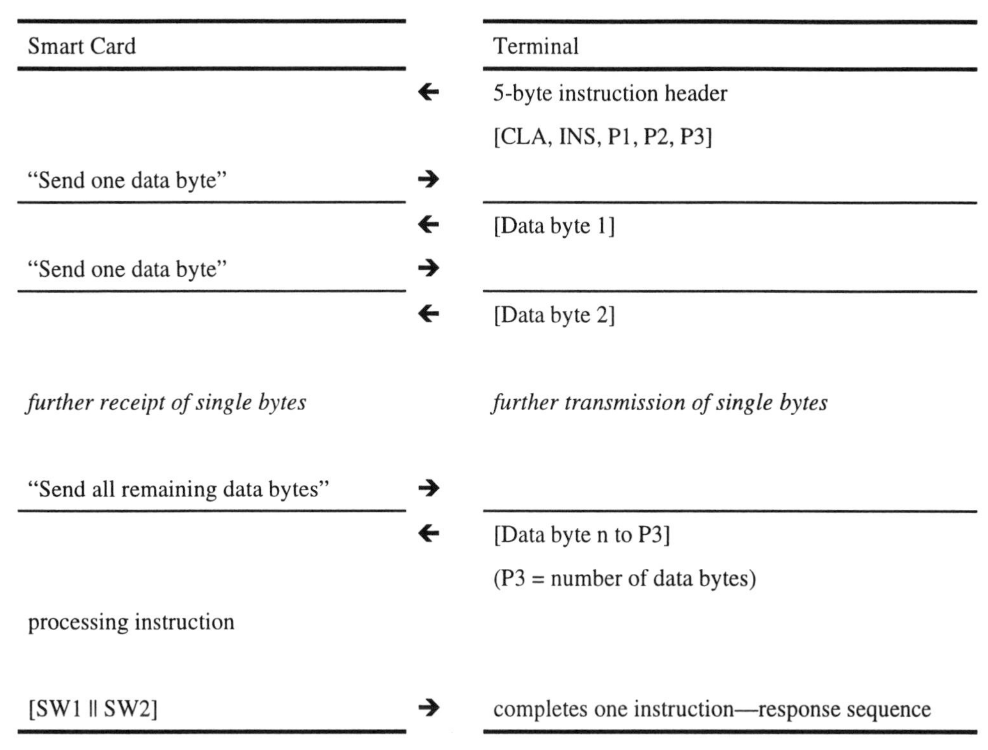

Figure 6.21 Receiving single bytes through T=0 (e.g. with UPDATE BINARY)

As far as the transmission protocol is concerned, at the end of the day the user is mainly interested in the data transmission speed as well as in the error detection and correction mechanisms.

Sending a byte, that is 8 bits, involves the transmission of 12 bits, including 1 start bit, 1 parity bit and 2 etus for the guard time. Assuming a 3.5712 MHz clock frequency and a 372 divider, the transmission of one byte takes 12 × 1 etu = 1.25 ms.

Figure 6.22 lists the transmission times for a few typical instructions.

Instruction	Number of data (bytes)	Number of protocol data (bytes)	Data transmission speed
READ BINARY	C: 5 R: 2 + 8	—	18.75 ms
UPDATE BINARY	C: 5 + 8 R: 2	—	18.75 ms
ENCRYPT	C: 5 + 8 R: 2 + 8	C: 5 R: 2	37.50 ms

Figure 6.22 T=0 transmission speed with some typical instructions (C — command, R — response). Assumptions: clock frequency = 3.5712 MHz, divider = 372, 2 stop bits, 8 bytes data per instruction

Of course, the transmission speed drops as soon as errors start occurring. Nevertheless, the single-byte repetition mechanism is very beneficial here, since only the incorrectly received byte needs to be retransmitted.

The T=0 protocol's error detection mechanism involves no more than a parity check at the end of each byte. This allows reliable recognition of 1-bit errors, but 2-bit errors are no longer detected. If a byte is lost during the transmission from terminal to card, this leads to an endless loop (deadlock) in the card, since it is waiting for a particular number of bytes and there is no possibility of a clock-controlled interrupt. The only practical solution for the terminal to exit from this communications dead end is to reset the card and completely rebuild the link.

In normal communication, the insufficient separation between the channel and transport layers does not lead to excessively severe repercussions. The efficient GSM application illustrates this best. However, secured data transmission very quickly does cause problems. With a partly encrypted header and a completely encrypted data section, the specified protocol routine can no longer support T=0 without great overheads. The reason is the need for an unencrypted instruction byte to be used by T=0 as a procedure byte.

Due to the absence of layer separation and the evident problems caused in the event of a poor connection, the T=0 protocol is often considered outdated. On the other hand, communication between the terminal and the card practically never gives rise to transmission errors. The good mean transmission speed, the minimal implementation overheads and its wide market penetration are this protocol's main advantages.

6.2.3 Transmission protocol T=1

The transmission protocol T=1 is an asynchronous half-duplex Smart Card protocol. It is based on the international standard ISO/IEC 7816-3 Amd. 1. This is a supplement to ISO/IEC 7816-3, and deals specifically with this protocol. T=1 belongs to the class of block protocols. In other words, the block is the smallest data unit which can be transmitted between the card and the terminal.

A feature of this protocol is its strict layer separation, and it can be classified in the OSI reference model as a data link layer. In this context, layer separation also means that data destined for higher layers, e.g. the application, can be processed completely transparently by the link layer. It is not necessary for layers other than the one involved to interpret or modify the contents of the transmitted data.

Secure messaging, in particular, requires strict adherence to layer separation. It is only then that encrypted user data can be transmitted across the interface without resorting to complex methods or various tricks. The T=1 protocol is currently the only international Smart Card protocol which allows secure data transmission in all its variations, without any problems or compromises.

The transmission protocol starts after the card had sent out the ATR, or after a successful PTS had been executed. The first block is sent by the terminal, the next one by the card. Communication then continues, with authority to transmit alternating between the two.

Incidentally, T=1 finds its application not only in Smart Card/terminal communications. It is often used to communicate between the terminal and its computer, for the transmission of user and control data.

The resulting transmission speed is, of course, of special interest in any protocol. Figure 6.3 lists the transmission times for a few typical instructions.

Instruction	Number of data (bytes)	Number of protocol data (bytes)	Transmission speed
READ BINARY	C: 5 R: 2 + 8	C: 4 R: 4	28.75 ms
UPDATE BINARY	C: 5 + 8 R: 2	C: 4 R: 4	23.00 ms
ENCRYPT	C: 5 + 8 R: 2 + 8	C: 4 R: 4	38.75 ms

Figure 6.23 T=1 transmission speed with some typical instructions (C — command, R — response). Assumption: clock frequency = 3.5712 MHz, divider = 372, EDC = XOR, 2 stop bits, 8 bytes data per instruction

Block structure

The transmitted blocks are essentially used for two different purposes: on one hand, for the transparent transmission of application-specific data, and on the other, for the transmission protocol's control data and for dealing with transmission errors.

The blocks consist of a leading prologue field, the information field and a final epilogue field. The prologue and epilogue fields are mandatory, and must be sent in every case. In contrast, the information field is optional and contains the data for the application layer. This is either an instruction APDU for the Smart Card, or a response APDU from the card.

One distinguishes between three fundamentally different block types in T=1: information blocks, reception acknowledgement blocks and system blocks.

- Information blocks (I-blocks) are used for the transparent exchange of application layer data.
- The reception acknowledgement block (R-block), which does not contain a data block, serves for positive or negative reception confirmation.
- System blocks (S-blocks) are used for control data which relate to the protocol itself. Depending on the specific control data, they may possess an information field.

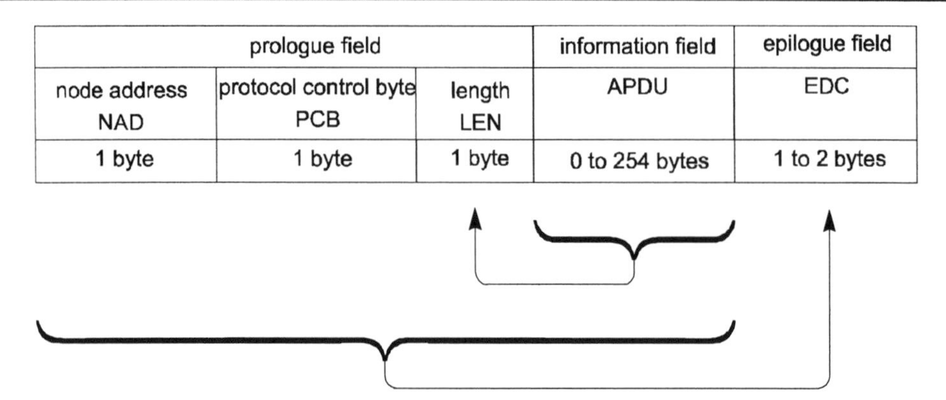

Figure 6.24 T=1 transmission block structure

Prologue field

The prologue field consists of the three subfields node address (NAD), protocol control byte (PCB) and length (LEN). It is three bytes long, and contains elementary control and pointing data for the actual transmission block.

Node address (NAD)

The first byte in the prologue field is designated a node address or NAD byte. It contains the block's target and source node addresses. These are each coded as 3 bits. If addressing is not used, the relevant bits are set to 0. Furthermore, for reasons of compatibility with older microprocessors, control is provided for EEPROM or EPROM programming voltage. However, this control no longer finds any practical application, since currently all Smart Card microprocessors have a charging pump on the chip.

b8	b7	b6	b5	b4	b3	b2	b1	Meaning
X	...	...	...	X	...	...	...	V_{pp} control
...	X	X	X	...	...	...	...	DAD (target node address)
...	...	...	...	...	X	X	X	SAD (source node address)

Figure 6.25 Node address (NAD field)

PCB field

The subfield following the node address is the protocol control byte, PCB for short. As the name suggests, it serves to control and supervise the transmission protocol. This increases the coding overheads. The PCB field encodes first and foremost the block type, as well as supplementary data needed in this context.

b8	b7	b6	b5	b4	b3	b2	b1	Meaning
0	...	...	...	...	...	...	...	Signals I-block
...	N(S)	...	...	...	...	...	...	Number in sequence
...	...	X	...	...	...	...	...	Supplementary data M
...	...	...	X	X	X	X	X	Reserved

Figure 6.26 PCB field in I-block

b8	b7	b6	b5	b4	b3	b2	b1	Meaning
1	0	...	...	...	...	...	...	Signals R-block
...	...	0	N(R)	0	0	0	0	No errors
...	...	0	N(R)	0	0	0	1	EDC/parity field
...	...	0	N(R)	0	0	1	0	Other errors

Figure 6.27 PCB field in R-block

b8	b7	b6	b5	b4	b3	b2	b1	Meaning
1	1	...	...	...	...	...	...	Signals S-block
...	...	0	0	0	0	0	0	Resynch request (only from terminal)
...	...	1	0	0	0	0	0	Resynch answer (only from Smart Card)
...	...	0	0	0	0	0	1	Request to modification of information field size
...	...	1	0	0	0	0	1	Answer to modification of information field size
...	...	0	0	0	0	1	0	Cancel request
...	...	1	0	0	0	1	0	Cancel answer
...	...	0	0	0	0	1	1	Request to modify waiting time (only from Smart Card)
...	...	1	0	0	0	1	1	Answer to modify waiting time (only from terminal)
...	...	1	0	0	1	0	0	V_{pp} error detection (only from Smart Card)

Figure 6.28 PCB field in S-block

LEN field

The 1-byte LEN field indicates in hexadecimal form the length of the information field. Its value can be between '00' and 'FE'. The code 'FF' is planned for future extensions, and currently should not be used.

Information field

In an I-block, the information field serves as a container for the application layer's data (OSI layer 7). The contents of this field are transmitted completely transparently. This means that the content is directly forwarded by the transmission protocol, without any analysis or evaluation.

In an S-block, this field is used for the transmission of data for the transmission protocol. This, though, is the only exceptional situation in which this field's contents are used by the transmission layer.

The size of the information field can range between '00' and 'FE' (254) bytes, as per ISO. The value 'FF' (255) is reserved by ISO for future use. Terminal and card may have different sizes of I-fields. The default value of terminal I-field length is 32 bytes (= IFSD — information field size for the interface device), and can be modified via a special S-field. This default value of 32 bytes also applies to the card (IFSC - information field size for the card), but it can be modified by an instruction in the ATR[4]. In practice, the I-field's size both in the terminal and in the card lies in the range 50 to 140 bytes.

Epilogue field

This field, transmitted at the end of the block, contains an error detection code which covers all previous bytes in the block. The calculation takes place either with an LRC (longitudinal redundancy check) or with a CRC (cyclic redundancy check). Which of these two error detection codes is used must be specified in the ATR's interface character. If it is not specified, the LRC method is implicitly used by convention. Otherwise, the CRC calculation is carried out as per ISO 3309. The dividing polynomial used there, $G(x) = x^{16} + x^{12} + x^5 + 1$, is identical with CCITT Recommendation V.41. Both error detection codes can only be used for recognition; they are unable to correct a block error.

The one-byte longitudinal redundancy check is computed via an XOR concatenation of all previous bytes in the block. This calculation can be executed very fast, and its implementation is not code-intensive. Normally it is performed on-the-fly during transmission or reception. It is present as standard in practically all T=1 implementations.

Utilization of the CRC procedure to generate an error detection code produces a far higher probability of error recognition than the relatively primitive XOR checksum. Nevertheless, in practice this approach is hardly exploited at the moment, since its implementation is very code-intensive and slow. In addition, the epilogue field needs to be extended to two bytes, which lowers transmission speed even further.

---

[4] Data elements in ATR: TAi (i>2).

Transmission sequence/reception sequence counter

Each information block in the T=1 protocol is provided with a transmission sequence number, consisting of only one bit and contained in the PCB byte. It is incremented modulo 2, i.e. it alternates between the values 0 and 1. The transmission sequence counter is also designated N(S), and its starting value at protocol initiation is 0. The terminal and Smart Card counters are incremented independently of each other.

The transmission sequence counter's purpose is mainly to require a faulty received block to be retransmitted, since the individual information blocks can each be unambiguously addressed via the N(S).

Waiting times

Various waiting times have been defined, in order to allow senders and recipients precisely specified minimum and maximum time intervals for different actions during data transmission, and in order to prevent the transmission protocol freezing in the event of errors through a defined interrupt. These are all covered in the standard by defined values, but in order to maximize transmission speed it is possible to modify them accordingly and to indicate this in the ATR's specific interface characters.

Character waiting time (CWT)

The character waiting time is defined as the maximum time interval between the rising slopes of two consecutive characters within a block. Thus, the recipient triggers a countdown clock at each rising slope. Its starting value is the character waiting time. If the clocks runs out and during this time no new character's rising slope was detected, the recipient assumes that the transmission block has been received in full. Therefore the "CWT reception criterion" can also be used generally for end-of-block recognition. It does, however, considerably reduce data transmission speed, as each block is extended by the CWT's length. Hence it is more appropriate to recognize the end-of-block by counting arriving bytes.

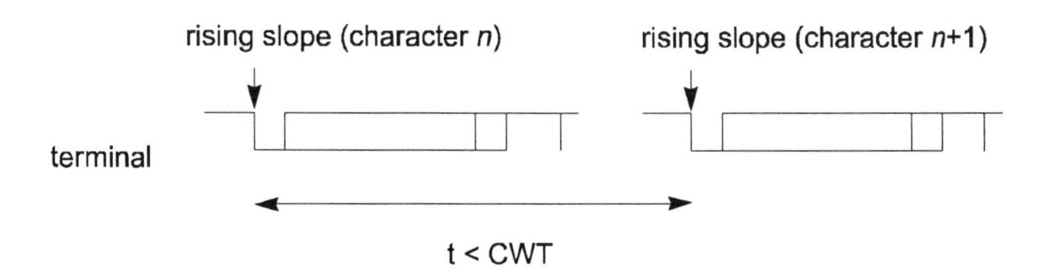

Figure 6.29 Definition of character waiting time (CWT)

The CWT is calculated according to the following formula, using the data element CWI contained in ATR:

$$CWT = (2^{CWI} + 11)\ \text{work etu}$$

The CWI's default value is 13, which gives us the following CWT:

$$CWT = (2^{13} + 11)\ \text{work etu} = 8203\ \text{work etu}$$

With a clock frequency of 3.5712 MHz and a 372 divider, this results in an interval of 0.85 seconds[5].

This interval, specified in the standard as the default setting, is too high for fast data transmission. In practice, the usual range for CWI is between 3 and 5. This means, in the case of a normal transmission sequence in which one character after the other crosses the interface without a time delay, that the recipient waits for the duration of an additional one to two bytes before the end of the block or a communications breakdown is detected.

Normally, the receiving routine recognizes the end of a block from the block length information in the LEN field. However, if the contents of this field are faulty, one uses the character waiting time as an additional means of reaching a defined interruption of reception. This problem only becomes acute, though, when the length information is too short, since in this case the recipient would wait for further characters which never arrive. This would freeze the transmission protocol, and this situation could only be overcome by a card reset. The character waiting time mechanism gets around this problem.

Block waiting time (BWT)

The block waiting time was set up in order to achieve a defined communications interruption in the event of a non-responding Smart Card. It is the maximum time interval between the rising slope of the last byte in a block sent to the card, and the rising slope of the first byte sent back by the card.

Seen in terms of a conventional T=1 block, this is the maximum time interval permitted between the rising slope of the XOR byte in the instruction block's epilogue field, and the rising slope of the NAD byte in the card's response.

If this waiting period expires without a response being sent from the card, the terminal may assume a faulty card and initiate appropriate correction mechanisms. This might be a card reset with a subsequent attempt at re-establishing communications.

---

[5] See also 15.4.2 Conversion table for ATR data elements.

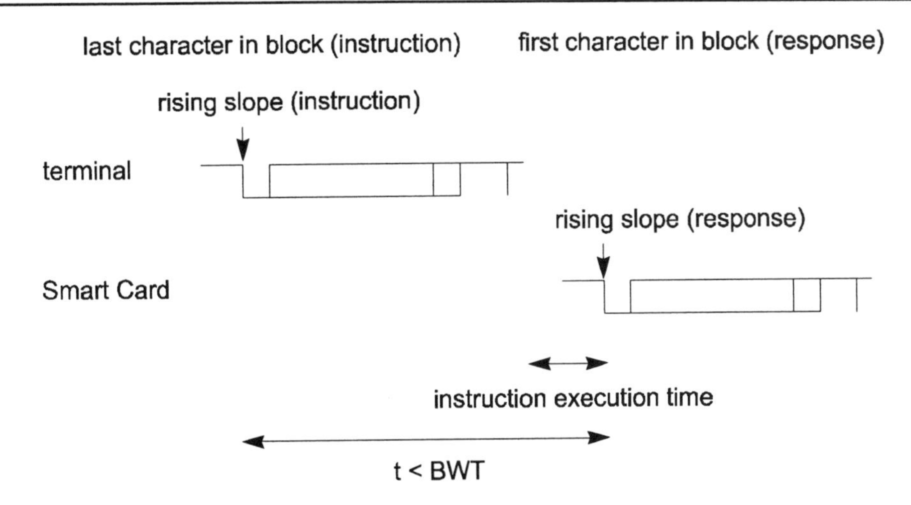

Figure 6.30 Definition of block waiting time (BWT)

In ATR, the interface characters specify a BWT value which is coded in the abbreviated form of a BWI:

$$BWT = 2^{BWI} \times 960 \times \frac{372}{f} s + 11 \text{ work etu}$$

If the ATR does not define a BWI value, one can use the default value BWI = 4. With 3.5712 MHz and a 372 divider, this gives 1.6 s as the value for the block waiting time:

$$BWT = 2^{4} \times 960 \times \frac{372}{3\ 571\ 200 \text{ Hz}} s + 11 \text{ work etu} = 2^{4} \times 0.1 s + 11 \text{ work etu} \approx 1.6 \text{ s}$$

As one can see, this value is quite generous. In practice, BWI is frequently taken to be 3, which results in a block waiting time of 0.8 s[6]. As a rule, typical instruction execution times in the card lie around 0.2 s. Hence a BWT of the above duration is a compromise between conventional instruction execution times and the fast detection of a Smart Card which no longer responds to instructions.

Block guard time (BGT)

The minimum time interval between the rising slope of the last byte and the rising slope of the first byte in the opposite direction, is defined as the block guard time. This is the

---

[6] See also 15.4.2 Conversion table for ATR data elements.

converse of the BWT, which is defined as the maximum time interval between the two rising slopes. Another difference is that the block guard time is mandatory for both transmission parties and must be observed, but the block waiting time is only significant for the Smart Card.

The block guard time's purpose is to allow the block sender a minimum time interval in which to switch over from sending to receiving.

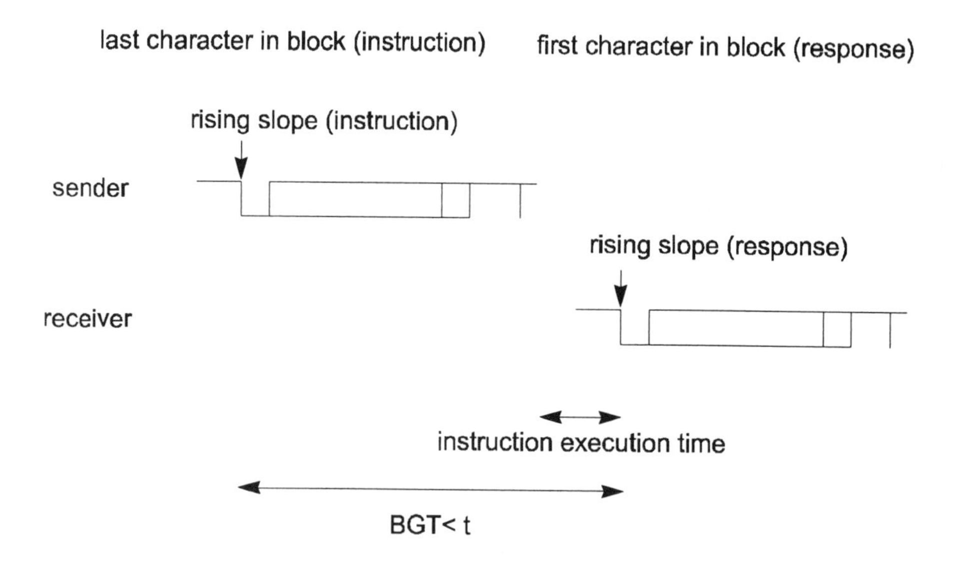

Figure 6.31 Definition of block guard time (BGT)

The block guard time's length is fixed, and is standardized at 22 etus. In a Smart Card driven at 3.5712 MHz with a 372 divider, this gives a value of *ca.* 2.3 ms.

Transmission protocol mechanisms

Waiting time extension

If the card needs somewhat longer to generate a response than the maximum allowed by the block waiting time (BWT), it can request from the terminal a waiting time extension. To achieve this, the Smart Card transmits a special S-block which requests an extension, and receives from the terminal a corresponding S-block in acknowledgement. The terminal cannot refuse this request.

The terminal is informed of the length of this extension by a byte in the information field. This byte, multiplied by the block waiting time, gives the new block waiting time, which is only valid for the most recently sent I-block.

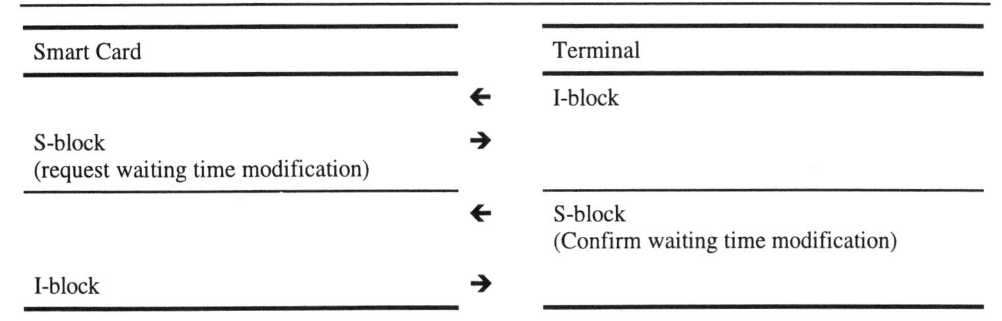

Figure 6.32 Progress of waiting time extension

Block chaining

One of the fundamental features of the T=1 protocol is the function of block chaining. It allows either of the parties in turn to transmit data blocks which are larger than the corresponding sending or receiving buffer. This is particularly useful with the limited storage space available in Smart Cards. Such chaining is only permitted in the case of information blocks, as they are the only block type which can display large data sizes. During this process, the user data is divided into individual blocks which are sent to the recipient one after the other.

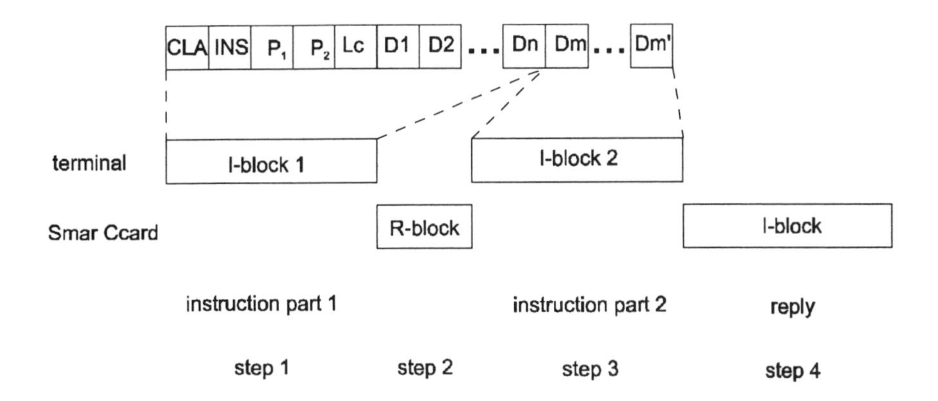

Figure 6.33 Example of block chaining when transmitting data blocks from terminal to Smart Card

The application layer data must be so divided that none of the segments generated are larger than the recipient's maximum block size. Then the first section is embedded in an information field as per the T=1 protocol, supplied with prologue and epilogue fields and sent to the recipient. The M-bit (more data bits) in the block's PCB field is set, to show the recipient that the block chaining function is being executed and that the following blocks contain chained data.

As soon as the recipient has successfully received this information block with the first part of the user data, it signals with an R-block — whose sequence number N(R) is the transmission sequence counter N(S) of the next I-block — that it is ready to receive the next chained I-block. The recipient is then sent this next block.

This back and forth of I- and R-blocks continues until the sender originates an I-block, whose M-bit in the PCB field indicates that this is the last block in the chain (M-bit = 0). On arrival of this block, all application layer data has been received and the recipient can process the block.

There is one restriction which applies to the block chaining routine. Within one instruction—response cycle, chaining may only proceed in one direction. So for example, if the terminal sends the chained blocks, the card may not send chained blocks in response.

One further restriction is not of a technical nature like the one mentioned above, but rather is due to the Smart Card's very limited memory. Implementation of the block chaining mechanism in software involves a certain amount of programming overheads, but its usefulness is very limited since instructions and responses tend not to be too long, and therefore normally do not require chaining. If the available size of the card's reception buffer held in RAM is insufficient to store all the data should block chaining take place, it is necessary to set up this buffer in the EEPROM. But this leads to a sharp decrease in transmission speed, since writing to EEPROM (in contrast to RAM) cannot take place at full processor speed.

As a result, many T=1 implementations possess no block chaining function, as the cost:benefit ratio often does not justify it. This illustrates the common fact that in practice, standards are often interpreted very generously. In this case, the interpretation assumes that block chaining is a supplementary option in T=1, and is not absolutely necessary.

Error processing

The T=1 protocol exhibits highly developed error detection and processing mechanisms. Should invalid blocks be received, the protocol attempts to apply precisely defined routines in order to re-establish error-free communications.

Seen from the terminal's perspective, this involves three synchronization steps. First, the sender of a faulty block receives an R-block which indicates an EDC/parity bit error or a general error. The recipient of this R-block must then retransmit the last block.

If it proves impossible to re-establish an error-free connection by using this mechanism, one proceeds to the second step. This means that the Smart Card receives from the terminal a resynchronization query in the form of an S-block. The terminal then expects a resynch response. Simultaneously, both terminal and card reset the transmission and reception counters to zero. This corresponds to the protocol status immediately following ATR. With this original state re-established, the terminal attempts to restart communications.

Steps 1 and 2 only affect the protocol layer. They produce no repercussions on the application itself. On the other hand, the third synchronization step does affect all Smart Card layers. If the terminal fails to establish an error-free connection via steps 1 and 2, it triggers a Smart Card reset through the reset channel. Unfortunately, this means that all the original session's data and conditions are lost. After this reset, communications must be completely re-established from the bottom up.

If even this procedure fails to provide a functional connection, the terminal deactivates the card after three attempts. Normally the user then receives an error message, viz. that the card is faulty.

Synchronization steps	Mechanism
Step 1	Retransmission of faulty block
Step 2	Resynchronization and retransmission of faulty block
Step 3	Reset Smart Card and re-establish connection

Figure 6.34 Steps in T=1 error detection and processing

6.2.4 Comparison of asynchronous transmission protocols

The previous sections described two international transmission protocols. In order to gain an overall view, Figure 6.35 summarizes their fundamental features as well as their advantages and shortcomings, and includes T=2 for comparison.

Criteria	T=0	T=1	T=2 (in preparation)
Data transmission	asynchronous, half-duplex, byte oriented	asynchronous, half-duplex, block oriented	asynchronous, full-duplex, block oriented
Standard	ISO/IEC 7816-3	ISO/IEC 7816-3 amd.1	ISO/IEC 10536-4
Divider	freely defined, common: 372	freely defined, common: 372	freely defined, common: 372
Block chaining	not possible	possible	possible
Error detection	parity bits	parity bits, EDC at end of block	parity bits, EDC at end of block
Memory needed for implementation	approx. 300 bytes	approx. 1100 bytes	approx. 1600 bytes

Figure 6.35 Comparison of asynchronous transmission protocols, conforming to international standards

It is appropriate to comment here briefly on the achievable transmission speed. One often tries to compare the two protocols T=0 and T=1, by calculating an effective transmission speed. This would only be valid for one particular instruction and in specific contexts. As soon as one attempts to generalize these calculations, however, they become meaningless and in fact misleading. Both protocols have their strengths and weaknesses as far as possible transmission speeds are concerned, but the latter depend on a multitude of individual factors. For example, these may include transmission error frequency, the size of the card's I/O buffer and the protocol's concrete implementation. Overall, it is possible to state that on average and in most applications, both protocols have approximately the same effective transmission speeds. Changing the protocol will achieve little in the way of increasing speed. It is more effective to lower the divider, as this approach would achieve considerably better results.

6.3 ANSWER TO RESET

After booting the power supply, the clock and the reset signal, the Smart Card sends out an answer to reset (ATR) at the I/O pin. This data string, up to 33 bytes long, is always sent with a 372 divider as per the relevant standard (ISO/IEC 7816-3), and contains various data relevant to the transmission and to the card. This divider should be selected even if the transmission protocol used after the ATR relies on a different divider (e.g. 512). This ensures that any card's ATR can be received independently of the protocol's parameters.

It is very rare for an ATR to reach maximum length. Much more often it is necessary to construct the ATR out of very few bytes. In particular, the ATR may not be too long in the case of applications where the card needs to be in use very quickly after the booting sequence. A typical example could be motorway toll payment by Smart Card electronic purses. Despite high vehicle speeds when driving through the data capture points, reliable logging must be possible in the short time available.

The ATR must be sent out between 400 and 40 000 clock cycles after the terminal's reset signal is sent. With a clock frequency of 3.5712 MHz, this corresponds to a time interval of 112 µs to 11.20 ms, and at 4.9152 MHz this is 81.38 µs to 8.14 ms[7]. If the terminal does not begin to receive the ATR within this interval, it repeats the booting sequence several times (usually three times), and attempts to detect an ATR. If all these attempts fail, the terminal assumes the card is faulty and rejects it.

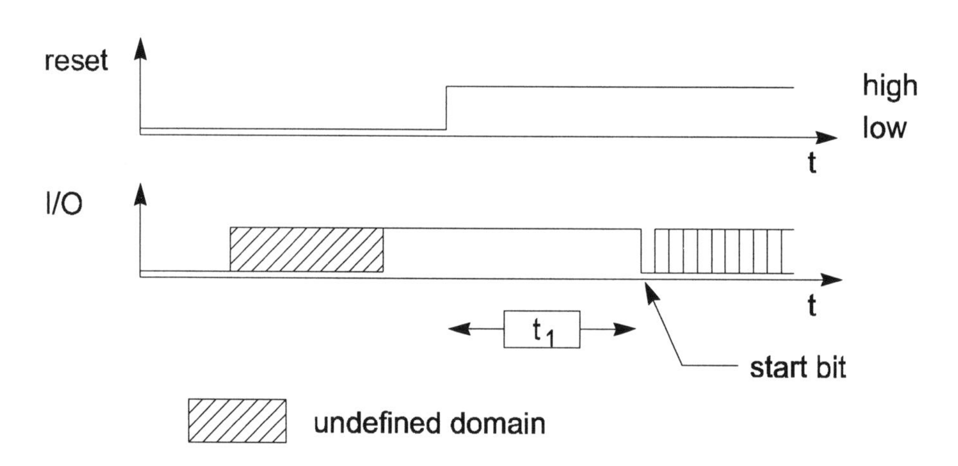

Figure 6.36 Time diagram of reset signal and ATR start in accordance with ISO/IEC 7816-3 (400 clock $\le t_1 \le$ 40 000 clock)

During the ATR, the rising slopes of individual bytes may be separated by a time interval of 9600 etus, according to ISO/IEC 7816-3. This period is designated the initial waiting time. With a clock frequency of 3.5712 MHz, it would be exactly one second. This means that the standard permits a one- second wait between the ATR's individual bytes, before

---

[7] See also 15.4.2 Conversion table for ATR data elements.

sending the next byte to the terminal. In some Smart Card operating systems, this time is utilized for internal calculations and writing to EEPROM.

The ATR's data string and data elements are defined and described in detail in ISO/IEC 7816-3. The basic ATR format is shown in Figure 6.37.

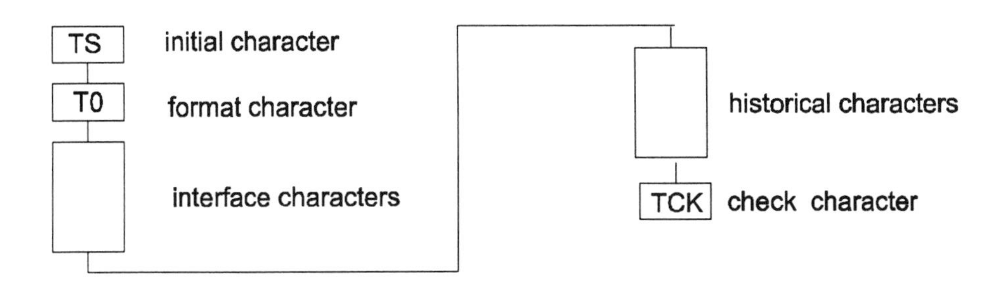

Figure 6.37 Basic ATR structure

The first two bytes, designated TS and T0, define various fundamental transmission parameters, as well as the presence of subsequent bytes. The interface characters specify special transmission characters and protocols important for the subsequent data transmission. The historical characters describe the extent of the Smart Card's basic functions. Depending on the transmission protocol, the last byte may consist of the check character, which is a checksum of the previous ATR bytes.

The initial character

This byte, denoted by TS, specifies the convention used for all the data in ATR as well as the subsequent communications routine. In addition, the TS contains a characteristic bit pattern which can be used by the terminal to identify the divider. The terminal measures the time between the first two falling slopes in the TS, divides it by three and the result is the duration of one etu. But since the divider to be used by the ATR is fixed at 372, the terminal does not normally evaluate the synchronization pattern. This first byte is a mandatory ATR component, and must be sent in every case; it can only be coded in two ways: '3B' in the direct convention or '3F' in the inverse convention.

The usual convention used in Germany is the direct one, but in France it is the inverse one. The convention does not affect transmission security. Of course, for historical reasons every operating system manufacturer prefers one or the other, but all terminals and many Smart Cards support both the direct and the inverse convention.

The format character

The second byte, T0, serves to indicate the subsequent "interface character", for which purpose a bit field is used. It further contains the number of subsequent "historical characters". Like TS, this byte is obligatory in every ATR.

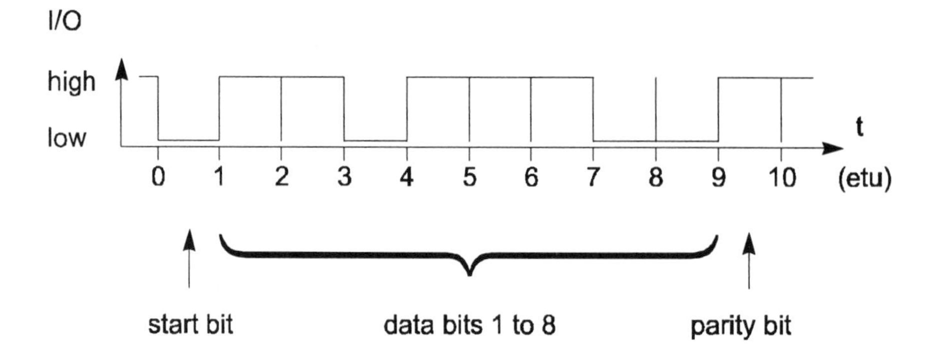

Figure 6.38 Time diagram of initial character TS with direct convention ('3B')

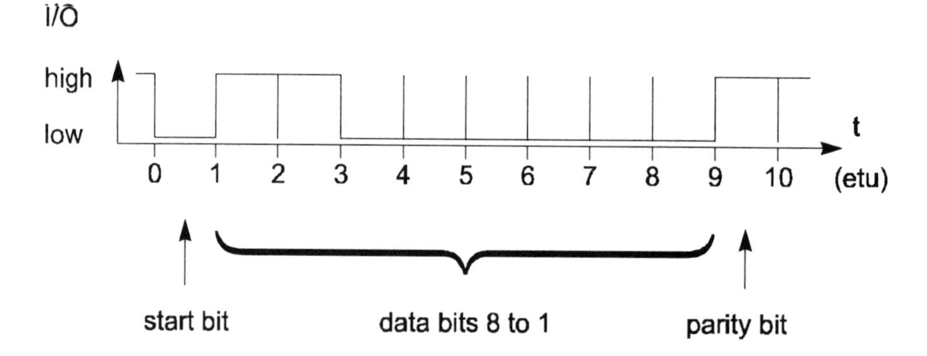

Figure 6.39 Time diagram of initial character TS with inverse convention ('3F')

b8	b7	b6	b5	b4	b3	b2	b1	Meaning
			'3B'					Direct convention
			'3F'					inverse convention

Figure 6.40 Initial character TS codes

b8	b7	b6	b5	b4	b3	b2	b1	Meaning
...	...	...	...	X	X	X	X	Number of historical characters (0 to 15)
...	...	...	1	...	...	...	...	TA1 transmitted
...	...	1	...	...	...	...	...	TB1 transmitted
...	1	...	...	...	...	...	...	TC1 transmitted
1	...	...	...	...	...	...	...	TD1 transmitted

Figure 6.41 Format character T0 codes

The interface characters

The interface characters specify all transmission parameters of the current protocol. They are constructed from the bytes TAi, TBi, TCi and TDi. However, the ATR transmission of these bytes is optional, and may be omitted if appropriate. Since fundamental values are defined for all of the protocol's transmission parameters, interface characters are often not required in the ATR in a conventional communications process.

Interface characters may be subdivided into global interface characters and specific interface characters. The former specify fundamental transmission protocol parameters, e.g. the divider, which are relevant for all subsequent protocols. The latter, on the other hand, lay down parameters for a very specific transmission protocol. The work waiting time for T=0 is a typical example.

The global interface characters apply in principle to all protocols. For historical reasons (originally only T=0 was included in ISO), several of these characters are only relevant to this protocol. If T=0 is not to be implemented, they can be omitted. Then the default values apply.

Figure 6.42 Interface character structure

The TDi byte's only purpose is to ensure the chaining of any subsequent interface characters. The most significant TDi half-byte indicates, bit-wise, the presence of the subsequent interface characters. This is analogous to the format character T0's encoding. The least significant TDi half-byte indicates the currently valid transmission protocol.

If no TDi byte is present, this means that TAi+1, TBi+1, TCi+1 and TDi+1 are not being transmitted.

b8	b7	b6	b5	b4	b3	b2	b1	Meaning
...	...	...	...		X			Transmission protocol number (0 to 15)
...	...	...	1		...			TAi+1 transmitted
...	...	1	...		...			TBi+1 transmitted
...	1	...	...		...			TCi+1 transmitted
1	...	...	...		...			TDi+1 transmitted

Figure 6.43 TDi byte codes

The other interface characters not used for chaining, TAi, TBi and TCi, specify the available transmission protocol(s). As described in ISO/IEC 7816-3, they are interpreted as follows.

Global interface character TA1

The divider (clock rate conversion factor) is encoded in the most significant half-byte as FI. The bit adjustment factor D is encoded in the least significant half-byte as DI.

b8	b7	b6	b5	b4	b3	b2	b1	IFSC
	X				...			FI
	...				X			DI

Figure 6.44 TA1 codes

F	internal clock		372	558	744	1116	1488	1860	RFU
FI	0000		0001	0010	0011	0100	0101	0110	0111
F	RFU	512	768	1024	1536	2048	RFU	RFU	—
FI	1000	1001	1010	1011	1100	1101	1110	1111	—

Figure 6.45 FI codes

D	RFU	1	2	4	8	16	RFU	RFU
DI	0000	0001	0010	0011	0100	0101	0110	0111
D	RFU	1/2	1/4	1/8	1/16	1/32	1/64	—
DI	1001	1010	1011	1100	1101	1110	1111	—

Figure 6.46 DI codes

The above-mentioned encoding of the divider F and transmission adjustment factor D allow the typical transmission speeds to be determined in line with the standard. This is summarized in the Appendix (15.4.2 Conversion table for ATR data elements).

The following interrelationships apply. The duration of one ATR and PTS bit, designated initial etu, is calculated as follows:

$$\text{initial etu} = \frac{372}{f_i} \text{s}$$

The transmission protocol which follows the ATR and PTS defines the duration of a bit independently of the ATR. It is known as work etu and defined as

$$\text{work etu} = \frac{1}{D} \times \frac{F}{fs}\,\text{s}$$

This makes it possible to modify and adjust the transmission speed individually through the two parameters, bit rate adjustment factor D and clock rate conversion factor F. The applied clock frequency is abbreviated as *f*, and is defined in hertz in the above formulae.

Global interface character TB1

This byte encodes a factor II for programming voltage in bits b7 and b6. Bits b5 to b1 define PI1. The highest bit, b8, is always set to 0, i.e. it is not used.

b8	b7	b6	b5	b4	b3	b2	b1	IFSC
0	X		...	...	...	...	...	II
0	...		X	X	X	X	X	PI1

Figure 6.47 TB1 codes

These parameters were needed in the first generation of Smart Cards, which had an EPROM for data storage instead of the currently usual EEPROM. The high voltages and currents required for EPROM programming had to be supplied by the terminal via the V_{PP} contact. However, since Smart Cards without internal charging pump no longer exist, we will skip the detailed encoding.

Thus, data elements PI1 and II always have the value 0, which signals that no external programming voltage is needed. If the data element TB1 is omitted in the ATR, the standard's default V_{PP} supply setting of 5 V at 50 mA applies.

Global interface element TC1

This encodes an extra guard time, designated N, as a hexadecimal and even byte without a preliminary character. Its value specifies the number of etus by which the guard time should be extended. TC1 is interpreted linearly, except that N = 'FF' means that a guard time which normally is two etus should be shortened to one etu. In practice, this means a speed increase of *ca.* 10%, since one bit fewer needs to be sent.

b8	b7	b6	b5	b4	b3	b2	b1	IFSC
				X				Additional guard time N

Figure 6.48 TC1 codes

Global interface TB2

This byte holds the value of PI2. The data element provides the external programming voltage in tenths of a volt. Hence, it is not normally used any longer in ATR for the same reason as TB1.

b8	b7	b6	b5	b4	b3	b2	b1	IFSC
				X				PI2

Figure 6.49 TB2 codes

The following specific interface character is additionally defined for transmission protocol T=0.

Specific interface character TC2

The last data element in the T=0 protocol encodes the work waiting time and is designated TC2. The work waiting time is the maximum time interval between the rising slopes of two consecutive bytes:

work waiting time = (960×D×WI) work etu

If the ATR does not contain a TC2, the default value of the work waiting time, WI = 10, is used.

b8	b7	b6	b5	b4	b3	b2	b1	Meaning
				X				WI

Figure 6.50 TC2 codes

The following additional bytes are defined for the T=1 transmission protocol in accordance with ISO/IEC 7816-3 Amd. 1. The interface characters prescribed for T=0 find application here only as and when required. Data elements — such as TC1 for extra guard time — may be omitted, as they would be irrelevant in T=1.

In this protocol, the data element index i must always be greater than 2. The specific interface characters TAi, TBi, TCi (i > 2) always apply to whichever transmission protocol is given by $TD_{(i-1)}$.

Specific interface character TAi (i > 2)

The TAi byte contains the maximum length of the information field that can be accepted by the card (IFSC). Its value must be in the range 1 to 254. IFSC's default value is 32 bytes.

b8	b7	b6	b5	b4	b3	b2	b1	Meaning
				X				IFSC

Figure 6.51 TAi (i > 2) codes

Specific interface character TBi (i > 2)

The least significant half-byte — the four bits b4 to b1 — contains the codes for the character waiting time CWT[8], so that the following applies:

$$CWT = (2^{CWI} + 11)\ \text{work etus}$$

The most significant half-byte holds the value BWI, with which the block waiting time BWT[9] can be calculated as follows:

$$BWT = 2^{BWI} \times 960 \times \frac{372}{f} s + 11\ \text{work etus}$$

b8	b7	b6	b5	b4	b3	b2	b1	Meaning
	...				X			CWI
	X				...			BWI

Figure 6.52 TBi (i > 2) codes

Specific interface character TCi (i > 2)

The bit b1 encodes the error detection code's calculation method.

Since the standard covering the ATR's data elements does not define all the possible transmission protocol parameters within the interface characters setup, different implementations can use additional interface characters.

A typical example is provided by the national German protocol T=14. Several additional ATR bytes are defined in it for specific use by this protocol. Decoding, though, is only possible by the users of this protocol, since it can only be executed by knowing the corresponding specification. But this is neither covered by a standard nor known outside the relevant application.

---

[8] See also 15.4.1 Time interval for ATR.

[9] See also 15.4.2 Conversion table for ATR data elements.

b8	b7	b6	b5	b4	b3	b2	b1	Meaning
...	...	...	...	...	...	...	0	LRC in use
...	...	...	...	...	...	...	1	CRC in use
0	0	0	0	0	0	0	...	Reserved for future applications

Figure 6.53 TCi (i > 2) codes

Global interface character TA2

This byte indicates the possible PTS modes. This is explained in more detail in Section 6.4, which deals with the PTS.

TA2 encoding is shown in Figure 6.54.

b8	b7	b6	b5	b4	b3	b2	b1	Meaning
0	...	...	...			...		Switching between negotiable mode and specific mode possible
1	...	...	...			...		Switching between negotiable mode and specific mode not possible
...	0	0	...			...		Reserved for later applications
...	...	...	0			...		Transmission parameters explicitly defined in the interface characters
...	...	...	1			...		Transmission parameters implicitly defined in the interface characters
...	...	...	...			X		Protocol T=X to be used

Figure 6.54 TA2 codes

The historical characters

For a long time the historical characters were not defined by any standard, as a result of which they contained the most diverse data, depending on the system manufacturers' preferences.

Many firms use the available bytes to specify the operating system as well as the ROM mask's version number. This is mostly encoded in ASCII, and is easy to interpret. The historical characters' presence within the ATR is not prescribed. Thus it is possible to leave them out altogether, which in some cases can be useful as it makes the ATR shorter and quicker to transmit.

The international standard ISO/IEC 7816-4 provides for an ATR file, in addition to the historical characters. This file, with the reserved FID '2F01', contains further data about the ATR. It is designed as an extension to the historical characters, which are limited to 15 bytes. The contents of this file, whose structure is not laid down by the standard, is ASN.1 encoded.

Figure 6.55 Structure of historical character

The data elements held in the ATR file or in the historical characters, may be multi-layered data concerning the Smart Card and the operating system in use. Thus, the concept involves the storage of card-supported file selection functions, implicit selection and data concerning the mechanism of the logic channels. Furthermore, additional data may be held there about the card issuer, the card and chip serial numbers and the ROM, mask, chip, and operating system version numbers. The coding of the relevant data object is covered in ISO/IEC 7816-5.

The check character

This last byte in the ATR is an XOR checksum from byte T0 to the last byte before the TCK. It can also apply a parity check to ensure correct ATR transmission. Nevertheless, despite the supposedly simple format and computation of this byte, there are several significant differences between the various transmission protocols.

Where only the T=0 transmission protocol is indicated in the ATR, the TCK checksum may not be present at the end of the ATR. In that case it is not transmitted at all, since byte-wise error detection through a parity check and repetition of the faulty byte are mandatory in T=0. In contrast, a TCK byte must be present in the case of T=1. Calculation starts with byte T0 and ends with the last interface character, or the last historical character if present.

6.4 PROTOCOL TYPE SELECTION

In the ATR's interface characters, the Smart Card displays various data transmission characters such as e.g. the transmission protocol and the character waiting time. If a terminal wants to modify one or more of these parameters, a protocol type selection (PTS) must be performed as per ISO/IEC 7816-3 and ISO/IEC 7816-3 Amd. 2, before actual execution of the protocol. If the card permits it, the terminal can use this to modify certain protocol parameters.

The PTS can be implemented in two different ways. In the negotiable mode, the standard values of the divider F and the transmission adjustment factor D remain unchanged until a PTS is successfully carried out. If the card uses the specific mode, then the values for F and D given in the ATR are mandatory also for the transmission of the PTS. The TA2 byte indicates which of these two modes is supported by the card.

Description	Value	Meaning	Comment
TS	'3B'	direct convention	
T0	'B5'	Y1 = 1011 = 'B' K = '5'	TA1, TB1, TC1 follow 15 historical characters
TA1	'11'	FI = 0001 = '1' DI = 0001 = '1'	F = 372 D = 1
TB1	'00'	II = 0 PI1 = 0000 = '0'	I = 0 V_{PP}-PIN not in use
TD1	'81'	Y2 = 1000 = '8' T = 1	TD2 follows
TD2	'31'	Y2 = 0011 = '3' T = 1	TA3 and TB3 follow
TA3	'46'	I/O buffer length = 70 bytes	Length of ICC I/O buffers (layer 7)
TB3	'15'	BWI = '1' CWI = '5'	BWT = 2011 etu CWT = 43 work etu
T1	'56'	"V"	"V 1.0"
T2	'20'	" "	
T3	'31'	"1"	
T4	'2E'	"."	
T5	'30'	"0"	
TCK	'1E'	The check character	XOR-checksum from T0 up to and including T5

Figure 6.56 Example of Smart Card ATR with transmission protocol T=1

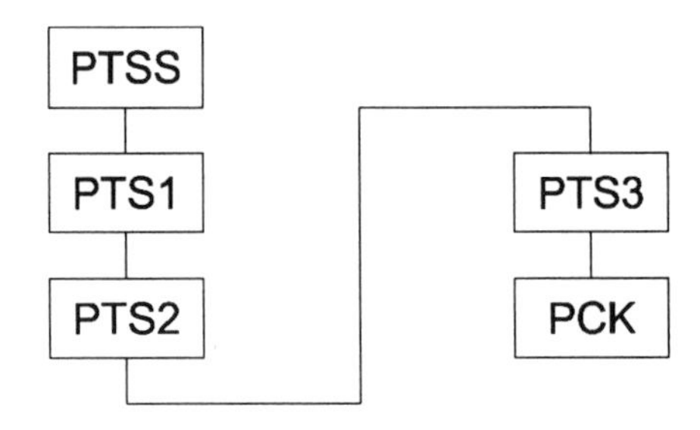

Figure 6.57 Basic PTS structure

The PTS request must be performed immediately on ATR receipt by the terminal. If the card permits the requested protocol parameter modification, it transmits the received PTS bytes back to the terminal. In principle, this is a reflection of the received data. In the other case, the card transmits nothing and the terminal must go through another reset sequence so that the card leaves this state. The PTS may only be carried out once immediately after the ATR; a repeat PTS transmission is forbidden by ISO/IEC 7816-3.

In practice, though, it is very rare for a PTS to have to be performed at all, since in Smart Cards currently in use the transmission parameters are designed to be in exact agreement with the terminal.

The first byte, the initial character PTSS, informs the card unambiguously that a PTS request will be made by the terminal immediately after the ATR. Therefore it always has the value 'FF' and its presence in a PTS is mandatory.

The data element following the PTSS, the format character PTS0, is also a required component of every PTS. This may optionally be followed by up to three bytes, called the parameter characters and designated PTS1, PTS2 and PTS3. They encode various parameters of the transmission protocol initiated after the PTS.

b8	b7	b6	b5	b4	b3	b2	b1	Meaning
...	...	...	...			X		Transmission protocol to be used
...	...	...	1			...		PTS1 available
...	...	1	...			...		PTS2 available
...	1	...	...			...		PTS3 available
0	...	...	...			...		Reserved for future functions

Figure 6.58 PTS0 codes

b8	b7	b6	b5	b4	b3	b2	b1	Meaning
	X				...			FI
	...				X			DI

Figure 6.59 PTS1 codes

b8	b7	b6	b5	b4	b3	b2	b1	Meaning
...	...	...	...	...	...	0	0	No additional guard time necessary
...	...	...	...	...	...	0	1	N = 255
...	...	...	...	...	...	1	0	Additional guard time of 12 etu
X	X	X	X	X	X	...	...	Reserved for future functions

Figure 6.60 PTS2 codes

The data element PTS3 is reserved for future applications, and hence cannot yet be described here.

The last byte in the PTS is named PCK, and is the XOR checksum of all previous bytes starting with PTSS. Just like PTSS and PTS0, it is a mandatory component of PTS, in contrast with all other data elements which are optional.

If the card can interpret the PTS and then modify the transmission protocol accordingly, it acknowledges this by transmitting the received PTS back to the terminal. If the PTS request contains items which the card cannot execute, it waits until the terminal performs a reset. The worst disadvantage of this procedure is the considerable waste of time until the actual transmission protocol is initiated.

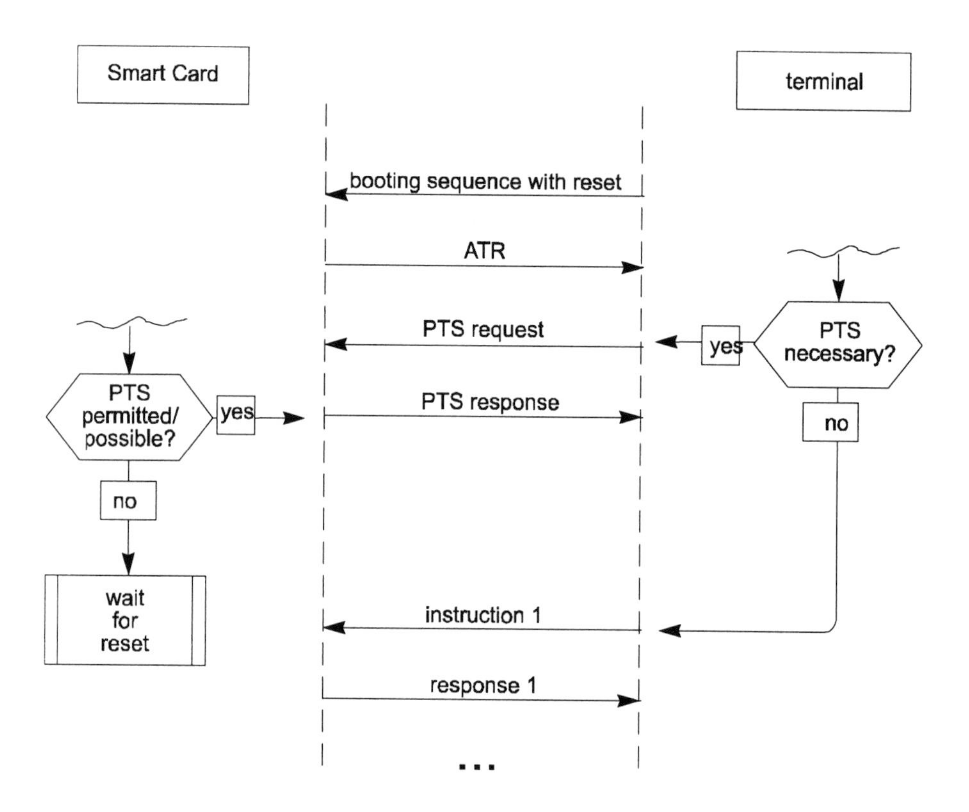

Figure 6.61 Typical PTS procedure in a GSM card

The PTS described above would not convert the protocol in the case of terminals which cannot execute a PTS, but have their own special transmission protocol. But this is exactly the case e.g. in the cardphones used in Germany. A special procedure was designed to allow the possibility of protocol conversion, the above restriction notwithstanding.

Since all terminals carry out a reset sequence if the ATR is unrecognized, it was decided that the Smart Card should switch the transmission protocol over after each reset. This can best be illustrated by an example: at the first reset, the card transmits the ATR for T=14 and is then ready to communicate in T=14. After the second reset it transmits an ATR for

T=1 and can communicate in T=1; after the third reset it can run T=14 once again. Technically this is not an optimal solution, since a device should always behave identically after every reset, but it does represent a practical solution for a heterogeneous terminal universe.

It is possible to minimize this disadvantage by having the Smart Card always respond with the same ATR after a power-on-reset (cold reset). A cold reset is always performed by the card directly after insertion in the card reader and the booting sequence. A reset over the card's reset line (warm reset), on the other hand, switches the transmission protocol over. Hence the card behaves identically after each "real" booting, and any additional triggering of a reset causes the transmission protocol to be switched.

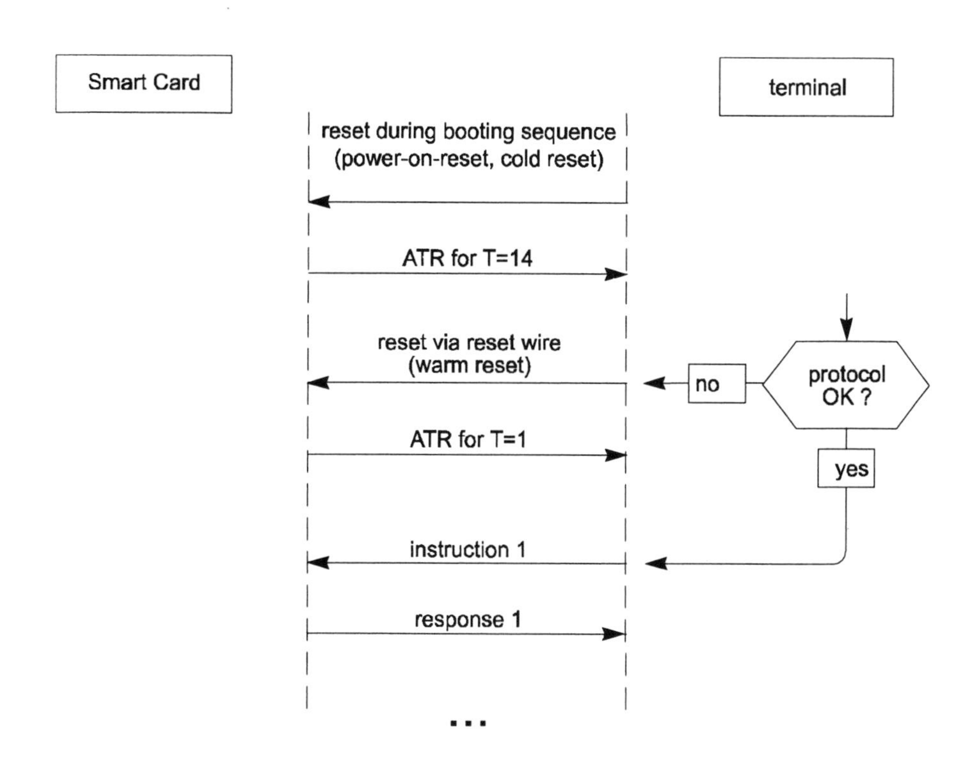

Figure 6.62 Typical PTS reset procedure

6.5 MESSAGE STRUCTURE

The entire data exchange between Smart Card and terminal takes place using APDUs. The term APDU is short for "application protocol data unit". It denotes internationally standardized data units in the application layer, which in the OSI model is layer 7. This is the layer which in Smart Cards is located directly above the transmission protocols. The protocol-dependent TPDUs are the data units in the layer directly below.

A distinction is made between instruction APDUs, which represent instructions to the card, and response APDUs, which are the card's replies. Stated simply, the APDUs are, so to speak, containers which hold a complete instruction to the card or a complete response from the card. They are carried by the transmission protocol transparently, that is, without modification or interpretation.

The APDUs which satisfy ISO/IEC 7816-4 are so structured that they are independent of the transmission protocol. Thus, an APDU must not be altered in its contents or format by differing transmission protocols. This applies above all to the two standardized protocols T=0 and T=1. The requirement of protocol independence influenced the structure of APDUs, since they must be capable of transparent transmission by both the byte protocol T=0 and the block protocol T=1.

6.5.1 Instruction APDU structure

An instruction APDU consists of a header and a body. The body may be of arbitrary length, or even be absent when the corresponding data field is empty.

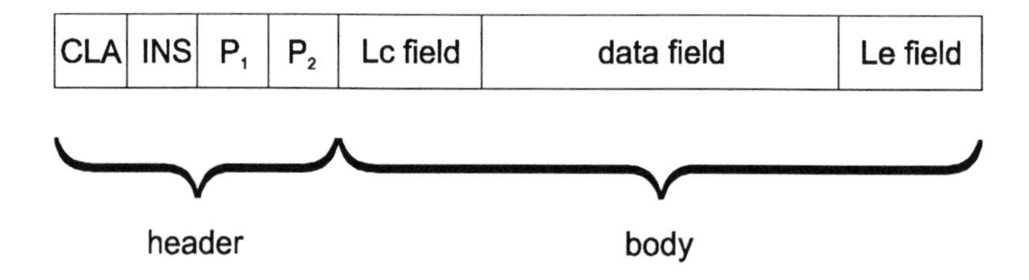

Figure 6.63 Structure of an APDU command

The header consists of the four elements class (CLA), instruction (INS) and parameters 1 and 2 (P1, P2). The class byte is currently still used to identify applications and their specific instruction sets. Thus, GSM uses the class byte 'A0', and the code '8X' is the one mostly used for manufacturer-specific instructions. In contrast, ISO-based instructions are encoded by class byte '0X'. The standard additionally specifies the class byte for identification of secure messaging and logical channels. Nevertheless, this is compatible with its use as an application identifier, as mentioned above[10].

The next byte in the APDU instruction is the instruction byte. Almost the entire address space of this byte can be exploited, the sole restriction being to even codes only. This is necessary since the T=0 protocol can activate the programming voltage by an instruction byte which is incremented by one, in the procedure byte. The instruction byte thus needs always to be even, to distinguish between these[11].

---

[10] See also 15.4.5 Table of class bytes used.

[11] See also 15.4.6 Table of the most important Smart Card instructions.

b8 to b5	b4	b3	b2	b1	Meaning
...	...	...	X	X	Number of the logical channel
...	0	0	...	...	No Secure Messaging
...	0	1	...	...	Secure Messaging not in accordance with ISO, own procedure used
...	1	0	...	...	Secure Messaging in accordance with ISO, header not authentic
...	1	1	...	...	Secure Messaging in accordance with ISO, header authentic
'0'	...	...	...	...	Structure and codes in accordance with ISO/IEC 7816-4
'8', '9'	...	...	...	...	Structure in accordance with ISO/IEC 7816-4, user specific codes and meanings of instructions and responses ("private use")
'A'	...	...	...	...	Structure and codes in accordance with ISO/IEC 7816-4, specified additional documents (e.g. GSM 11.11)
'F'	1	1	1	1	Reserved for PTS

Figure 6.64 The most important codes of the class byte (CLA)

The two parameter bytes are used, in the first place, to describe more closely the instruction selected by the instruction byte. Therefore they serve mainly as switches, to select various instruction options. Thus they are used to choose various options in SELECT FILE or to specify the offset in READ BINARY.

The section after the header, i.e. the body, which can be omitted bar a length specification, fulfils a double role. First, it fixes the length of the data section sent to the card (Lc field[12]), and also specifies the length of the data section to be sent back from the card (Le field[13]). Secondly, it contains the instruction-related data which are sent to the card. If the Le field's value is '00', then the terminal expects the card to send the maximum amount of data available for this instruction. This is the only exception, as otherwise the length is described numerically.

Normally, Le and Lc fields are one byte long. It is possible to construct from this a three-byte Le/Lc field. This can be used to represent lengths up to 65 536, since the first byte would be encoded as an escape sequence 'FF'. The standard already defines this three-byte length specification for future applications, but due to currently available memory sizes this cannot yet be implemented.

By combining the previously described parts of the instruction APDU, the four general cases shown in Figure 6.66 can be distinguished.

---

[12] Length command.

[13] Length expected.

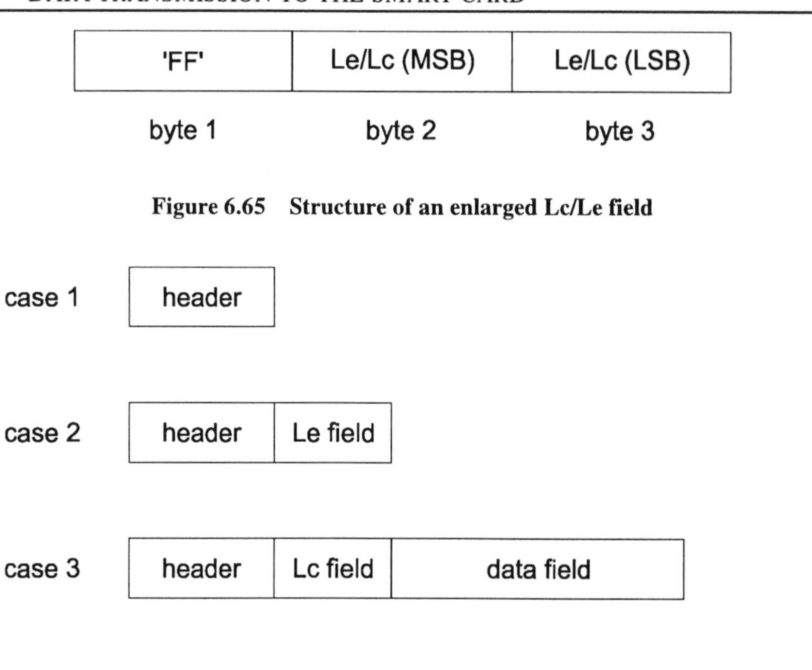

Figure 6.65 Structure of an enlarged Lc/Le field

Figure 6.66 Command APDU cases

6.5.2 Response APDU structure

The response APDU, sent by the card in reply to an instruction APDU, consists of an optional body and a mandatory trailer (Figure 6.67).

The body consists of the data field, whose length was determined in the previous instruction APDU's Le byte. This instruction notwithstanding, the length can be nil if the Smart Card has interrupted instruction processing due to an error or wrong parameters. This is then indicated in the two single-byte status words 1 and 2 in the trailer.

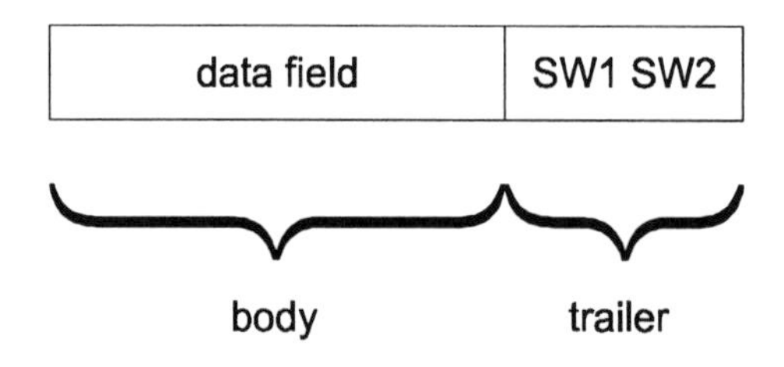

Figure 6.67 Structure of response APDU

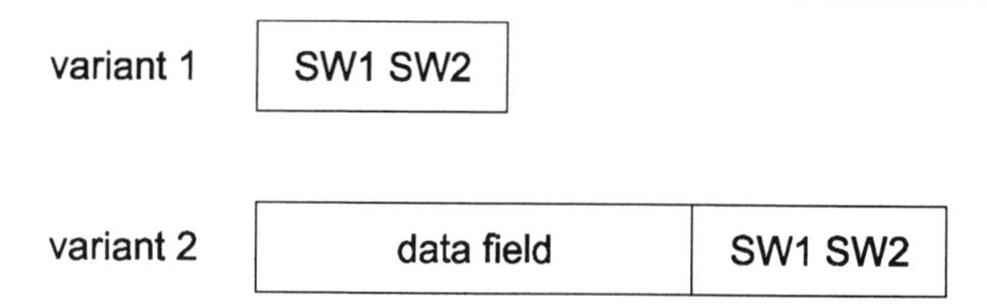

Figure 6.68 Variations on response APDUs

The trailer must, in every case, be transmitted by the card in response to an instruction. The two bytes SW1 and SW2, also designated the return code, contain the response to the instruction. For example, the return code '9000' means that an instruction was executed successfully and completely. The system under which the more than 50 different codes are organized, is visualized in Figure 6.69[14].

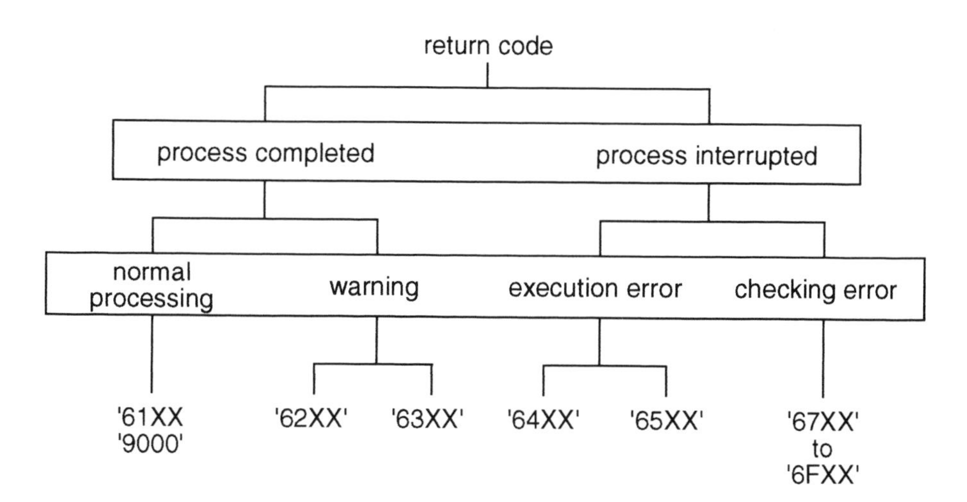

Figure 6.69 System of return codes in accordance with ISO/IEC 7816-4

If the return code '63XX' or '65XX' is sent back after an instruction is executed, it means that the card's non-volatile memory has undergone some change. If, on the other hand, the code '62XX' or '64XX' is returned, then the instruction was interrupted without a change to the non-volatile memory.

It may be noted that although a standard for return codes exists, many applications use codes which deviate from it. The only exception is the code '9000', which all but universally designates successful execution. With all other codes it is necessary to inspect the underlying specification in order to discover their meaning unambiguously.

---

[14] See also 15.4.8 Important Smart Card return codes.

6.6 SECURE MESSAGING

The entire data exchange between terminal and Smart Card takes place via digital electric pulses across the card's I/O channel. It is theoretically and technically simple to solder a wire to the I/O contact field, record the whole communication and analyse it at some later time. This provides information about all data transmitted in both directions.

It would be somewhat more expensive to isolate the I/O contact field electrically, and install on it a dummy contact. It would be further necessary to use a thin wire to connect the dummy contact as well as the original contact electrically to a computer. With this arrangement it is straightforward to permit only certain instructions to reach the card, or to interpolate one's own instructions.

Both these typical methods of attack would only be successful if secret data travelled unprotected over the I/O channel. In principle, therefore, data transmission should be so designed that an attacker, even though always able to tap into a transmission and also to insert extraneous transmission blocks into the protocol, would still be incapable of gaining any advantage therefrom.

Various mechanisms and procedures exist for protection against these or even more sophisticated attacks. Together they are known as secure messaging. They are not specific to Smart Cards, as they have been in use for quite some time in data communications. The specific features characterizing the Smart Card domain are that neither the computational power of the communicating parties, nor the transmission speed, are particularly high. Therefore, the standard procedures which are in general use have been scaled down to accord with Smart Card capabilities, whilst at the same time the security offered by these methods was not permitted to suffer.

The purpose of secure messaging, then, is to ensure the authenticity, and if necessary the confidentiality, of all or some of the transmitted data. A variety of security mechanisms are used to meet this requirement. A security mechanism is defined as a function which satisfies the following: a cryptographic algorithm, a key, an argument, and initial data if needed. A global condition needs also to be satisfied. All security mechanisms must behave completely transparently vis-à-vis the relevant protocol layers, so that procedures already standardized are not compromised by secured messaging. This applies particularly to the two transmission protocols T=0 and T=1, as well as to the generally used and standardized Smart Card instructions.

Before executing the procedure, both communicating parties must agree on the cryptographic algorithm to be used and on a common secret key. According to Kerckhoff, the whole security of the method relies on this key. If it were revealed, the secure messaging approach would become nothing more than a generally known, additional checksum which reduces data transmission speed and with which one could at most correct transmission errors.

Several variations of secure messaging have been around for some years. They are all relatively rigid, and tailor-made for a particular application. Most of them cannot be faulted as far as security is concerned. Nevertheless, none of them have become internationally established or were sufficiently flexible to be adopted as a standard.

The criteria of transparency to existing instructions, two fundamentally different transmission protocols and maximum adaptability, have resulted in the standardization of a very flexible but therefore also expensive and complex secure messaging method, under ISO/IEC 7816-4. It is based on the embedding of all user data in TLV-encoded data objects. Three different types of such objects are defined:

- Data objects for plaintext — contain data in plaintext (e.g. an APDU data section)
- Data objects for security mechanisms — contain the outputs of security mechanisms (e.g. MAC)
- Data objects for auxiliary functions — contain control data for secure messaging (e.g. the padding method used).

The class byte is used to indicate whether the instruction uses secure messaging. The two available bytes can encode whether the procedure specified in ISO/IEC 7816-4 is utilized, and whether the header is included in the cryptographic checksum (CCS)[15]. If the header is included in the computation then it is authentic, as it cannot be changed unobtrusively during transmission.

Plaintext data objects

According to the standard, all data which is not BER-TLV encoded must be embedded, i.e. encapsulated, in data objects. Various identification characters exist, and they are listed in Figure 6.70. Bit 1 of the ID determines whether a data object is included in the computation of the cryptographic checksum. Thus, a bit which is not set (e.g. 'B0') means that the data object is not part of the calculation, and a set bit (e.g. 'B1') that it is.

Tag	Meaning
'B0', 'B1'	BER-TLV encoded and contains data objects connected with secure messaging
'B2', 'B3'	BER-TLV encoded and contains data objects not connected with secure messaging
'80', '81'	No BER-TLV encoded data
'99'	Secure messaging information

Figure 6.70 Tags for simple text data objects

Data objects for security mechanisms

The data objects used in security mechanisms are divided into those used for authenticity, and those used for confidentiality. The ID characters defined for this purpose are listed in Figures 6.71 and 6.72. The concept of authenticity covers all data objects which are to do with cryptographic checksums and digital signatures. Data encryption and the identification needed to this end during secure messaging come under the heading of confidentiality. Depending on the particular procedure used, the relevant ID should be looked up in both tables and used during secure transmission.

---

[15] Encoding of class byte, see 6.5.1 Instruction APDU structure.

Tag	Meaning
'8E'	Cryptographic checksum
'9A', 'BA'	Entry value for digital signature
'9E'	Digital signature

Figure 6.71 Tags for data objects for authentication

Tag	Meaning
'82', '83'	Cryptogram tag: simple text is BER-TLV encoded and locks data objects for secure messaging
'84', '85'	Cryptogram tag: simple text is BER-TLV encoded and does not lock data objects for secure messaging
'86', '87'	Tag for padding method used '01' — padding with '80 00 ... ' '02' — without padding

Figure 6.72 Tags for confidential data objects

Data objects for auxiliary functions

The data objects for auxiliary functions are used to determine global conditions during secure messaging. The two parties use them to exchange data about the cryptographic algorithms and keys used, starting data and similar fundamental information. Theoretically, these may differ between transmitted APDUs or even between instruction and response. In practice, though, these data objects are very rarely used, since all secure messaging global conditions are defined implicitly, and do not have to be specifically agreed during the course of communication.

Using the options offered by secure messaging under ISO/IEC 7816-4 and only indicated above in outline, we describe below two fundamental procedures. The description is kept deliberately simple, in order to make understanding of the complex mechanisms as easy as possible. Due to the very high flexibility offered, there are many additional and also more complex combinations of security mechanisms. The two procedures shown here represent a synthesis between simplicity and security.

The authentic procedure protects the user data, i.e. the APDUs, with a cryptographic checksum (CCS, MAC) against manipulation during transmission. In contrast, the combined procedure is used to encrypt the user data completely, so that an attacker cannot draw any conclusions about the data contents of the transmitted instructions and the responses to them. A transmission sequence counter is only seen in combination with one of these two procedures. This counter, whose initial value is a random number, is incremented with each instruction and each response. This allows both parties to discover whether an instruction or a response was lost or inserted. Used together with the combined procedure, a transmission sequence counter can also make equal APDUs appear to be different, which concept is also known as "diversity".

6.6.1 The authentic procedure

The authentic procedure guarantees an authentic, that is non-falsifiable, APDU transmission. The recipient of an APDU, i.e. an instruction or a response, can determine whether it was altered during transmission. This makes it impossible for an attacker to modify data within an APDU without this fact being noticed by the recipient.

The implementation of this procedure is indicated by a bit in the class byte, so that the recipient can act accordingly and check the arriving APDU for authenticity. The APDUs themselves are transmitted in plaintext and are not encrypted. That is, the transmitted data is public both before and after transmission, and can be received and evaluated by the attacker, assuming the transmission channel is being tapped into. This is not necessarily a disadvantage, since under data protection legislation it is preferable not to transmit confidential data publicly. This offers the user at least the theoretical possibility of finding out which data is exchanged between his or her Smart Card and the terminal.

In principle, any block encryption algorithm can be used to compute the cryptographic checksum. The following descriptions always assume, for practical reasons, a DEA with a fixed 8-byte block length. Hence the individual data objects need to be made up to a multiple of 8 bytes, known as padding. Data objects whose length is already a multiple of 8 bytes are nevertheless extended by one block. After padding, a cryptographic checksum (CCS) of the entire APDU is computed with the DES algorithm in CBC mode. This 8-byte checksum is attached directly to the APDU as a TLV-encoded data object with the four least significant bytes omitted; all padding bytes are removed after calculation of the checksum. The APDU, reformatted through the above procedure, is then sent across the interface. This procedure extends APDU length by 8 bytes which only lowers the transmission speed by a fraction, assuming conventional transmission block sizes.

The data objects for the control structures can indicate explicitly which algorithm and which padding method are used. Here we assume, for simplicity, that the Smart Card and the terminal implicitly know all parameters of the secure messaging system being used.

When this protected APDU arrives at the recipient, the latter extends it again to a multiple of 8 bytes and computes in turn a MAC on the APDU. By comparing it with the received MAC, calculated by the sender, the recipient can determine whether the APDU was changed during transmission.

It is a precondition for computing a cryptographic checksum that a secret DEA key exists and is known to both parties. The key must be secret, since otherwise an attacker would be able to breach communications security; an APDU could be intercepted, changed as desired and a new and correct MAC computed. This new MAC would then replace the original one, and the newly generated APDU transmitted onward.

In order to improve protection of the MAC formation key against attacks through known plaintext—ciphertext pairs, it is usual to employ dynamic keys. These are generated by encrypting a random number previously exchanged between the terminal and the card, using a common key known to both parties.

The additional steps necessary for the transmission and reception of an APDU which is protected by the authentic procedure naturally reduce the effective transmission speed. On average, one may regard an approximate halving of transmission speed, compared with plaintext, as normal.

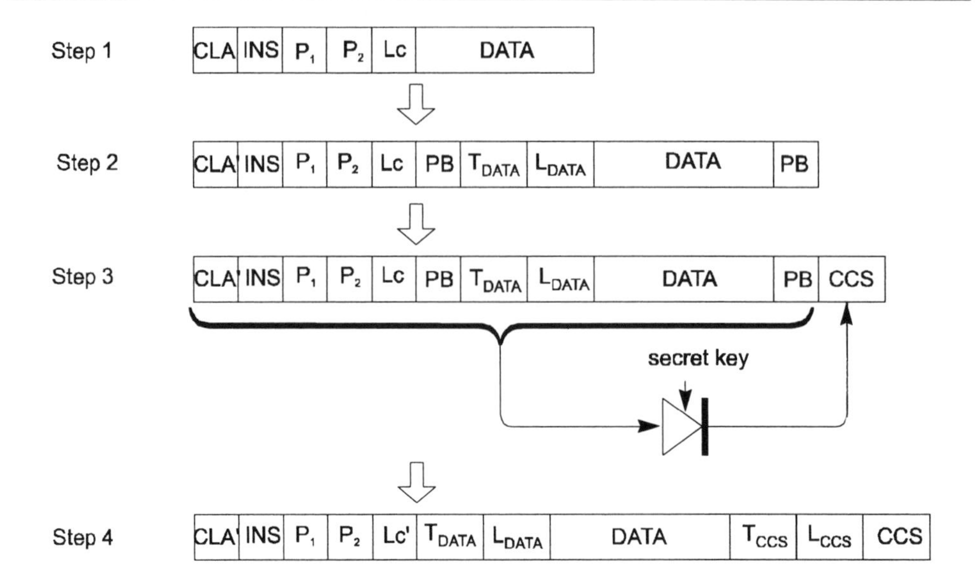

Figure 6.73 Creation of APDU instruction in authentication procedure. A case 3 instruction (e.g. UPDATE BINARY) is used, the header of which is included in the cryptographic checksum (CCS) (the creation of an APDU response is analog, PB — padding bytes).

Step 1 APDU format
Step 2 Transformation of data sections into TLV encoded data and padding of data objects by a multiple of eight
Step 3 Calculation of CCS
Step 4 Supplementing APDU with a TLV encoded data object containing the CCS

6.6.2 The combined procedure

The combined procedure represents the next step in improving security, compared with the authentic procedure. Transmission of the APDU's data section no longer takes place in plaintext, but rather in encrypted form.

The procedure builds on the authentic one. The data objects, which as in the previous method are protected by a cryptographic checksum, are again extended to a multiple of 8 bytes (padding) and encrypted with the DEA in CBC mode. The header is omitted from this process, which is necessary for reasons of compatibility with the T=0 protocol. If it is preferred to encrypt the header as well so as to make the instruction which is being sent to the card unrecognizable, then the T=0 instruction ENVELOPE is necessary. One bit in the class byte serves to indicate secure messaging. Encryption is followed by transmission across the interface. Since the recipient knows the secret key that was used in the encryption operation, he can decrypt the APDU. The decryption is then checked for correctness by recalculating the attached cryptographic checksum in the same level of the transmission layer.

If this method is employed, an attacker listening in on the I/O channel cannot discover which data is exchanged between card and terminal in the instruction and response. Nor is it possible to replace one of the encrypted blocks within the APDU, since they are linked to each other through the DEA operating in CBC mode. This means that any replacement would be noticed by the recipient at once.

As far as the cryptographic algorithm is concerned, the same is true here as was already noted when discussing the authentic procedure. In principle, any block encryption algorithm can be employed. Just as in the previous method, the keys should be of a temporary nature only, so that each session utilizes its own derived key.

With regard to the security benefits, blanket usage of the combined procedure would be recommended for all APDUs due to the increase in security, but for a considerable and significant drop in data transmission speed.

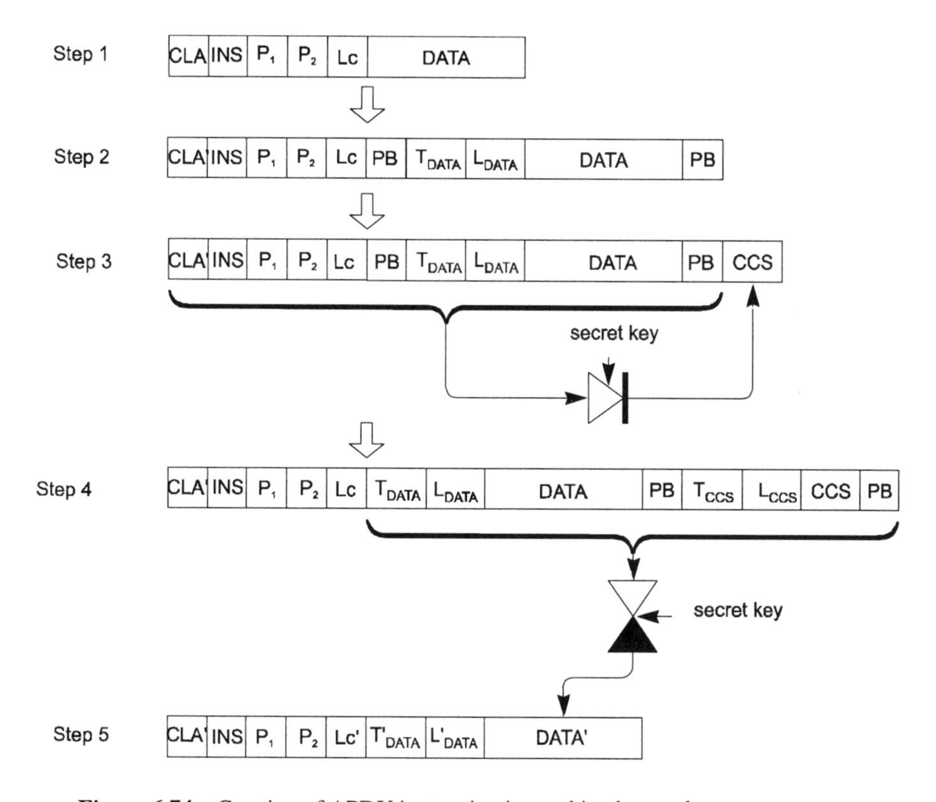

Figure 6.74 Creation of APDU instruction in combined procedure.
A case 3 instruction (e.g. UPDATE BINARY) is used, the header of which is included in the cryptographic checksum (CCS) (the creation of an APDU response is analog, PB — padding bytes).

Step 1 APDU format
Step 2 Transformation of data sections into TLV encoded data and padding of data objects by a multiple of eight
Step 3 Calculation of CCS
Step 4 Supplementing APDU with a TLV encoded data object containing the CCS, padding of data objects by a multiple of eight
Step 5 Encoding of APDU data section

The speed reduction is considerable. The time ratio between an unprotected APDU and one transmitted with the combined procedure is a factor of approximately four. Even the difference between the authentic and combined procedures comes to a factor of two. Therefore it is necessary to weigh carefully in each case which data should be transmitted in such an expensive but secure fashion.

6.6.3 Transmission sequence counter

Secured messaging using the mechanism of transmission sequence counter is not an independent security method. It can only be implemented sensibly in combination with either the authentic or the combined procedure, otherwise the counter could be altered by an attacker and remain undetected.

The operating principle is to allocate to each APDU a sequence number, which is a function of its transmission time. This would make the removal or insertion of an APDU during protocol execution immediately noticeable, and result in appropriate steps (interruption of communications) being taken by the recipient.

The function is based on a counter whose initial value is a random number, and which at the start of communications is sent by the card to the terminal at its request. The counter is incremented every time an APDU is transmitted. The counter's length should be neither too short, nor too long because of the additional transmission overheads. In the following comments the usual figure of two bytes is assumed, but in practice it may be longer.

There are two basic variations on the inclusion of the sequence counter in instruction and response APDUs. It can be added directly as a numerical value in a data object in the respective APDU. Alternatively, the counter may be XOR-ed with the same amount of data in the APDU. Then the cryptographic checksum is computed, and finally the altered data is restored. The recipient of this APDU is aware of the counter's expected value, and can alter the APDU in the same way as the sender. Finally he would compute the cryptographic checksum and test the APDU for correctness.

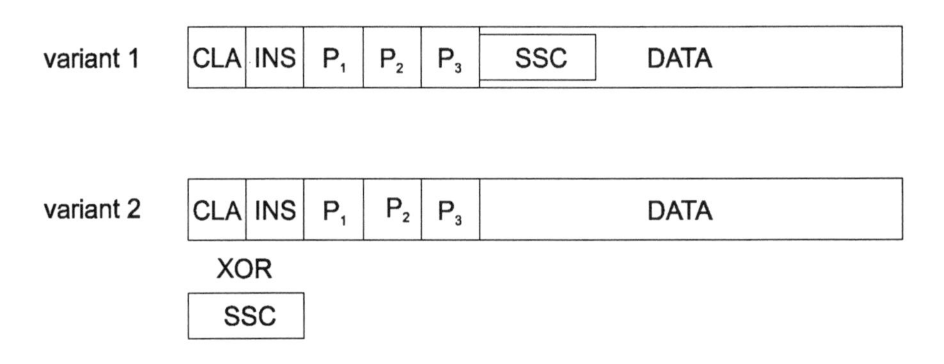

Figure 6.75 The two types of send sequence counters within an instruction APDU.
Type 1: send sequence counter as TLV-encoded data object in data section.
Type 2: send sequence counter only for CCS calculation through XOR tied to the APDU data
(SSC – send sequence counter)

The following process takes place during communications: first the terminal requests from the Smart Card an initial counter value. The latter sends back to the terminal a two-byte random number. The terminal transmits to the card the first of the secured instructions, together with the transmission sequence counter. Either the authentic or the combined procedure can be employed to protect the counter and the body. The card receives the protected APDU, checks first whether the authentic or combined procedure indicates that tampering has taken place, and then compares the counter. If it agrees with the one stored in the card, no APDU was inserted or removed during transmission.

It is apparent that the use of a transmission sequence counter is not only of interest where several instructions are to be executed in a particular order, but even in the case of a single instruction, since the session is individualized by the counter. The counter protects above all against reinsertion of APDUs already sent, and against APDU interception.

If the sequence counter is used together with the combined procedure, each encrypted block is different, a situation known as "diversity". This relies on the fact that the counter is incremented at every exchange of one APDU, and in an efficient encryption algorithm the alteration of one bit of plaintext modifies the appearance of the entire ciphertext block.

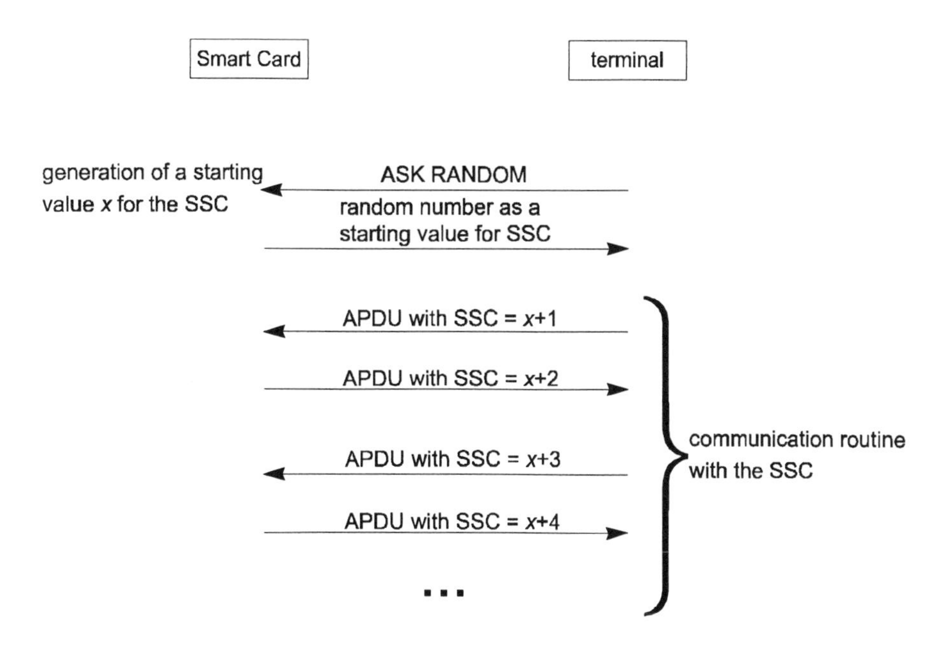

Figure 6.76 Transmission of APDUs with SSC (SSC — send sequence counter)

6.7 LOGICAL CHANNELS

In future Smart Cards, those containing many independent applications, it is intended to make the latter addressable over so-called logical channels. These would make it possible for up to four applications on one card to exchange data with a terminal in parallel with each other. As before, only the one existing serial interface would be used, but at the logic level the applications would be addressed individually.

The decision as to which instruction belongs to which application will be based on two bits in the class byte (bit 1 and bit 2)[16]. This makes up to four logical channels available, and up to four card applications sessions can proceed simultaneously in parallel. However, one restriction still applies to communicating with the four applications. The processes which access the card from outside must be synchronized among themselves and may not exhibit cross-overs, since the card's response APDU contains no information about the original logical channel. That is, externally it is impossible to determine which response has sent back which return code. The result of this absent channel specification is that after each instruction is sent to the card, it is necessary to wait for the relevant response.

The field of application of this very powerful mechanism is, above all, the simultaneous running of several applications. For example, a user conducts a telephone conversation, for which he uses the GSM application in the multi-application Smart Card. In order to confirm an appointment with his opposite party, he needs briefly to consult his personal organizer which is also on the card. Using a second logical channel, and in parallel to the GSM application, he searches for the date in a file within the organizer application, and tells his stressed manager whether he can attend the business meeting or not. This is a typical application of logical channels. Another example is the secure transfer of electronic money between two payment accounts on the same card.

However useful logical channels appear, their management by the card's operating system is fraught with difficulties. Each logical channel means, in effect, no less than a complete and free-standing Smart Card with all its states and conditions. At the end of the day, from the system's perspective this means the management of all data relating to several sessions running in parallel. The complexity involved is enormous, and only possible with large and expensive microprocessors. If, in addition, one requires individual secure messaging and all the possible variations on authentication for every separate logical channel, the demands on memory increase to a point which cannot be met by any currently available Smart Card microprocessor.

---

[16] See also 6.5.1 Instruction APDU structure.

7 The instruction set

The communications routine between a terminal and a Smart Card is always based on the so-called request–response procedure. This means that the terminal sends a request (i.e. an instruction) to the card, the card immediately processes it, produces a result and returns it to the terminal as its response. The card, therefore, never sends out data without first having received an instruction from the terminal. Even the ATR is no exception to this rule, since it is an answer to a reset, which in a way also represents a type of instruction to the card.

Communication itself always proceeds within the framework of a transmission protocol, namely T=0 or T=1. These fairly uncomplicated protocols meet the special requirements posed by Smart Card applications, and are optimized for this role. Deviations from these very precisely specified protocols during application run-time are not permitted. The protocols make it possible to send data to the card and back again, in a way which is completely transparent as far as the transmission layer is concerned. The data is embedded in a sort of container, called the application protocol data unit (APDU). The APDUs sent by the terminal to the card are the instructions to the card. The replies produced as a result – the responses – also arrive at the terminal in an APDU embedded in the transmission protocol. These mechanisms form the basis for a series of instructions to the card, which trigger actions in it, such as read or write instructions relating to the files stored on the card.

In Smart Card applications, the files are used either to store data, to store authorizations or both. This has led to the development of instruction sets specially optimized for these requirements, and transmission protocols, which are instruction sets only used in the Smart Card industry. Due to the severely restricted memory size in Smart Cards, and the economical pressure to keep it down so as to remain competitive, the instruction set is usually tailored to a particular application. All the instructions which are not needed in that application are removed in a rigorous optimization process. Only very few operating systems have a complete instruction set which is not reduced in this manner for a particular application. These cards, however, are correspondingly more expensive and therefore cannot be produced in large numbers. They are only used during the prototype and testing phases of large projects.

Another phenomenon typical to Smart Card instruction sets, as is common in new technologies, is that of diversification. Each manufacturer active in this field attempts to invent unique instructions for his operating system or for the application being developed. Often this is forced on them by the fact that no functionally comparable instructions exist in the standards. Or else manufacturers attempt, quite deliberately, to steal a march on the competition by producing advanced instructions which are optimized with regard to functionality and memory usage, or even to use these products to deny other manufacturers a foothold in the market. At the end of the day, the choice must be made by the system distributor. Nevertheless, a decision in favour of instructions based on existing standards is always one for an open, more easily expandable and also proven system, one in which many applications could eventually be incorporated in one card. On the other hand, there are plenty of examples where the introduction of Smart Cards was only made possible in the first place through the use of highly optimized special instructions.

There are four international standards plus more or less finalized draft standards which define typical Smart Card instruction sets. These standards describe rather more than 50 instructions and their corresponding execution parameters. The instructions defined in these four separate standards are largely compatible in coding and functionality.

The standard developed for use in the telecommunications industry, GSM 11.11 (prETS 300608) forms the normative basis for the GSM card. In practical terms, this standard is now beyond modification due to the millions of cards produced and currently in circulation. Some very limited additions and extensions are still possible. Hence this standard effectively applies to Smart Cards worldwide, and in principle forms the basis for most Smart Card operating systems.

European standard EN 726-3, which like GSM 11.11 finds its application in the telecommunications field, is a superset of GSM 11.11. It defines many additional general instructions, which in contrast to GSM 11.11 are not tailored specially for one application. This standard also contains a large number of instructions for application management, which is mainly of interest in the domain of multi-application Smart Cards.

In turn, a large proportion of the instructions in the two above-mentioned standards are contained in ISO/IEC 7816-4, which is a general and international industry standard for Smart Card instruction sets. It is not dedicated to a particular field, such as telecommunications, but attempts to cover all Smart Card areas of application.

Special instructions which are only employed in a limited field are not covered by these standards, and therefore must be specified individually. One example of this situation is the instruction set for a universal electronic purse, currently defined in the preliminary CEN standard prEN 1546. When finalized, this will become a European standard covering all the instructions necessary for an electronic purse and all its execution parameters. Such standards, restricted to a single application, are only created in areas which are of particular interest to the authorities or to particular branches of industry, since the cost would otherwise be prohibitive.

The instructions in the standards described above can be classified by function as follows:

- File selection
- File reading and writing
- File searching
- File operations

- Identification
- Authentication
- Execution of cryptographic algorithms
- File management
- Instructions for electronic purses/credit cards
- Completion of operating systems
- Smart Card hardware testing
- Special instructions only implemented in a single application
- Instructions supporting transmission protocols

In current, real-life Smart Card operating systems, only a subset of all the instructions listed above is implemented. Depending on the manufacturer, there may be greater or smaller deviations from the standards in terms of functionality and coding. However, the basic functions are always present, though they may be severely restricted due to memory or cost limitations. Whenever a new application is at the drawing board stage, the precise coding and functional specifications of each instruction should be obtained from the appropriate operating system manufacturer.

The following sections describe the most important and common Smart Card instructions, selected from the standards GSM 11.11, EN 726-3, ISO/IEC 7816-4, prEN 1546-3 and EMV-2.

It is impossible, however, to purchase even a single Smart Card anywhere in the world which contains all those instructions. As a conservative estimate, the memory required for full implementation would exceed by a factor of three to four the whole of the largest memory currently available in any Smart Card microprocessor. It is, moreover, unnecessary for a card to be able to execute all those instructions. Depending on the planned area of application and on the operating system, some instruction classes can be supported more comprehensively than others.

Multi-application cards have an additional advantage in that additional applications can be installed on them after they have been personalized. For example, a card designed for applications involving encryption would contain the full spectrum of cryptographic instructions along with their various algorithms, always assuming that sufficient memory is available. Each type of implementation requires a different selection of instructions from the classes listed above.

In order to obtain an overall view of the instructions available to the designers, each description given below specifies the standard in which the instruction is covered. If the source is not quoted, the instruction is one which is used internally by Smart Card manufacturers but cannot be assigned to any of the standards. Some of these instructions, though, are very useful, and will no doubt be standardized in the future. That is why they are listed here and their basic functions described.

Under each instruction, the response listed is the one obtained in the event of successful execution. Otherwise, if an operation is prohibited or a card fault occurs, the terminal only receives a 2-byte return code. Several of the described instructions also possess parameters for the selection of additional functions. Often these variations only exist theoretically in the standards but not in real operating systems, since they are too expensive to implement or pointless in practice. Therefore, this chapter does not list or explain every variant covered by the standards, since our aim is to concentrate on functions relevant to real cards. When describing the instructions, we usually followed the standard in which the instruction's functional options are the widest.

7.1 FILE SELECTION

Without exception, the file managers in all modern Smart Card operating systems are object oriented. This means, *inter alia*, that before any action is performed on an object (which corresponds to a file), it must first be selected. Only then does the system know which file is being operated on, and all subsequent file-specific instructions only apply to this one file. Of course, file access conditions still need to be checked within the operating system, i.e. whether the relevant instructions are even permitted or indeed possible.

The file manager is always implicitly selected after a card reset, and does not need to be specifically identified. Any subsequent file selection is executed with the instruction SELECT FILE. Addressing is either via the 2-byte FID (file identifier) or – additionally in the case of DFs – with the 5- to 16-byte AID (application identifier). The latter can also be quoted in partial form, so that its lowest bytes can be omitted. An additional parameter allows the card to select the first, last, next or previous DF to which the abbreviated AID relates.

In the older GSM 11.11 definition, file selection is only possible with the 2-byte FID. In contrast, the ISO instruction set also supports a functional extension in the form of file selection through a path specification to the relevant file.

Only successful selection of a new file results in deselection of the previous one. Where selection failed to execute, e.g. because the file does not exist, the older selection remains in force. This ensures that a file is always selected, even in the event of errors.

After successful selection, the terminal may, if appropriate, request data about the new file. This request, and the quantity of data requested, are sent to the card as part of the SELECT FILE instruction. The standards define the exact contents of these data. They may cover, *inter alia*, the new file's structure, size and available memory. The quantity of the data may also depend on the file's type.

Figure 7.1 shows the explicit file selection options permitted by ISO/IEC 7816-4 via the SELECT FILE insrtuction, and Figure 7.2 represents a typical file selection process.

In addition to this explicit file selection through a FID, AID or a path in the SELECT FILE instruction, it is possible to select the file implicitly. However, this is only allowed via the standard read and write instructions. By specifying an additional parameter, namely a FID abbreviated to its lowest 5 bits, a file can be selected before actual execution of the instruction. The file must be an EF, and be contained within the current DF. The advantage of this procedure lies in a simplified execution of the instruction and an increase in execution speed, since an explicit SELECT FILE instruction no longer needs to be transmitted to the card[1].

---

[1] See also: impl. short-FID.

SELECT FILE		
Instruction	•	FID (if EF, DF, MF) *or* AID (if DF) *or* path from DF to file *or* path from MF to file *or* *switch*: replaced by DF *or* first, last, next, orefious DF (in case of partial AID)
	•	*switch*: sending back information on chosen files
Response	•	information on chosen file (if selected)
	•	return code

Figure 7.1 Functions of SELECT FILE covered by ISO/IEC 7816-4

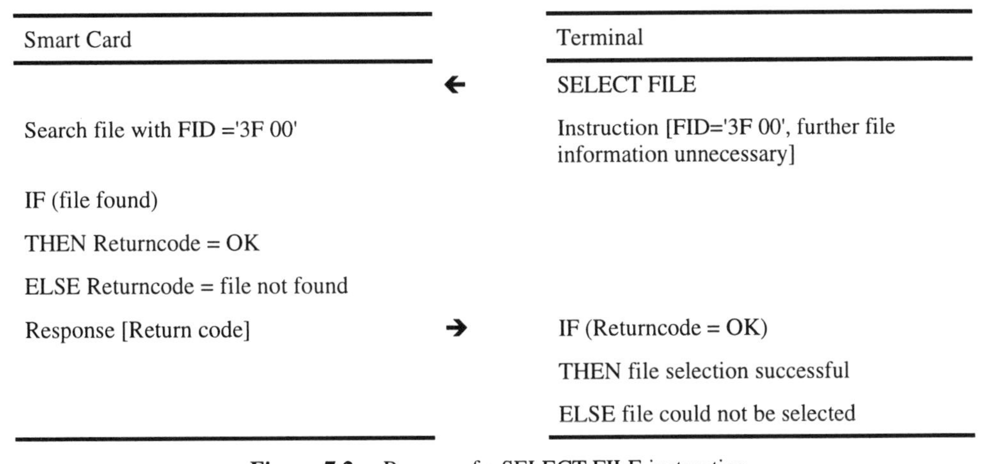

Smart Card		Terminal
	←	SELECT FILE
Search file with FID ='3F 00'		Instruction [FID='3F 00', further file information unnecessary]
IF (file found)		
THEN Returncode = OK		
ELSE Returncode = file not found		
Response [Return code]	→	IF (Returncode = OK)
		THEN file selection successful
		ELSE file could not be selected

Figure 7.2 Process of a SELECT FILE instruction

GSM 11.11 defines the instruction STATUS, which returns the same data to the terminal as does successful file selection with SELECT. This data is concerned with the type and structure of the file currently selected, its size, FID, access conditions and whether the file status is "locked". This instruction is rarely used, and its purpose is mainly to allow the terminal to query, during a session, which file is currently selected and which global conditions apply.

STATUS		
Instruction	•	—
Response	•	about the selected file
	•	return code

Figure 7.3 Function of STATUS covered by GSM 11.11

EN 726-3 contains an instruction which supplements SELECT FILE and STATUS, and which is used to close applications. The CLOSE APPLICATION instruction supplies the FID of the application to be closed. This instruction is mainly useful when a terminal needs to ensure that a state reached by the card is reset. If the card's operating system does not support such an instruction, this situation can only be achieved by a card reset.

CLOSE APPLICATION		
Instruction	•	FID (of the current DF)
Response	•	return code

Figure 7.4 Function of CLOSE APPLICATION in accordance with EN 726-3

7.2 READ AND WRITE INSTRUCTIONS

The read and write class of instructions supports the utilization of Smart Cards for secure data storage. By using these instructions it is possible to write data to an EF, and later to read them. To the extent that the EFs possess particular access conditions, the reading of these files is only possible by authorized users. The information is thus stored safely against unauthorized access to the card.

Since different EF data structures exist, there are varying read and write instructions for these files. Unfortunately, this arrangement does not quite correspond to an object-oriented file management structure. In pure object orientation, the operating system would have to be designed so that an object itself can determine its access mechanisms. This is not the case in Smart Card file managers. The situation can be traced back to the history of the Smart Card industry, where instructions which were developed independently were later adopted in the standards. The precursors of Smart Cards, namely memory cards, only possess one memory area, which can be read and written to by specifying the offset and length. Viewed from the outside, this memory can be regarded as a single file with a transparent structure. In the first Smart Cards, the same type of formatting was used; the definitions for read and write access to transparent files originated during that period. In the case of file structures defined later on, namely those based on a record structure, other, specially adapted instructions were employed. This is how the two different access instruction types came to coexist.

Thus, it is necessary to subdivide this instruction class into those which operate on EFs with transparent structure, and those for the remaining file structures, namely cyclic, linear fixed and linear variable. However, certain standards (e.g. electronic purse as per CEN prEN 1546) specify explicitly that it is permitted to use read instructions designed for transparently structured files for reading other types as well. In this event, it is possible to obtain additional data about the file's internal structure.

EFs with the transparent logical structure are amorphous, that is, they possess no internal structure. They represent a linearly addressable memory, which can be accessed byte-wise. The instruction READ BINARY is used for reading, and the two instructions WRITE BINARY and UPDATE BINARY for writing.

The fundamental difference between the writing instructions WRITE and UPDATE relates to the ground state of the card's EEPROM. The ground state is the logical value of EEPROM bits in which the memory cells have reached their minimum energy state; i.e., when the EEPROM cells, which are small capacitors, contain no charge. Normally this is the logic state 0. In order to set a bit from state 0 back to state 1, it must be cleared. This recharges the capacitor.

The WRITE instruction can only be used to set the bits from the non-ground state, in our example 1, to the ground state, i.e. in this case 0. The WRITE instruction in our example is therefore a logical OR between the input data and the file contents. If the chip's ground state is represented by the value 1, the WRITE instruction must effect a logical AND between instruction data and the file data. The logical interaction between the instruction data and the file data is such that a WRITE instruction always achieves the EEPROM's ground state. In addition, the WRITE instruction may support a WORM write (write once, read many), if this is appropriate for the file.

The UPDATE instruction, in contrast, is a genuine write to the file. The previously held data has no effect on the file's contents after the UPDATE instruction. UPDATE can therefore be regarded as an initial file clear, followed by a WRITE.

These two instructions can be exploited to construct physically secure Smart Card counters. The principle involves a data field, in which set bits each play the part of a monetary unit. During payment, the counter is decremented bit by bit with WRITE instructions, through the OR operation. After authentication, the counter can be reincremented with the UPDATE instruction. The great advantage of this design is that the EEPROM cannot be forced to increase the counter values, e.g. through thermal interference, since the bit field's ground state represents the value 0.

As the name suggests, READ BINARY is a read instruction. Here, and with the WRITE/UPDATE BINARY instructions, file access is always implemented using a length specification and an offset to the first addressed byte. The most up-to-date operating systems also permit, before actual file access, implicit file selection by inputting a short-FID, but this is not yet provided for in all the standards or operating systems.

READ BINARY	
Instruction	• number of bytes to be read
	• nffset to first byte to be read
	• *optional:* short-FID for implicit selection
Response	• data read from the file
	• return code

Figure 7.5 Function of READ BINARY in accordance with ISO/IEC 7816-4

WRITE BINARY		
Instruction	•	number of bytes to be written
	•	offset to first byte to be written
	•	*optional:* Short-FID for implicit selection
Response	•	return code

Figure 7.6 Function of WRITE BINARY in accordance with ISO/IEC 7816-4

UPDATE BINARY		
Instruction	•	number of bytes to be overwritten
	•	offset to first byte to be overwritten
	•	*optional:* Short-FID for implicit selection
Response	•	return code

Figure 7.7 Function of UPDATE BINARY in accordance with ISO/IEC 7816-4

Figure 7.9 illustrates a typical sequence consisting of the READ BINARY instruction, a subsequent WRITE BINARY and finally UPDATE BINARY. All the steps and data assume that selection of the file, shown in Figure 7.8, has been successful, and that the access conditions have been satisfied.

ERASE BINARY is an exception among the instructions operating on transparent EFs. It cannot be used to write data to a file, but only to erase its contents starting from a given offset. If a second offset, indicating the end of the section to be erased, is not given, the instruction erases all the data to the end of the selected file. In the context of the ERASE BINARY instruction, erasing means that the data section specified in the instruction is set to the logical cleared status. This state must be defined individually in each operating system, since it may not be identical to the physically cleared state of the semiconductor memory.

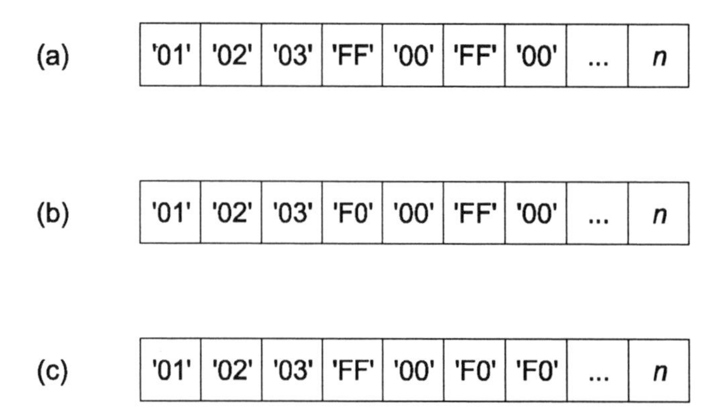

Figure 7.8 Example of EF sequence with transparent structure: a) File contents with READ BINARY; b) File contents after WRITE BINARY; c) File contents after UPDATE BINARY

Smart Card		Terminal
	←	READ BINARY
		Instruction [Offset = 2 bytes, number of bytes to be read = 5]
data requested := '03' ll 'FF' ll '00' ll 'FF' ll '00'		
Response [data requested ll Return code]	→	IF (Returncode = OK)
		THEN READ BINARY successful
		ELSE cancel
	←	WRITE BINARY
		Instruction [Offset = 3 bytes, number of bytes to be written = 2, data = 'F0 F0']
Response [Return code]	→	IF (Return code = OK)
		THEN WRITE BINARY successful
		ELSE cancel
	←	UPDATE BINARY
		Instruction [Offset = 5 Bytes, number of bytes to be written = 2, data = 'F0 F0']
Response [Return code]	→	IF (Return code = OK)
		THEN UPDATE BINARY successful
		ELSE cancel

Figure 7.9 Accessing file with transparent structure

ERASE BINARY		
Instruction	•	offset to first byte to be erased
		optional: offset to last byte to be erased
	•	*optional:* short-FID for implicit selection
Response	•	data read from the file
	•	return code

Figure 7.10 Function of ERASE BINARY in accordance with ISO/IEC 7816-4

Because linear fixed, linear variable and cyclic EFs have fundamentally different structures from that of transparent files, there are, in addition to the instructions described above, special instructons for accessing these particular data structures. All these files are record oriented. The smallest unit in the data field that can be written to is a single record. During reading, either the entire record or some part of it may be read, starting with the first byte. Compared with transparent structures, these file structures, which transform a linear, one-dimensional memory into a two-dimensionally addressable one, offer significantly more complex types of access. In principle, all data structures can be emulated by transparent ones; but in certain cases this may prove considerably more costly than using higher-level data structures.

After selection of an EF with a record-oriented structure, the card's operating system generates a record pointer whose initial value is undefined, and which can be set through READ/WRITE/UPDATE RECORD or SEEK. The pointer for the current file, however, remains fixed for as long as it is selected. Following successful explicit or implicit selection of another file, the record pointer once again becomes undefined.

All the instructions for record-oriented files can specify the type of access to the file contents. This is done through a parameter byte. The most basic method is direct access, using the required record's absolute number. This type of access does not alter the record pointer. Thus, the card is notified of the record number and returns its contents.

By supplying the parameter "first", the system sets the record pointer to the first record in the file, and it is read or written to as relevant. Other similar parameters are "last", "next" and "previous". Finally, the parameter "current" can be used to address the record indicated by the record pointer.

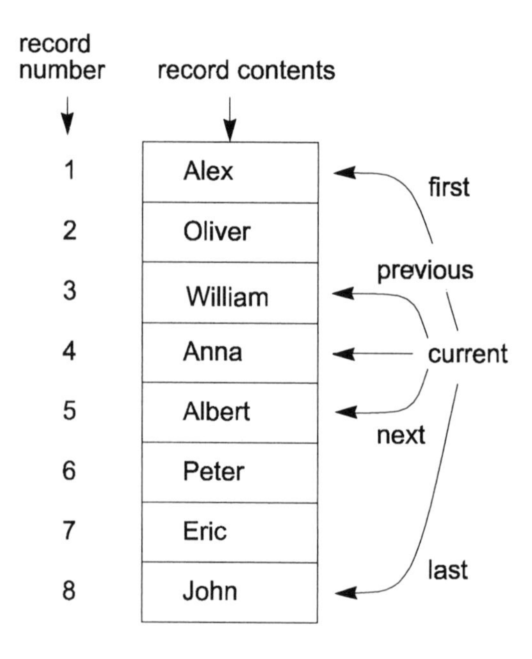

Figure 7.11 Access to record-oriented files

These methods of accessing records originated from the need to structure such large collections of data as telephone directories. Consider a record whose front section contains a surname and first name, followed (still within the same record) by the corresponding telephone number. Using a READ RECORD instruction and the parameters described above, it is possible to examine the list of phone numbers by "turning pages" forwards and backwards within this EF, or by jumping to the first or last entry. It is also possible to alter the record pointer with the search instruction SEEK, described in the next section.

name	telephone number
Alex	01603 660 668
James	01603 618 213
William	01603 632 650
Anna	01603 632 744
Albert	01502 586 213
Peter	01692 407 383
Eric	01603 860 007
Mark	01842 761 245
Paul	01493 440 942
John	01362 668 033

Figure 7.12 Example of telephone directory in fixed linear file

ISO/IEC 7816-4 also provides the option of reading all records from the first one up to a specified record number, or from a specified number up to the last one, using READ RECORD in a single instruction–response cycle. In large files, though, these instructions can very quickly reach the I/O buffer's limits.

READ RECORD		
Instruction	•	number of records to be read *or* mode (first, last, next, previous record) *or* read all records from n to the last record *or* read all records from the first to n
	•	*optional:* short-FID for implicit selection
Response	•	data read in file
	•	return code

Figure 7.13 Function of READ RECORD in accordance with ISO/IEC 7816-4

WRITE RECORD		
Instruction	•	record to be written
	•	number of record to be written *or* mode (first, last, next, previous record)
	•	*optional:* short-FID for implicit selection
Response	•	return code

Figure 7.14 Function of WRITE RECORD in accordance with ISO/IEC 7816-4

UPDATE RECORD		
Instruction	•	record to be overwritten
	•	number of record to be overwritten *or* mode (first, last, next, previous record)
	•	*optional:* short-FID for implicit selection
Response	•	return code

Figure 7.15 Function of UPDATE RECORD in accordance with ISO/IEC 7816-4

Figure 7.16 illustrates the execution of several read and write operations on the file shown in Figure 7.12.

Functionally, the instruction APPEND RECORD could equally be considered part of the File Management section. It is used to add records to existing record-oriented files. The instruction itself also provides the data for the entire new record. A relatively complex memory manager, in Smart Card terms, is necessary for this instruction to be available with all its functions. The manager's task is to set up a link between the new record and the ones already present. It is then possible, within the limits of the memory at our disposal, to add an arbitrary number of new records to the file. Often this is restricted to a fixed number of new records, which simplifies matters somewhat. During the generation of record-oriented files, memory is reserved as required for the records yet to be appended. This space is filled up by APPEND RECORD as required. Once this free memory is used up, no further APPEND RECORD instructions are possible.

If APPEND RECORD is used in conjunction with a linear fixed or linear variable file, the new record is always added at the end of the file. If the structure is cyclic, however, the new record is always numbered 1, which corresponds to the currently written record in files of this type.

The instruction can be used for various purposes. We have already mentioned telephone directories. Another example is that of a protocol file, in which the data to be logged is written to the card immediately, as the new record is generated.

Smart Card		Terminal
	←	READ RECORD *Instruction* [Record number = "2"]
Processing instruction *Response* ["Oliver" ll Return code]	→	IF (Returncode = OK) THEN READ RECORD successful ELSE cancel
	←	UPDATE RECORD *Instruction* [first, "Wolfgang"]
Processing instruction *Response* [Return code]	→	IF (Returncode = OK) THEN UPDATE RECORD successful ELSE cancel
	←	UPDATE RECORD *Instruction* [next, "Alex"]
Processing instruction *Response* [Return code]	→	IF (Returncode = OK) THEN UPDATE RECORD successful ELSE cancel
	←	READ RECORD *Instruction* [Record number = 2]
Processing instruction *Response* ["Alex" ll Return code]	→	IF (Returncode = OK) THEN READ RECORD successful ELSE cancel

Figure 7.16 Example of read and write operatons on record-oriented files

APPEND RECORD		
Instruction	•	record to be written
	•	*optional:* short-FID for implicit selection
Response	•	return code

Figure 7.17 Function of APPEND RECORD in accordance with ISO/IEC 7816-4

In order to complement file-based read and write instructions, two instructions exist which are designed for accessing data objects directly. Depending on the selected DF, certain data can be written to or read from files or into internal operating system structures, bypassing the file-oriented access mechanisms. Data objects can be written to with PUT DATA, or read with GET DATA. In both these instructions, the exact structure of the TLV-encoded data objects must be supplied, specifying whether the object's encoding method is application-specific or standardized. This information is important for the operating system, so that the data format can be recognized by the objects. The appropriate access conditions must also be satisfied in both these instructions.

GET DATA		
Instruction	•	number of data objects
	•	codes of data objects to be read
Response	•	data objects read
	•	return code

Figure 7.18 Function of GET DATA in accordance with ISO/IEC 7816-4

PUT DATA		
Instruction	•	structure of data object to be written
Response	•	return code

Figure 7.19 Function of PUT DATA in accordance with ISO/IEC 7816-4

7.3 SEARCH INSTRUCTIONS

Record-oriented structures offer the option of storing related data of the same format in one file. The typical example is a telephone directory which lists names and phone numbers. A search instruction can be used to avoid the need to read the entire directory, record by record, when looking for a particular name.

The SEEK instruction with an offset can be used to search for a specified character string in a record-oriented data structure. The string's length can be variable. The operating system must be informed of the search direction, i.e. whether it should proceed forwards (from low to high record numbers) or backwards (high to low). The start position must also be supplied, or the first, last or current record may be specified as the start position. If the character string is found, the system sets the record pointer to the record containing this item and sends the result of the search to the terminal.

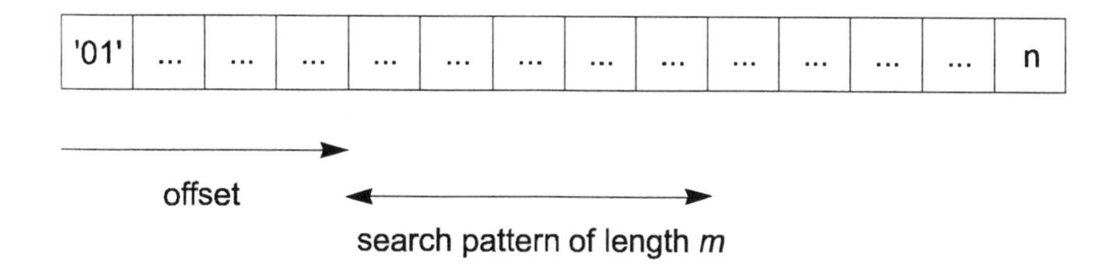

Figure 7.20 Searching within a record-oriented file

SEEK	
Instruction	• length of character string
	• character string
	• offset
	• mode (from the beginning, forwards, or backwards from the end, forwards from the next point, backwards from the previous point)
	• *record pointer:* number of found records sent back
Response	• record number (if selected)
	• return code

Figure 7.21 SEEK function specified in EN 726-3

Figure 7.22 illustrates several possible ways of using a SEEK instruction on the linear fixed file shown in Figure 7.12.

7.4 FILE OPERATIONS

A number of instructions exist which allow file contents to be modified by means other than simple writing. The main representatives of this class are the instructions INCREASE and DECREASE. They increment or decrement, by a specified value, a counter structured in the form of data held in a cyclic file.

The reason for specifying a cyclic file structure is that the prescribed protocol function assumes this to be the case. These instructions are most extensively used in simple low-value electronic purses and counters.

Smart Card		Terminal
	←	SEEK *Instruction* [character string = "Hans" ‖ search direction = "from beginning to end" ‖ send record no.]
Response [Record No. = 8 ‖ Return code]	→	IF (Returncode = OK) THEN "Hans" found ELSE "Hans" not found
	←	SEEK *Instruction* [character string = "Alex" ‖ search direction = "from end to beginning" ‖ send record no.]
Instruction [Record no. = 1 ‖ Returncode]	→	IF (Return code = OK) THEN "Alex" found ELSE "Alex" not found

Figure 7.22 Process of SEEK instruction

DECREASE	
Instruction	• value to be decremented
Response	• value decremented • new value of record • return code

Figure 7.23 DECREASE function in accordance with EN 726-3

INCREASE	
Instruction	• value to be incremented
Response	• incremented value • new value of record • return code

Figure 7.24 INCREASE function in accordance with EN 726-3

In Figure 7.25, we consider for the sake of simplicity a cyclic file holding just one record whose starting value is 10. The file's contents are restored to this value at the end of the run.

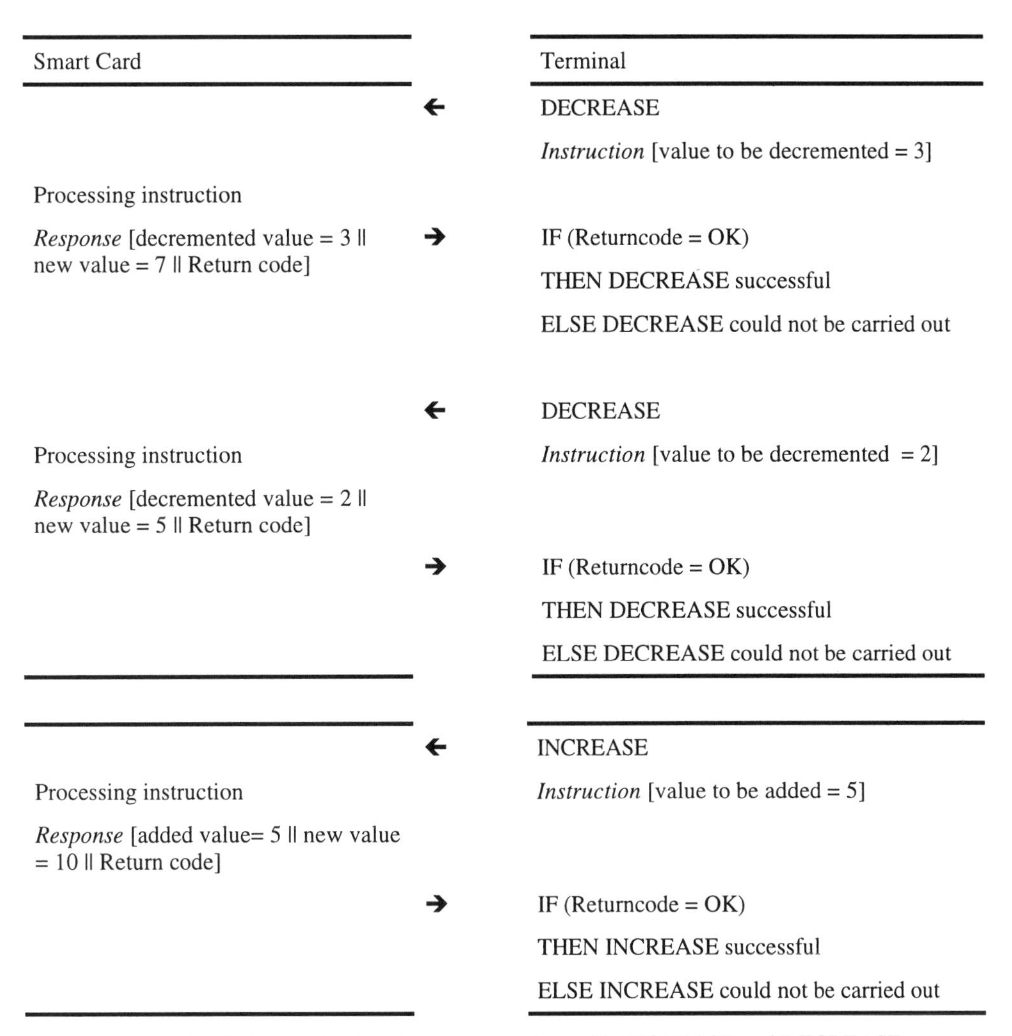

Smart Card		Terminal
	←	DECREASE *Instruction* [value to be decremented = 3]
Processing instruction *Response* [decremented value = 3 ‖ new value = 7 ‖ Return code]	→	IF (Returncode = OK) THEN DECREASE successful ELSE DECREASE could not be carried out
	←	DECREASE
Processing instruction *Response* [decremented value = 2 ‖ new value = 5 ‖ Return code]		*Instruction* [value to be decremented = 2]
	→	IF (Returncode = OK) THEN DECREASE successful ELSE DECREASE could not be carried out
	←	INCREASE
Processing instruction *Response* [added value= 5 ‖ new value = 10 ‖ Return code]		*Instruction* [value to be added = 5]
	→	IF (Returncode = OK) THEN INCREASE successful ELSE INCREASE could not be carried out

Figure 7.25 Example of instruction procedure with INCREASE and DECREASE

The instruction EXECUTE is one which is used to start program runs in executable EFs. The instruction can cause the executable program to receive data from the terminal and send back generated data in response.

The instruction and the related file structure are controversial, since in certain circumstances they may compromise the Smart Card's entire security system.

EXECUTE	
Instruction	• data sent to executable file
Response	• data withheld from executable file • return code

Figure 7.26 EXECUTE function in accordance with EN 726-3

7.5 IDENTIFICATION INSTRUCTIONS

In addition to their use as secure data carriers, Smart Cards also serve to identify individuals. The usual procedure involves identification by entering a password (Personal Identification Number or PIN) which is only known to the user and to the card.

PIN testing is a procedure known to most people from their own everyday experience. You key in the PIN at the terminal, and shortly thereafter the display shows whether it was correct or how many attempts are still permissible. The Smart Card receives from the terminal the instruction VERIFY PIN which contains the usual 4-figure PIN, and compares it with the one stored in EEPROM. If the two are identical, the card's state automaton is updated and the terminal receives a message confirming a positive outcome. The error counter is also reset to 0. If the keyed PIN and the PIN stored in the card are different, the counter is incremented. Once it reaches its predefined maximum value, the card is locked against any further attempts.

In many Smart Card operating systems it is possible to employ several PINs. It is then necessary to enter the relevant PIN's addressing number when using instructions which involve identification. As a rule, though, card issuers attach great importance to the existence of a single PIN per card, even where several are technically possible.

The PIN is often redesignated CHV in the telecommunications industry, CHV being short for "chip holder verification". Since the instructions described below originated in the telecommunications industry, their names include the abbreviation CHV.

VERIFY CHV	
Instruction	• PIN • PIN number
Response	• return code

Figure 7.27 VERIFY CHV function in accordance with EN 726-3

PINs have steadily proliferated since their introduction as ID numbers for card users. Currently, an average card user is expected to memorize perhaps 10 or 20 different PINs relating to various cards and other authorizations. This unrealistic expectation is illustrated by the number of people who jot down their PIN on the card itself. The use of Smart Cards allows the user to choose a PIN at will, and in fact to have the same PIN for all cards.

Although this facility may conflict with security, since the discovery of one PIN by an attacker makes them all known, it is still better than writing it on the card for all to see.

The instruction CHANGE CHV implements PIN modification. If the PIN currently stored in the card is known, it is possible to replace it with a new one. If the current PIN is given incorrectly, the operating system increments the error counter as a security measure against a possible attack. Once the current PIN is correctly entered into the card, it stores the chosen new one in the appropriate memory section and resets the error counter.

CHANGE CHV	
Instruction	• old PIN • new PIN • PIN number
Response	• return code

Figure 7.28 CHANGE CHV function in accordance with EN 726-3

If a PIN's error counter has reached its maximum value, it can be reset with the instruction UNBLOCK CHV and a second PIN, known as PUK (personal unblocking key). Normally the PUK is longer (e.g. 8 digits) than the usual 4-figure PIN. The user does not need to memorize it, since it only needs to be known in the event of the PIN being forgotten. It then suffices to look it up in one's records back home. However, it will be of little use to reset the error counter if it doesn't help one to rediscover the forgotten PIN. Hence the instruction UNBLOCK CHV must also provide the card with a new PIN.

In hybrid cards, which carry a magnetic stripe as well as a chip, it should not be possible to alter the PIN on the chip, since then the stripe would carry a different PIN from the one on the chip: this would lead to severe problems. In this event, therefore, the error counter is simply reset to its original value and the customer is sent a letter which specifies the old PIN.

UNBLOCK CHV	
Instruction	• PUK • new PIN
Response	• return code

Figure 7.29 UNBLOCK CHV function in accordance with EN 726-3

GSM contains two further instructions which can affect PIN interrogation. These are DISABLE CHV and ENABLE CHV, which serve to switch the PIN check off and on. In the OFF position, all file access conditions which previously required a PIN check are disabled. Both instructions are very popular in the field of mobile communications, since the PIN does not need to be entered afresh every time the mobile phone is switched on. It is possible to object to both these instructions on security grounds, since they can remove the protection afforded by the PIN against unauthorized use. On the other hand, of course, the user could also have used CHANGE CHV to choose a trivial PIN such as '0000', whose degree of protection is just as minimal.

DISABLE CHV	
Instruction	• PIN • PIN number
Response	• return code

Figure 7.30 DISABLE CHV function in accordance with EN 726-3

ENABLE CHV	
Instruction	• PIN • PIN number
Response	• return code

Figure 7.31 ENABLE CHV function in accordance with EN 726-3

For obvious reasons, the above-mentioned PIN checking procedures are vulnerable. It might be possible, given the right circumstances, using a found or stolen card and the right PIN, to obtain a large financial gain. All the instructions associated with PIN or PUK comparison must be secured against analysis of the card's electrical or time-related behaviour. For example, power consumption during VERIFY PIN must not vary as a function of the PIN's correctness or otherwise. Equally, the time taken to execute PIN instructions must not depend on the PIN being correct or false. Different run times may have fatal repercussions on the card's security, and thus eventually on that of the entire system, since this would permit a very easy discovery of the correct PIN, and all the PINs in the system would be useless as user identifiers.

7.6 AUTHENTICATION INSTRUCTIONS

In addition to identifying the card holder, there is a further series of instructions for authenticating the terminal and the card. Since each of these two parties is equipped with a complete computer, the procedures used can be made much more complex than the PIN checks, and thus also more secure.

In PIN checks, the card receives a secret code in plaintext (the PIN) across the interface, and only needs to compare it with the one held in memory. As a result, any tapping into the transmission would have fatal consequences. Modern authentication procedures are so designed that such an attack becomes impossible.

In principle, authentication involves checking a code known to both parties without having to send it across the interface. The procedures are designed to ensure that tapping into the transmission would not compromise authentication security[2].

---

[2] See also 8.2 Authentication.

Depending on the operating system, various instructions exist for the authentication of the card and/or terminal. For the sake of clarity, this section and the rest of the chapter refer to authentication between card and terminal. In IT terms, though, the "rest of the world" is authenticated to an application on the card. It is not the card as a whole which is checked for authenticity, but rather whether the incorporated microprocessor shares a common secret with the external world. This fact needs to be taken into account in certain applications.

In many operating systems, the keys used in authentication are protected by an error counter. If a terminal attempts a faulty authentication process once too often, the card locks the relevant key against further attempts. This procedure cannot be faulted as far as system security is concerned, but it does suffer from a disadvantage. Resetting the authentication key's error counter to its ground state frequently leads to very complex, logistically difficult and expensive administrative procedures. Therefore, there are many systems which do not have such an error counter.

For security reasons, only keys individual to the card should be used for authentication. Key generation can be based on a unique feature of each card. Serial number or chip manufacture number are very suitable for this purpose. These public, non-confidential numbers can be read from the card with the appropriate instruction. The procedure is not yet covered by a standard, so we refer to it here as a GET CHIP NUMBER instruction. The name varies from one operating system to the next, as does the nature of the exchanged data. However, we are only interested here in the functionality. The instruction GET CHIP NUMBER obtains a unique serial number from the card, which in view of the DES algorithms, should preferably be 8 bytes long. It is used for unambiguous identification of the chip, and for calculating keys unique to the card.

GET CHIP NUMBER		
Instruction	•	—
Response	•	chip number
	•	return code

Figure 7.32 GET CHIP NUMBER function

One further instruction is necessary for authentication. This is ASK RANDOM, as specified in EN 726-3. This requests a random number from the card to be used later during authentication. This instruction exists in ISO/IEC 7816-4 under the name GET REQUEST, and has the same function. In DEA authentication the length is typically 8 bytes, but this may differ for other cryptographic algorithms.

In order to illustrate the following run-time examples in outline, and to avoid complicating them unnecessarily, we shall omit the derivation of the unique key. This is a vital component, however, which is necessary for security reasons.

ASK RANDOM / GET CHALLENGE		
Instruction	•	—
Response	•	random number
	•	return code

Figure 7.33 ASK RANDOM function in accordance with EN 726-3 and GET CHALLENGE function in accordance with ISO/IEC 7816-4

The instruction INTERNAL AUTHENTICATE serves to authenticate the card by the terminal, or in the case of multi-application cards, to authenticate one application. The instruction checks that the card is genuine. The card receives a random number which it encrypts, using a key known only to it and to the terminal and a cryptographic algorithm such as DEA. The result of this operation is returned to the terminal. It now performs the same encryption as the card, and compares the result with that contained in the card's response. If the two are identical, it follows that the card also knows the secret authentication key and so must be genuine. The card has thus been authenticated.

The instruction EXTERNAL AUTHENTICATE is used by the terminal to demonstrate to the card that it is connected to a genuine terminal. The instruction must be initiated by the terminal, since the communications procedure must always run within the command–response framework. However, the card can force terminal authentication by blocking access to certain files until the terminal is successfully authenticated.

INTERNAL AUTHENTICATE		
Instruction	•	random number
	•	algorithm number
	•	authentication key
Response	•	enc (key; random number)
	•	return code

Figure 7.34 INTERNAL AUTHENTICATE function in accordance with IEO/IEC 7816-4

Smartcard		Terminal
	←	INTERNAL AUTHENTICATE
X := enc (key; random number)		*Instruction* [random number, key number]
		X' := enc (key; random number)
Response [X ‖ Return code]	→	IF (Returncode = OK) AND (X = X') THEN instruction successful, Smartcard authenticated ELSE no authorization given

Figure 7.35 INTERNAL AUTHENTICATE instruction process

This proceeds as follows: first the terminal requests from the card a random number (via the instruction ASK RANDOM), which it then encrypts using the secret key. In the second instruction, EXTERNAL AUTHENTICATE, the terminal returns the encrypted random number to the card. The card performs the same operation with the secret key which it also knows, and compares the computed result with the one obtained from the card. If they are identical, then the terminal must also be in possession of the secret authentication key and is thus genuine.

After successful terminal authentication, the operating system alters the state of the card's state automaton. This allows the terminal to access certain files for reading or writing. For the user, this is the recognizable result of an external authentication.

EXTERNAL AUTHENTICATE	
Instruction	• enc (key; random number) • algorithm number • authentication key
Response	• return code

Figure 7.36 EXTERNAL AUTHENTICATE function in accordance with IEO/IEC 7816-4

Smart Card		Terminal
	←	ASK RANDOM *Instruction* []
Response [random number ‖ Return code]	→	IF (Returncode = OK) THEN instruction successful ELSE cancel
		X := enc (key; random number)
	←	EXTERNAL AUTHENTICATE *Instruction* [X, key number]
X' := enc (key; random number) IF (X = X') THEN terminal authorized ELSE no authorization given		
Response [Return code]	→	IF (Returncode = OK) THEN instruction successful, terminal authorized ELSE no authorization given

Figure 7.37 EXTERNAL AUTHENTICATE procedure

If the instructions INTERNAL and EXTERNAL AUTHENTICATE are executed one after the other, the communicating parties are reciprocally authenticated. Thus they both know that the other party is genuine. However, this requires a total of three complete instruction sequences. In order to simplify this expensive and time-consuming procedure, the three instruction sequences and their transmission data have been linked together and the two authentication instructions integrated into one, known as MUTUAL AUTHENTICATE. It is not currently covered by any Smart Card standard, but ISO/IEC 9798-2 suggests a future standardization. This single authentication instruction also increases security, since it is no longer possible to insert instructions between the two one-way authentications. A further security improvement for the procedure results from the fact that it is impossible to obtain plaintext–ciphertext pairs by tapping into terminal–card communications, such insertion being an ideal method of attack.

MUTUAL AUTHENTICATE	
Instruction	• enc (key; random number terminal, random number Smart Card, chip number) • algorithm number • authorization key number
Response	• enc (key; random number Smart Card, random number terminal) • return code

Figure 7.38 MUTUAL AUTHENTICATE function

Mutual authentication proceeds as follows: first the terminal requests the card's chip number (GET CHIP NUMBER). Now the terminal can compute the card's individual key. The terminal then obtains a random number (ASK RANDOM) from the card and generates a random number.

Once the terminal has received the card's random number, it forms a block consisting of the two random numbers and the chip number, which it encrypts with the secret authentication key and a cryptographic algorithm in CBC mode. It sends the resulting ciphertext block to the card, which decrypts it and compares the chip number and the random number with those previously transmitted. If the two are genuine, the terminal has been authenticated.

Now the card swaps the two random numbers round, omits the chip number and encrypts the whole block again with the secret key. After receiving this block and decrypting it, the terminal can compare the known random numbers and determine whether the card was in possession of the secret authentication key. If this is the case, the card is also authenticated.

7.7 INSTRUCTIONS FOR CRYPTOGRAPHIC ALGORITHMS

The following instructions, all based directly on a cryptographic algorithm, are not currently laid down in any standard. Nevertheless, they are of great importance for many applications, since they can be used to make Smart Cards into cheap encryption and decryption devices.

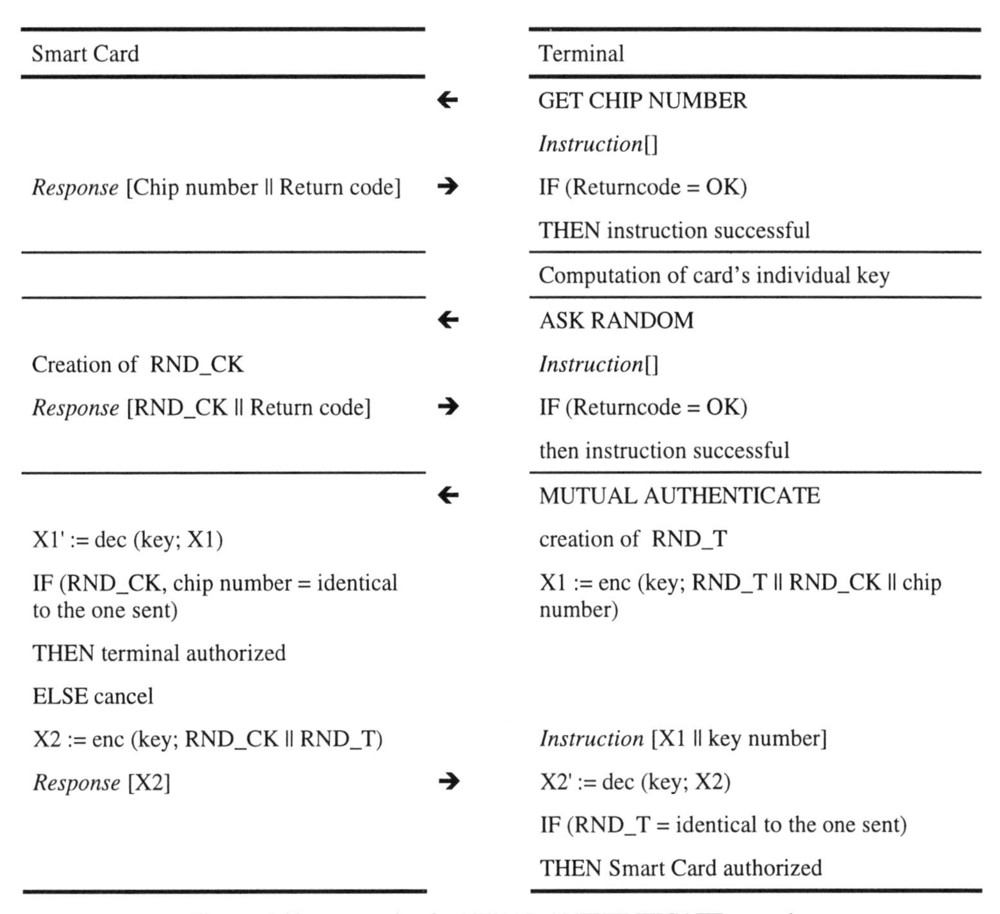

Smart Card		Terminal
	←	GET CHIP NUMBER
		Instruction[]
Response [Chip number ll Return code]	→	IF (Returncode = OK)
		THEN instruction successful
		Computation of card's individual key
	←	ASK RANDOM
Creation of RND_CK		*Instruction*[]
Response [RND_CK ll Return code]	→	IF (Returncode = OK)
		then instruction successful
	←	MUTUAL AUTHENTICATE
X1' := dec (key; X1)		creation of RND_T
IF (RND_CK, chip number = identical to the one sent)		X1 := enc (key; RND_T ll RND_CK ll chip number)
THEN terminal authorized		
ELSE cancel		
X2 := enc (key; RND_CK ll RND_T)		*Instruction* [X1 ll key number]
Response [X2]	→	X2' := dec (key; X2)
		IF (RND_T = identical to the one sent)
		THEN Smart Card authorized

Figure 7.39 Example of MUTUAL AUTHENTICATE procedure

The ENCRYPT instruction works on specified data. There is usually no choice of which encryption algorithm will be employed, as Smart Card memory is generally too small to hold two different algorithms. The algorithm's mode of operation, though, can be selected via certain parameters, so that in the case of a block encryption algorithm it is possible to choose between ECB and CBC modes. It is often also possible to calculate a MAC over the input data, instead of encrypting the whole block. This would be the simplest form of a signature.

Since the length of the data sent to the card does not necessarily have to be a multiple of the algorithm's block length, the padding method needs to be chosen via a further parameter. It is also necessary to address a key which is held in the card and required by the algorithm.

The inverse function to ENCRYPT is called DECRYPT. This instruction can decrypt the specified data, using the same modes as ENCRYPT. The card needs to know the relevant key, whose value is supplied in the instruction.

ENCRYPT	
Instruction	• data to be encrypted • mode (ECB, CBC, MAC) • padding method • authorization key number
Response	• encrypted data *or* MAC • return code

Figure 7.40 ENCRYPT function

A MAC computing function is not necessary, since checking a MAC involves two sets of calculations performed independently of each other and the two results compared. However, the padding method needs to be specified, in order to separate the padding data within the card from the user data. Fully symmetrical implementation of ENCRYPT and DECRYPT is then possible.

DECRYPT	
Instruction	• encrypted data • mode (ECB, CBC) • padding method • authorization key number
Response	• decrypted data • return code

Figure 7.41 DECRYPT function

With the introduction of public key algorithms into the Smart Card industry, it became necessary to use appropriate instructions to exploit the new functions. Smart Cards are superbly suited to digital signatures, since the secret key for the signature algorithm can be kept securely in memory. Currently, there are various incompatible corporate standards and no national or international ones, but here we present two possible instructions.

The first instruction, called SIGN DATA for our purposes, can be employed for data signatures. The data string, usually converted to a hash value, is sent to the card. The hash value's length should correspond to the public key algorithm's input value, in order to make padding unnecessary. After selecting the secret key the card executes the relevant algorithm.

SIGN DATA	
Instruction	• data to be signed (i.e. its hash value) • authorization key number
Response	• signature • return code

Figure 7.42 SIGN DATA function

Any sufficiently fast computer can be used to check the digital signature produced by SIGN DATA, since the key is public. The time required by Smart Cards for this computation is significantly shorter than on most PCs, so the operation can be performed in the card using VERIFY SIGNATURE. All that is required is knowledge of the digital signature, a key number and the hash value of the data. After checking, the card sends back the appropriate response to the terminal.

VERIFY SIGNATURE	
Instruction	• data to be verified (i.e. its hash value) • signature • authorization key number
Response	• return code

Figure 7.43 VERIFY SIGNATURE function

The instructions SIGN DATA and VERIFY SIGNATURE could be used to greater effect if the required hash value were created at the same time as the digital signature was computed. This would make it unnecessary to calculate the data's hash value externally, but it cannot be done owing to the very slow data transmission speed (in absolute terms) between terminal and card. It is so slow that transmission times for texts produced by conventional word-processing will take several minutes at least. With such waiting times, user acceptance would tend to be nil.

7.8 FILE MANAGEMENT

Most modern Smart Card operating systems permit the expansion of files, the creation of new files, deletion, locking and the execution of other management functions on files, all within the limits imposed by specific security provisions.

Nevertheless, and particularly in the case of cards with only a single application, most or all management functions are omitted, since in general they are very large in terms of size and thus increase the demands in terms of memory and, consequently, cost. In multi-application cards it is almost impossible to avoid supporting certain management functions, so that not all the applications have to be loaded into the card simultaneously when it is personalized.

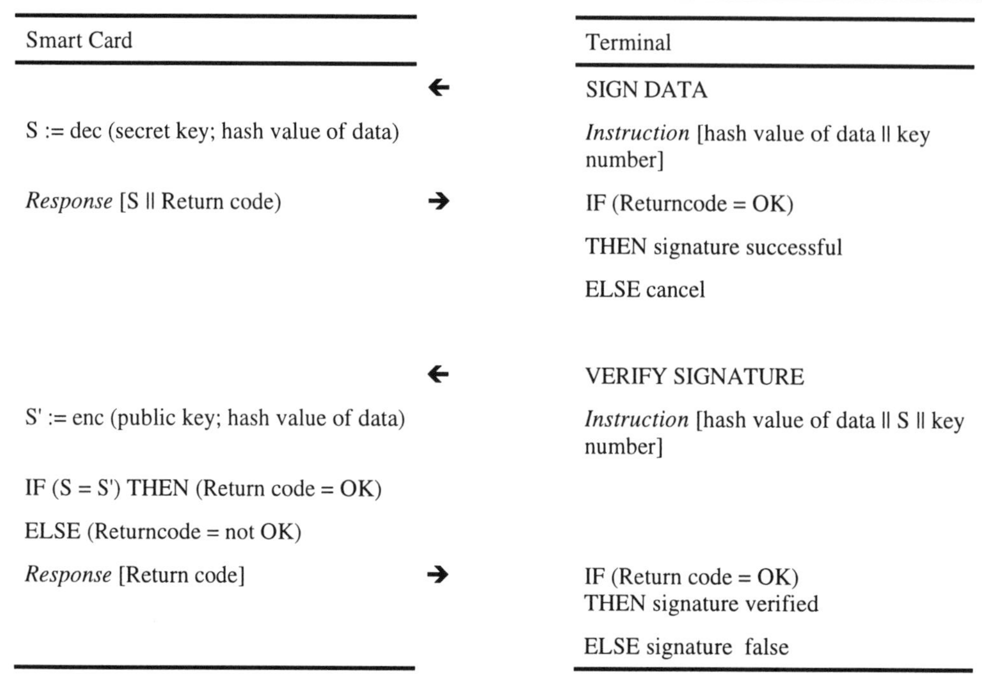

Figure 7.44 SIGN DATA and VERIFY SIGNATURE procedures

From a security perspective, management functions should only be executed after reciprocal authentication since they are an ideal lever for a hacker.

Imagine a case in which a file which stores confidential data is deleted by an unauthorized person who then generates a new file bearing the same name. The replacement file is not subject to any access conditions in terms of reading data. As far as the terminal is concerned, the file continues to exist as before; the terminal therefore continues to write secret data to it but now it can be read by anyone. This type of attack is by no means new – it has been around for many years in a somewhat modified form. The point, though, is that when applied to file management it succeeds time and time again.

Another weak link is represented by the execution of management functions in publicly accessible terminals, which by definition are insecure. Here, data transmission must be protected by secure messaging. It is only then that an application provider has the option at his disposal of securely loading files and applications into the previously issued card, for instance via public cardphones.

Before the individual applications can be generated, and particularly in the case of multi-application Smart Cards which can be used by several applications providers, the available memory needs to be subdivided and authorization keys allocated for file creation. This prevents a single application provider from using up the entire memory for his application, leaving no space for other applications. One solution to this problem exists in the form of a procedure with which an application can be pre-allocated memory space, and at the same time a card- and application-specific key for file generation is entered into the card. The instruction used is the (non-standardized) REGISTER. Subsequently, and with knowledge of the key, new files can be generated. The result of this procedure is strict separation between memory space allocation and the introduction of new files into the card. A multi-application card issuer can now sell memory space to several customers without having to worry about memory piracy.

REGISTER	
Instruction	• AID of new DF
	• maximum memory space for new application
	• key to generate new files (i.e. for CREATE FILE)
Response	• return code

Figure 7.45 REGISTER function

The CREATE FILE instruction permits the generation of a DF or EF after Smart Card completion. Here it is important to ensure that a particular logic state must be reached, for instance, after successful reciprocal authentication. Depending on the environment in which a CREATE FILE instruction is executed, data transmission needs to be protected by secure messaging.

After a file is generated together with all its access conditions, attributes and any other properties, it can be selected with SELECT FILE and then accessed. The operating system must prevent a semi-formed file, resulting from the interruption of the generating procedure, from being used for tampering purposes. Furthermore, it must be impossible to read the contents of old data in memory when these have only been partially overwritten by new files.

CREATE FILE	
Instruction	• new file type
	IF (file type = DF) THEN [
	• AID of new file]
	IF (file type = EF) THEN [
	• FID of new file
	• access conditions
	• file properties]
	IF (file structure = transparent) THEN [
	• file size]
	IF (file structure = linear fixed) OR (file structure = cyclic) THEN [
	• number of records
	• length of records]
	IF (file structure = linear variable) THEN [
	• number of records
	• length of individual records]
Response	• return code

Figure 7.46 CREATE FILE function in accordance with EN 726-3

The file header contains complete access conditions for the file. For example, it stores data concerning the status by means of which a READ or UPDATE of file contents may proceed. At the very point when the card is being personalized, or during extended management procedures in the file tree, it is of great benefit to be able to access files directly without object protection. This is possible by using the non-standardized instruction CHANGE ATTRIBUTES to modify the access conditions to a previously selected file. It is a vital precondition that the instruction may only be executed after reciprocal authentication and in a secure environment.

CHANGE ATTRIBUTES		
Instruction	•	new access conditions
Response	•	return code

Figure 7.47 CHANGE ATTRIBUTES function

The addition of a new application to a Smart Card's file tree proceeds as shown in Figure 7.49.

If the REGISTER instruction is available, it is used by the card issuer to determine the maximum memory space which is available for the subsequent application. A temporary AID (application identifier) is also chosen for the application's DF, and a key for CREATE FILE inserted. Knowing this key makes subsequent file creation possible.

The user receives the card which has been prepared in this way. If necessary, an application provider can load an additional application on the user's card, for cardphones, for instance. The application provider needs to know the secret loading key, which of course is only revealed by the issuer if a contractual relationship exists between them.

After successful reciprocal authentication between card and terminal, the application provider may create its files in the allocated DF. This procedure may take place either on the provider's premises or through a public cardphone. Finally, the provider fills the EFs with the necessary data and keys, and sets the access attributes for the files. When the process has been completed, the application is operational, and the user can make use of the new functions.

Currently, the steps listed above are mostly in the future. The technical problems involved can be regarded as solved, by and large, but implementation requires far more than that. In the long term, though, the procedure described is the one most likely to be followed.

The INVALIDATE instruction permits the terminal to lock reversibly a previously selected file. Read and write access to the file are then prohibited. Only selection is possible.

INVALIDATE		
Instruction	•	—
Response	•	return code

Figure 7.48 INVALIDATE function in accordance with EN 726-3

Smart Card		Terminal
mutual authorization between Smart Card and terminal		
	←	REGISTER
		Instruction [AID ‖ memory space ‖ key]
Processing instruction		
Response [Return code]	→	IF (Return code = OK)
		THEN instruction successfully carried out
		ELSE instruction failed
mutual authorization between Smart Card and terminal		
	←	CREATE FILE
		Instruction [...]
Processing instruction		
Response [Return code]	→	IF (Return code = OK)
		THEN instruction successfully carried out
		ELSE instruction failed
		numerous attempts:
	←	UPDATE BINARY / RECORD
		Instruction [...]
Processing instruction		
Response [Return code]	→	IF (Return code = OK)
		THEN instruction successfully carried out
		ELSE instruction failed
		numerous attempts:
	←	CHANGE ATTRIBUTES
		Instruction [...]
Processing instruction		
Response [Return code]	→	IF (Return code = OK)
		THEN instruction successfully carried out
		ELSE instruction failed

Figure 7.49 Example of addition of new application

The inverse function to INVALIDATE is REHABILITATE, with which a locked file can be unlocked, after first being selected. It goes without saying that these two instructions can only be permitted when a predetermined security status obtains, otherwise anyone would be able to lock or unlock files at will.

The instruction LOCK is the irreversible variant of INVALIDATE. A file locked with LOCK cannot be unlocked. This step is completely irrevocable. One way of using it is to lock an application upon expiry. After execution of a LOCK instruction the file can only be selected; any other instruction is prohibited by the operating system.

REHABILITATE	
Instruction	• —
Response	• return code

Figure 7.48 REHABILITATE function in accordance with EN 726-3

Smart Card		Terminal
mutual authorization between Smart Card and terminal		
	←	SELECT FILE
Processing instruction		*Instruction* [FID]
Response [Return code]	→	IF (Return code = OK)
		THEN file selection successful
		ELSE file could not be selected
	←	INVALIDATE
Processing instruction		*Instruction* []
Response [Return code]	→	IF (Return code = OK)
		THEN file locked
		ELSE file could not be locked
	←	REHABILITATE
Processing instruction		*Instruction*[]
Response [Return code]	→	IF (Return code = OK)
		THEN instruction successful
		ELSE file could not be unlocked

Figure 7.49 Example of failure with INVALIDATE and REHABILITATE

LOCK	
Instruction	• —
Response	• return code

Figure 7.50 LOCK function in accordance with EN 726-3

The final and irreversible locking of files suffers from a great disadvantage, namely that valuable card memory space is blocked for ever. When files are no longer needed it is considerably more elegant to clear the memory space they occupy, and make other or new applications accessible.

It is important to ensure that besides removing the file from the file tree, the entire memory used by the file is physically emptied. This is the only way of ensuring that the file's entire contents, which may still be secret and worth protecting, are overwritten and cannot be accessed by anyone. When the memory released through clearing is to be allocated to new files, file deletion becomes complicated and expensive, so only very few operating systems implement this process in its entirety. Completely free memory management usually requires more space than is available in a Smart Card for this purpose.

In principle, the instruction DELETE FILE is treated just like the ones described above for file locking and unlocking. A file selected using this instruction can be completely removed from the card's memory. Whether this released memory space can or cannot be used by other files depends on the operating system. As a rule, no free memory manager is available and the released memory is lost for good once the instruction is executed.

DELETE FILE	
Instruction	• FID
Response	• return code

Figure 7.51 DELETE FILE function in accordance with EN 726-3

The instructions defined for the above-mentioned functions depend on the operating system's manufacturer and the system version. A future standard will probably be agreed, but in practice all these management functions are currently incompatible and independent of each other. It is only the overall functionality which is similar.

7.9 INSTRUCTIONS FOR ELECTRONIC PURSES

The European standard for universal electronic purses, prEN 1546 part 3, defines a total of six instructions for electronic purses and twelve for the security module in the terminal which itself may be a Smart Card. This section describes the basic format of the four most important instructions used by a smart electronic purse[3]. These instructions can be exploited

[3] See also 12.3.1 CEN standard prEN 1546, describing the routines and general system structures.

to run an application on a Smart Card which makes it possible to purchase services by means of a pre-paid card, and also to re-credit it. Error recovery instructions and those used for currency conversion, parameter amendment and reversal, as well as all the security module instructions, are not described further.

The following descriptions are just as appropriate to the section on application-specific instructions, since the instructions illustrated are defined precisely for this one application. They cannot be used for any other purpose, as they have been optimized for electronic purses. However, here we have chosen to dedicate to them a whole section, since in future electronic purses will constitute one of the main applications for Smart Cards, alongside those involving telecommunications.

All electronic purse transactions are divided into three, in accordance with prEN 1546. In the first section, the card is initialized (INITIALIZE IEP for Load/for Purchase). The second instruction executes the actual transaction (i.e. credit the card or make a payment). The transaction just performed is acknowledged in the optional third section. All the accounting instructions are designed to directly access the files in the card's payment application, for both writing and reading. These are files which hold various parameters, the financial balance and the protocol records. The separate steps involved in a transaction are executed through the following instructions. The above-mentioned standard specifies in detail the precise internal routine for each instruction, as well as its functional details and the sequence of the separate steps, so that all implementations have at least the same overall configuration in common.

INITIALIZE IEP for Load	
Instruction	• —
	• loading amount (M_{LDA})
	• currency code ($CURR_{LDA}$)
	• PPSAM description (PPSAM)
	• random number (R)
	• user specific data (DD)
Response	• describes purse provider (PP_{IEP})
	• describes IEP
	• cryptographic algorithm used (ALG_{IEP})
	• expiry data ($DEXP_{IEP}$)
	• purse balance (BALIEP)
	• IEP transaction number (NT_{IEP})
	• key information (IK_{IEP})
	• signature S_1
	• return code (CC_{IEP})

Figure 7.52 INITIALIZE IEP for Load function in accordance with prEN 1546-3

The instruction INITIALIZE IEP can serve several purposes. One parameter is used to switch between the initialization of card-crediting transactions, payment transactions and other types.

The instruction INITIALIZE IEP for Load is the first step in loading the account into the card. The entered data, such as currency code and amount, are checked in the card to ensure they correspond with previous values in the parameter files. In addition, freely definable data (user-determined data) can be stored in a protocol file. Finally, the transaction counter is incremented and a signature S_1 formed from various data (e.g. actual balance, due date), so it can be transmitted to the terminal, secure from interference.

In step two, the card essentially receives data concerning the key to be used and a signature, S_2. This originates in the terminal's security module and allows the card to authenticate the security module, besides protecting the data. The card itself is already authenticated by the terminal's security module, through the previous INITIALIZE IEP for Load instruction. After successfully testing S_2, the card increases the account balance, completes the current record in the protocol file and generates a signature, S_3, for confirmation. This is then used by the terminal's security module as acknowledgement of the correct entry of the amount paid in.

CREDIT IEP	
Instruction	• key information (IK_{PPSAM})
	• signature S_2
	• user specific data (DD)
Response	• signature S_3
	• return code (CC_{IEP})

Figure 7.53 CREDIT function in accordance with prEN 1546-3

The second procedure described here, with its relevant instructions, demonstrates how payment is made using the Smart Card. The first instruction used here to initialize the transaction is also INITIALIZE IEP, but with the 'for Load' option. This instruction does not provide the card with data, but rather increments the transaction counter. Finally, a signature S_1 is formed from data such as due date, transaction counter and IEP identifier. The data, thus secured for transmission by the signature, are returned to the terminal together with a few other data.

The actual payment transaction is executed by the next instruction, designated DEBIT IEP. It sends data concerning the amount to be debited, the current key versions and another signature to the electronic purse held in the card. This allows the terminal's authenticity to be checked, as was the case when crediting the card. If the check is successful, the account and the payment transactions protocol file are updated and a further signature, S_3, are created to acknowledge the entire transaction. This is a response to the DEBIT IEP instruction. The S_3 signature serves as confirmation to the terminal's security module that the amount was properly debited to the card's account.

INITIALIZE IEP for Purchase	
Instruction	• —
Response	• describes purse provider (PP_{IEP}) • describes IEP • cryptographic algorithm used (ALG_{IEP}) • expiry date ($DEXP_{IEP}$) • purse balance (BAL_{IEP}) • currency code ($CURR_{IEP}$) • authentication mode (AM_{IEP}) • IEP transaction number (NT_{IEP}) • key information (IK_{IEP}) • signature S_1 • return code (CC_{IEP})

Figure 7.54 INITIALIZE IEP for Purchase function in accordance with prEN 1546-3

DEBIT IEP	
Instruction	• describes PSAM • PSAM transaction number (NT_{PSAM}) • amount to be debited (M_{PDA}) • currency code ($CURR_{PDA}$) • key information (IK_{PSAM}) • signature S_2 • user specific data (DD)
Response	• signature S_3 • return code (CC_{IEP})

Figure 7.55 DEBIT function in accordance with prEN 1546-3

This is just a brief overview of the four most important electronic purse instructions, as described in prEN 1546. The standard is very extensive and covers many options and variations for system design, which may affect the instruction formats[4].

---

[4] See also 12.3.1 CEN standard prEN 1546.

7.10 CREDIT CARD INSTRUCTIONS

Part two of the joint specification for credit cards incorporating chips covering Europay, Mastercard and Visa describes two instructions specific to this application. A whole section has been devoted to this instruction since it is credit cards incorporating chips which are expected to be produced in particularly large numbers in the future, and are therefore of great importance.

GET PROCESSING OPTIONS	
Instruction	• processing options data object list
Response	• application interchange profile
	• application file locator
	• return code

Figure 7.56 GET PROCESSING OPTIONS function in accordance with EMV-2

The instruction GET PROCESSING OPTIONS is employed to initiate payment. It is used to input the terminal's TLV-encoded data into the card, for processing the rest of the payment routine (processing options data object list). This could include the amount of the transaction. In response, the card sends back a BER-TLV encoded data object containing the functions supported by the card (application interchange profile), and an application file locator.

The second payment instruction in a smart credit card is called the GENERATE APPLICATION CRYPTOGRAM. In this instruction, all data are TLV-encoded in both the instruction APDU and the response APDU. The instruction sends all the data necessary for a payment transaction to the card, as well as the requested application certificate. The card now computes the rest of the payment routine on the basis of the transmitted and stored data. In its response it sends to the terminal an application certificate, which in the simplest case can be the transaction certificate. This concludes the payment procedure.

The application certificate returned by the card may, however, contain an authorization query instead of the application certificate. If, in computing the rest of the transaction routine, the card obtains the result that an on-line query is required, then the application certificate response data will include a query to the authorization centre in charge of the terminal. Once it has been processed by the centre, the card obtains the relevant data in a second GENERATE APPLICATION CRYPTOGRAM instruction, and can then produce the payment transaction certificate and send it to the terminal[5].

---

[5] See also 12.4 Chip-containing credit cards.

GENERATE APPLICATION CRYPTOGRAM	
Instruction	• application cryptogram required • transaction-related data
Response	• cryptogram information data • application transaction counter • application cryptogram • return code

Figure 7.57 GENERATE APPLICATION CRYPTOGRAM function in accordance with EMV-2

7.11 COMPLETING THE OPERATING SYSTEM

During Smart Card microprocessor manufacture, only the ROM is programmed but the EEPROM's memory remains empty apart from a chip number and an individual key. Once the card is assembled, the operating system in ROM needs to be supplemented and completed by the EEPROM's sections. Only then does the card contain a full operating system with all its functions.

In order to write these parts of the EEPROM, a relatively simple loading program is present in the ROM code which enables data to be written to the EEPROM after previous key-checking. The EEPROM memory is addressed linearly, byte-wise or section-wise. Once all necessary data has been input to the EEPROM in this way, the card's operating system is switched over from pure ROM operation. The processes and routines can now also run in EEPROM. This conversion can be performed by an instruction whose execute condition has been satisfied through comparison of a checksum of the completed EEPROM data. The checksum ensures that all data are correctly stored in the EEPROM. This completion does not rely on excessively complex functions or authorization procedures since it needs to be supported by the ROM part of the operating system, and even the smallest errors here would make it impossible to complete the card. This would be a very time-consuming error and ultimately a very expensive one.

The following sections provide examples of the three instructions needed to complete a Smart Card operating system. They vary greatly from one operating system to the next, and also between manufacturers. We can only illustrate the operating functions. However, practically all Smart Cards use either this method or a similar one. The instruction COMPARE KEY checks a key sent to the card against one stored in the EEPROM during the manufacturing stage. The key is unique to the card and is quite long, about 32 bytes. If the comparison is successful, the subsequent loading instructions are permitted. If not, an error counter is incremented which from a predefined value (usually 3) onwards locks the card against further access. The card should then be disposed of, as no other functions are present apart from the ATR.

After the loading key has been successfully checked via COMPARE KEY, all necessary data can be written to the EEPROM using the instruction WRITE DATA. It is then possible to address the entire EEPROM and to write to it byte-wise. This means that besides operating system data, complete applications may also be incorporated. This, incidentally,

COMPARE KEY		
Instruction	•	key
Response	•	return code

Figure 7.58 COMPARE KEY function

WRITE DATA		
Instruction	•	data
	•	EEPROM address
Response	•	return code

Figure 7.59 WRITE DATA function

is the normal method used for cards with a very small memory which lack space for complex CREATE FILE instructions with the corresponding state automata.

If the available ROM in the card is so small that even EEPROM test instructions cannot be stored in it, then it is possible to simulate its basic functionality using this instruction. A particular memory location is addressed so many times that the card returns a writing error. If the number of writing steps is counted, then the number of possible write/clear cycles is known. This is the main statement expected from an EEPROM test instruction as part of quality assurance.

Once all data have been written to EEPROM with one or a number of consecutive WRITE DATA instructions, the EEPROM's contents are checked and the completion procedure is terminated. The instruction used here is COMPLETION END. After successful execution of this instruction, it is usual to trigger a card reset to initialize the operating system again and allow the new state to be created.

The sequence of completion instructions listed above is illustrated in Figure 7.63. It is controlled by an automaton during the completion process so that only the instructions described can be executed in the precisely defined order.

COMPLETION END		
Instruction	•	EEPROM checksum
Response	•	return code

Figure 7.60 COMPLETION END function

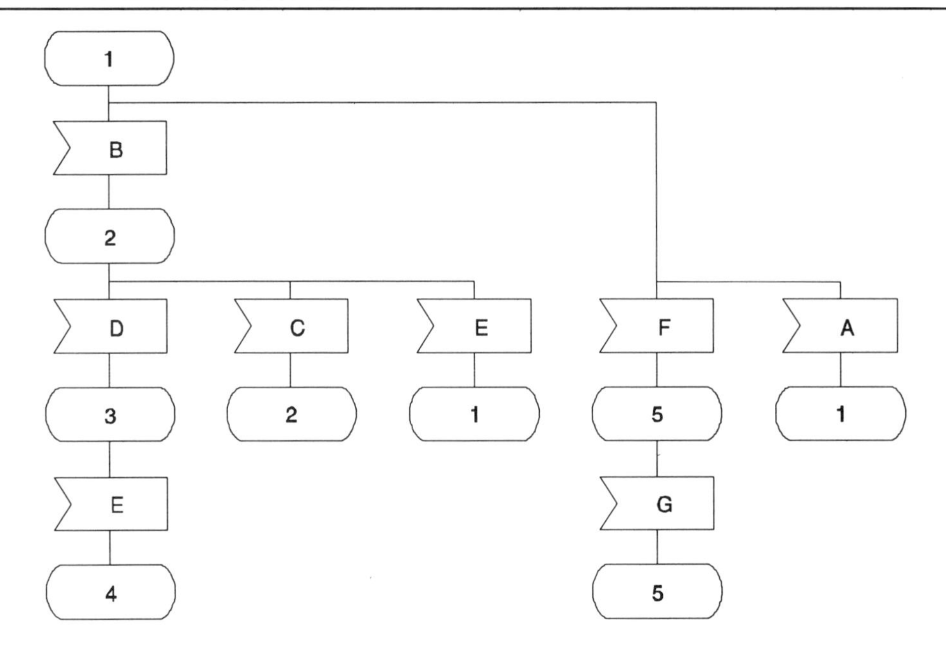

Figure 7.61 Completion instructions.
Conditions: 1 Basic condition after Smart Card reset
2 Smart Card ready for data entry into EEPROM
3 Smart Card completed
4 Basic condition of Smart Card after completion
5 Smart Card irreversibly locked
Transfer A all commands except COMPARE KEY
B COMPARE KEY (successfully carried out)
C WRITE DATA
D COMPLETION END
E Reset Smart Card
F COMPARE KEY (3-times unsuccessful)
G all instructions and reset

7.12 HARDWARE TESTING INSTRUCTIONS

During initialization, a Smart Card's operating system tests various hardware components both implicitly and explicitly. The instructions described here, however, go far beyond the self-test routines integrated directly into the operating system. During the course of production quality assurance the microprocessor's critical parts need to be subjected to special testing. These tests focus particularly on the EEPROM, since experience shows that most problems crop up here. The processor's functions are tested implicitly, the terminal having received the correct ATR.

Since no standard exists for test instructions, their functions and coding depend on the system manufacturer, or sometimes even on the system itself.

Smart Card		Terminal
	←	COMPARE KEY
Processing instructions		*Instruction* [key for completion]
Response [Return code]	→	IF (Return code = OK)
		THEN instruction successfully carried out
		ELSE instruction unsuccessful
		numerous attempts
	←	WRITE DATA
Processing instructions		*Instruction* [data ll address]
Response [Return code]	→	IF (Return code = OK)
		THEN instruction successfully carried out
		ELSE instruction unsuccessful
	←	COMPLETION END
Processing instructions		*Instruction* [EEPROM checksum]
Response [Return code]	→	IF (Return code = OK)
		THEN instruction successfully carried out
		ELSE instruction unsuccessful

Figure 7.62 Example of typical completion sequence

The test instructions may be fixed in ROM for security reasons. However, it is quite usual to load them into the EEPROM via the completion instructions, and to execute them there. Obviously, this may lead to problems in the case of an EEPROM which is not fully functional. The advantage lies naturally in the larger space then available in ROM. For a secure operating system, the installation of test instructions must be restricted to the phase preceding completion. This means that all test instructions in an initialized or even personalized card are locked. The exceptions are the RAM test and the ROM or EEPROM checksums, since these instructions do not affect security.

The following are several instructions which may be used for extensive hardware tests. A faulty RAM would lead to a complete system crash even before the ATR, but it is possible for only isolated RAM bits or bytes to be faulty. This would only have repercussions for special functions or parts thereof.

The instruction TEST RAM sends an appropriate response to the terminal concerning the entire condition of the RAM. During the test itself, all available bytes must be written to and be read using a variety of test patterns. A typical test involves alternate writing with '55' and 'AA', since both these hexadecimal values form a chequered pattern at the bit level. Another effective method is a waveform test in which the RAM is written to once from the lowest to the highest address and then read, and vice versa. The precise implementation of this test depends on the operating system. Sometimes this test is omitted.

TEST RAM		
Instruction	•	—
Response	•	return code

Figure 7.63 TEST RAM function

CALCULATE EDC is a very simple test which computes an EDC checksum over the whole ROM or over the EEPROM sections to be defined, and returns it to the terminal. This method is used to determine whether the ROM mask has remained unaltered and whether EEPROM cells have flipped over. The EEPROM test only relates to static areas which cannot be changed deliberately during the card's lifetime. The terminal compares the checksum received from the card with a reference value, and decides whether the memory contents are still consistent.

CALCULATE EDC		
Instruction	•	*switch:* ROM / EEPROM
Response	•	checksum
	•	return code

Figure 7.64 CALCULATE EDC function

The EEPROM's continuous check TEST EEPROM relates to sections which are to be written to. The card receives two patterns, which it writes alternately in an area of memory. The area's extent as well as the number of writing attempts can be specified in the instruction, within certain limits. After each writing step the operating system checks the memory contents for errors.

The fact that a card has been supplied with the number of writing cycles allows this instruction to be performed internally. This speeds up execution considerably compared with individual instructions. If the card discovers a write error, it sends the number of previously executed cycles to the terminal together with the faulty address.

The continuous EEPROM test is not restricted to destructive testing, as it can also be used for writing to freely selectable EEPROM areas. The number of cycles would then be set to one.

TEST EEPROM		
Instruction	•	pattern 1
	•	pattern 2
	•	number of writing cycles
	•	start address in EEPROM
	•	end address in EEPROM
Response	•	number of writing cycles executed (in case of faults)
	•	faulty address (in case of faults)
	•	return code

Figure 7.65 TEST EEPROM function

The COMPARE EEPROM test is used to check whether the written pattern is still present. This combination of instructions is used mainly for EEPROM data testing at various temperatures. This is done by writing a pattern to several sections of memory, placing the card in a temperature-controlled cabinet and checking after a given interval whether the EEPROM's contents have been retained.

COMPARE EEPROM	
Instruction	• comparative value
	• address
Response	• return code

Figure 7.66 COMPARE EEPROM function

The instruction TEST EEPROM can be used in principle to clear the entire EEPROM, when the start and end addresses are specified accordingly. However, many operating systems have an instruction which we will call DELETE EEPROM. It is used to clear the entire EEPROM in one step, by overwriting all the addresses.

This instruction is employed for two purposes. After various other tests the EEPROM can thus be set, with little difficulty, to a predefined state. The second application is used before initialization and personalization and is designed to speed them up. A cleared EEPROM can be written to faster, since the obligatory pre-write clearing time is omitted.

DELETE EEPROM	
Instruction	• —
Response	• return code

Figure 7.67 DELETE EEPROM function

7.13 APPLICATION-SPECIFIC INSTRUCTIONS

There are many instructions which are tailored to a particular application. These are mainly designed to save memory space or to minimize execution time. In the main, these instructions are so specific that they are not covered by standards, or else the standards restrict their use to a particular field of application.

It would go beyond the scope of this chapter to list all the application-specific instructions. We are therefore restricting ourselves to one typical example, the only instruction specific to GSM 11.11, namely RUN GSM ALGORITHM. It is used for simultaneous generation of a dynamic, card-specific key, and for card authentication against the GSM background system. This function is so specific to the GSM application that it makes little sense to include it in a general Smart Card standard. It relies on a cryptographic algorithm only used in GSM, and the two initial values generated by the random number received would be useless in any other application.

RUN GSM ALGORITHM	
Instruction	• random number
Response	• dynamic key
	• enc (key; random number)
	• return code

Figure 7.68 RUN GSM ALGORITHM function

7.14 TRANSMISSION PROTOCOL INSTRUCTIONS

In principle, transmission protocols should be so structured as to be wholly independent of the application layer's data and instructions. This is also provided for in the OSI layer model. Unfortunately, theory and practice clash in this area. There are two instructions whose only purpose is to execute transmission protocol mechanisms at application level, namely GET RESPONSE and ENVELOPE. The instruction MANAGE CHANNEL is yet another function used not only by the application layer.

In the T=0 protocol it is impossible, during an instruction–response cycle, to send a block of data to the Smart Card as well as to receive one from it[6]. Thus, the protocol does not support case 4 instructions. However, since such instructions are used in practice, T=0 is obliged to resort to an auxiliary construct. Its function is simple. The case 4 instruction is sent to the card, and, if successful, then a special return code is received. This signals to the terminal that the instruction has generated obtainable data. The terminal then sends a GET RESPONSE instruction to the card and receives this data. This completes the instruction–response cycle for the first instruction. As long as no instruction other than GET RESPONSE has been sent to the card, the original data can always be requested.

GET RESPONSE	
Instruction	• amount of data to be sent
Response	• data
	• return code

Figure 7.69 GET RESPONSE function in accordance with ISO/IEC 7816-4

Where instructions are completely encrypted as part of secure messaging, transmission problems can occur during the T=0 protocol since it requires both an unencrypted instruction byte and an unencrypted Le byte. The ENVELOPE instruction is used to get around this restriction, by embedding a complete APDU with header and data section into the ENVELOPE APDU data section. This can then be encrypted without any restrictions

[6] See also 6.2.2 Transmission protocol T=0.

and transmitted by any protocol. The same applies to the response generated by the card, which is also embedded in the ENVELOPE APDU.

ENVELOPE		
Instruction	•	instruction-APDU
Response	•	response-APDU
	•	return code

Figure 7.70 ENVELOPE function in accordance with ISO/IEC 7816-4

Logical channels allow up to four applications on a single Smart Card to be addressed independently of each other[7]. The linking of instruction and application takes place via two bits in the class byte. Before using a new logical channel, the card must be explicitly informed via the instruction MANAGE CHANNEL. This signals to the card that an additional channel is needed. Its number can be specified explicitly by the terminal, or after a query to the card the latter can supply the number of a free channel. When a new logical channel is opened from the standard channel numbered zero, the card behaves vis-à-vis this channel as after a reset, i.e. the MF is selected and the secure state has not yet been reached. When a new logical channel is opened from another which is not the one numbered zero, the currently selected DF as well as the secure state are retained. After closing a logical channel, the relevant file selection and the secure state are cleared.

MANAGE CHANNEL		
Instruction	•	switch: logical channel open/closed
		IF (certain channel required) THEN
	•	number of logical channel
Response	•	number of logical channel (if new logical channel available)
	•	return code

Figure 7.71 MANAGE CHANNEL function in accordance with ISO/IEC 7816-4

[7] See also 6.7 Logical channels.

Security methods

8

One of the main advantages of Smart Cards, as compared with other media such as magnetic cards or disks, is the protection and confidential storage of data. This requires made-to-measure and optimized chip hardware and various encryption procedures. However, this is not merely a matter of the chip and the algorithms used in the software. The security of the system itself and the design principles used by its development team are also of fundamental importance. This chapter summarizes the essential principles, procedures and strategies utilized en route to secure Smart Cards.

8.1 USER IDENTIFICATION

The unambiguous identification of individuals has occupied the human mind since the dawn of history. The simplest form of identification is a card bearing a photograph, or a signature written in the presence of witnesses. The photograph attached to the ID card is compared with the actual person, and the outcome of this comparison is a statement about his or her true identity.

However, in the field of information technology the comparison is not so simple, since it must be made not by humans but by a computer. That is why the established method is one which uses passwords entered via a keyboard, and compared by the computer with stored reference values. The computer decides if the operator is genuine or not, depending on the result of this comparison. This procedure is further restricted in the case of Smart Cards, in which not all alphanumeric keys can be used as ID codes but only the digits 0 to 9. This code is usually known as the PIN (Personal Identification Number).

In principle, three different options can be used to identify a user:

- confidential information (e.g. the PIN input)
- possession of a physical object (e.g. key to a door)
- measurement of biological characteristics (e.g. fingerprints)

The first two methods suffer from the disadvantage that the person to be identified needs to memorize something or carry the item on his or her person. This is unnecessary in the third option, but in most cases the measurement is technically complex since very simple biological features, such as weight or height, are ruled out.

One further point is very important in this context. In many cases the input and checking of a PIN are not merely an identification of the user, but also represent a declaration of acceptance by the user, who by entering the PIN agrees to a particular course of action. For example, consider a PIN input to a cash dispenser. The user is identified by knowing the secret PIN, but he or she also agrees to have a particular amount of cash dispensed. This is a serious problem if biological characteristics are to be used for identification, since these can be measured without the person's explicit agreement.

The input of PINs to various automated equipment has by now become commonplace. The sharp increase in PIN checks for various purposes has put a heavy strain on the ordinary citizen's memory. It is very dangerous to jot down PINs on the card or somewhere close to it, as this endangers the whole security of the system. However, how many people can memorize 20 different PINs? That is why in recent years the emphasis has shifted towards research into other identification methods. Biological features by which a machine can uniquely identify a person are ideal for this purpose.

8.1.1 Input of secret numbers

The most common method of user identification is by entering a secret number, which is generally abbreviated as PIN. This is compared inside the card with a reference value, and the result is sent to the terminal. Normally PINs consist of four digits, which may be any number between 0 and 9. However, ISO 9564-1 recommends that a PIN should have between four and twelve alphanumeric characters, to reduce the probability of PIN discovery merely through trial and error. The input of non-numeric symbols is impossible at many terminals on technical grounds, as a numeric keyboard is often the only one available. Nor does the number of characters in a PIN depend solely on the required level of security, but also very much on the average user's memory. People have become used to the idea of memorizing four figures over the years, and if the code were increased to five or even more digits, there is likely to be quite strong resistance. The same applies to the very reasonable desire to change the PIN periodically, although this may be acceptable in high-security applications with a small number of users. Mass applications, however, demand maximum simplicity; there should be no need to have an exceptional memory in order to remember PINs.

The PIN is used to check the user's identity for purposes of access. Yet on the other hand, the user may be just as anxious to ascertain whether the terminal is genuine. There have been cases in which dummy keypads were attached to real cash dispensers. Criminals used this method to discover the PINs of unsuspecting customers who had typed them in on the bogus keypad. The criminals subsequently stole the card itself, and were subsequently able to obtain money by using the PIN. This only happened because it was impossible for the card-holder to check the identity of the terminal or of the keypad connected to it.

Yet there are simple procedures which can deal with this type of risk. A secret code known only to the user is stored in a file on the card. This may be a name or a number which can only be altered by him or her. When the card is inserted in the terminal, reciprocal authentication of these two items takes place. Once each party recognizes the other as genuine, the card permits access to the file containing the user's secret code, and the code is then displayed on the screen. The user can see the code and knows that the terminal is genuine, otherwise it would not have been able to read the file. The PIN can now be input safely. This simple procedure prevents PINs being entered onto dummy or sabotaged terminals. The code may be an arbitrary word or number, and the user should be able to change it at will, so as to ensure that nobody else can discover it. The method can be modified or extended for other applications.

8.1.2 Biometric methods

The increasing use of passwords and PINs is leading to ever-intensifying user resistance to this type of identification. Memorizing a few frequently used number (or letter) combinations may not present a problem for most people; but if a PIN is only used rarely, for example where a customer only needs to use a cash dispenser every few weeks, the average person may find the code difficult to remember. This is all the more so because the user is under a subtle form of stress, as the dispenser may swallow the card if the wrong number is entered three times!

This must surely be one of the main reasons why biometric methods are finding increasing favour in more and more areas. They may not be any faster or safer than PINs, but they can simplify things greatly for the user. Once the security offered by them is similar to that of PINs, system operators will be prepared to consider them. After all, biometric features cannot be transferred to other people as PINs can: one identifies the actual person and not a secret code common to the user and the system operator.

8.1.2.1 Basics

A biometric identification procedure is one which can identify a person unambiguously by unique, individual and biological properties. One may distinguish between physiological and behavioural characteristics. If the features tested by the procedure are directly associated with a human body and do not depend on conscious behavioural patterns, then these are referred to as physiological features. On the other hand, behavioural patterns use certain elements which can be changed consciously within certain limits, yet are typical of that person.

The problem of user acceptance is only one aspect of biometric feature checks. The greater the similarity a procedure is to existing, known identification methods, the more willing users are to accept a procedure and use it. A typical example is the handwritten signature, which has been used in almost all cultures for generations as a means of identification and to signify agreement or consent. There are also socio-psychological aspects to the implementation of certain procedures. In most countries, fingerprinting is employed mainly by the police and security forces. This may adversely affect acceptance of biometric procedures which employ fingerprinting or similar techniques.

Another point to bear in mind is the user's concern over the medical and hygienic aspects of biometric identification. There may be anxiety about the potential for transmitting disease. For instance, if retinal examination is used, there may be fear of damage to the eye caused by the laser used. Even where such fears are wholly irrational and groundless, they may strongly influence user behaviour and acceptance. This should be considered before any biometric identification procedure is implemented.

PIN entry not only checks the user's knowledge of the secret code; it is also a legally valid statement signifying "I consent". This relationship is very important where procedures other than PINs are to be used. It can be stated categorically that the examination of the fovea of the human retina from a distance of three metres cannot be interpreted as a token of consent by the individual concerned to any course of action, however it might be worded. Incidentally, this method is not yet a technically realistic proposition anyway. In almost all countries, the user must perform a conscious action for consent to be assumed. For instance, breaking the seal on a cardboard box containing software is a definite consent to accept the licence conditions printed on it. Biometric methods in which a person is subject to a totally passive examination must therefore be supplemented by appropriate explanations which elicit an element of consent.

Naturally, not all biological features are suitable for personal identification. The following points constitute the minimum criteria to be satisfied before a physical feature can be used for automated identification purposes:

- it must be capable of being uniquely assigned to a particular individual
- tampering must be impossible
- natural changes through ageing are not too significant
- it is measurable effectively (method, duration and cost)
- the amount of data generated is small (no more than several hundred bytes)
- the method and the feature are both acceptable to future users.

In any kind of measuring technique, the result does not always turn out to be identical. This effect may be present even in the simplest cases. For example, if one measures the length of a sheet of paper several times, the results are often different. There are many reasons for this and it may not present a problem, since the mean of all the measurements will be close to the actual value.

Experience shows that in principle, variations between measurements are proportionate to the complexity of the process used. For instance, there is a huge difference between weighing a bar of chocolate and finding the distance to the moon. Measurements of human beings are particularly problematic due to the wide range of possible results.

Figure 8.1 illustrates the measurement of a biological feature, for example the length of a finger. The range of measurements is recorded on the abscissa (x-axis). The ordinate (y-axis) displays the probability of correct identification based on the measured feature. If this were an ideal feature and an ideal method, there would be no variation between individual measurements, and the curve would be reduced to a vertical line. However, a real feature together with a real method produces the Gaussian bell-shaped curve shown. If a result deviates from the reference value, one can no longer be certain that the person has been correctly identified.

In order to check a biological feature, one first needs to capture the human data. This can be achieved by repeated measurement of the same feature and calculation of the mean. The result is a reference value which can then be stored in the Smart Card. The Smart Card performs the identification checks when required to ascertain whether the reference pattern agrees with the current measured value. Depending on the method used, it may be necessary to first use a powerful computer to manipulate the current values in order to facilitate their use by the card for comparison purposes. This is very similar to a PIN check. However, in the method under discussion, identification does not ensue with 100% certainty, so that a threshold value is needed in order to determine whether the individual can be accepted or not. The threshold value needs to be specified for each procedure and application.

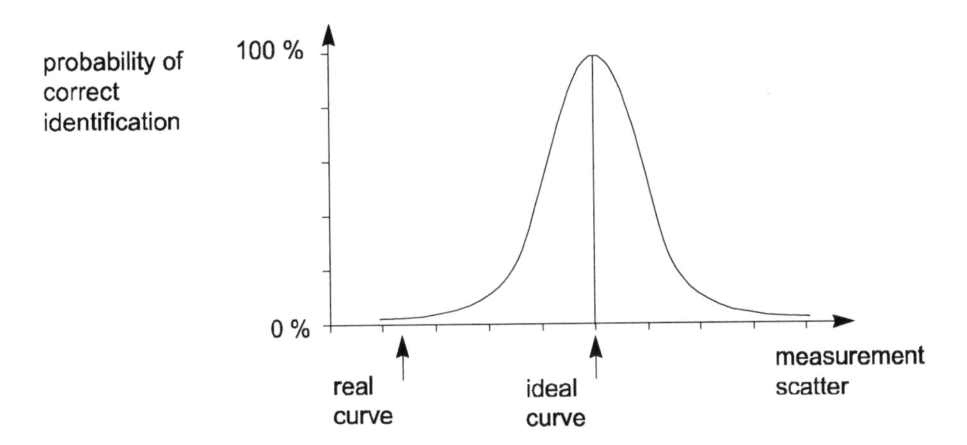

Figure 8.1 Probability distribution with repeated measurements of biological features of a person

If another person's curve is plotted in the probability distribution diagram next to the one for the person to be identified, Figure 8.2 is obtained. The additional curve represents an arbitrary individual whose biometric feature is close enough to that of the first individual for it to exert some influence over the identity decision. Since both curves tend to the horizontal asymptotically, they possess a joint point of intersection. At this point the probability that the person is genuine is the same as the probability of him or her being false. That is why biometric ID systems also use an adjustable threshold value, which determines from which probability onward correct identification can be assumed. The marked threshold value divides the two curves into four regions. They indicate the decision taken regarding the individual's identity after measurement of the feature.

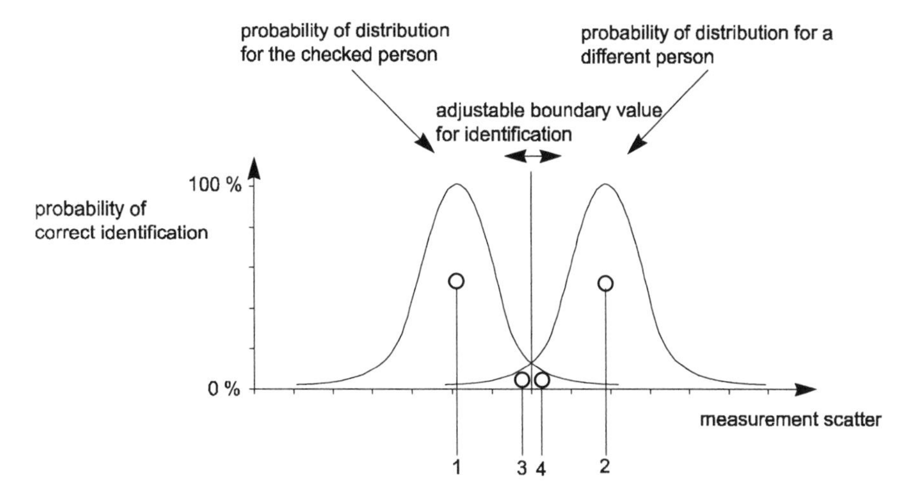

Figure 8.2 Probability distribution and evaluation when measuring biometric features
1 – true acceptance,
2 – true rejection,
3 – false acceptance,
4 – false rejection

Essentially, the diagram demonstrates that there is no such thing as an absolutely positive identification. It is only possible to assume, with a high degree of probability, that this is the right person. The magnitude of this probability can be adjusted by means of the threshold value. The position of this value cannot be freely selected in practice, since a strict criterion for correct identification also means a large number of false negatives.

PIN checking does not require the card to contain complex algorithms, since the only comparison needed is between the input and the stored PIN. Unfortunately, matters are not quite so simple in the case of biometric features. The reference value is stored in the card, but comparison with the current measurement cannot, as a rule, be made in the card itself, due to the very high computing power required. The routines used often require special signal processors or even minicomputing power. Of course, such performance cannot be expected of a Smart Card. Therefore, the processing of the raw data is often performed externally. Compressed data is fed to the card which compares them with the reference value by using simple algorithms.

8.1.2.2 Physiological features

Some physiological characteristics which cannot be consciously altered, also change little over time. The typical patterns in fingerprints never change during an individual's lifetime, nor do those of the blood vessels in the retina. This is not true of the face, which does not change fundamentally but can be transformed remarkably through hairstyle, beard, etc. Basically, though, it is possible to say that in the case of biometric features which rely on adult physiology, no continuous adjustment of the reference pattern is necessary since changes are too small or non-existent.

It would be a simple matter to list several additional features to the ones described below, but space prohibits a mention of more than the most important and commonest used in this field.

Facial features

Ordinary experience suggests that the human face is suitable for use as a biometric feature. Implementing this assumption in a technical context, however, is fraught with difficulty. Faces can change greatly over a short period of time, and their appearance depends very much on such external factors as glasses, facial hair, make-up, illumination and angle of view.

If one photographs a face in visible light and processes the captured data appropriately, it is possible in many cases to make a statement about the identity of the person behind the face. In terms of technical equipment, this complex task requires *inter alia* very powerful computers, fuzzy logic and neural networks. This method might be predicted to become a very interesting one for biometric purposes, but currently it cannot produce enough reliably positive identification to enable it to be widely used.

Retinal features

Human retinas are unique to each individual, due to the pattern of blood vessel nodes and capillaries. Data capture is performed by means of an infrared beam directed through the pupil. The reflected radiation is intercepted by a CCD camera, which sends the image to a computer for analysis.

Retinal imaging is one of the very best biometric methods, since it allows a person to be identified with a very high degree of probability. Nevertheless, the procedure has met with considerable opposition from users, since the eyes need to be brought very close to the apparatus. This results in fear of infection and of the infrared beam. Another problem is caused by certain types of contact lens which screen out a high proportion of waves in the infrared range and cause the measurement to fail.

Iris

The iris is a variable diaphragm which cuts down the rays reaching the retina. It is a biological feature which, like the retina, is unique to each individual. It is possible to keep the measuring apparatus at a greater distance than is the case with retinal examination, since the procedure is simpler. The surface of the eye is recorded by a CCD camera in the visible range. Processing is similar to that of retinal images. Again, contact lenses may greatly affect the results and lead to problems.

Hand geometry

Identification systems based on three-dimensional measurements of the hand or part thereof have been in use since the 1970s. Measurable features include finger length, finger diameter and fingertip radius. A unique identifier can be determined by using only a very few measurement points (e.g. five). From the technical point of view, data capture itself can be achieved very simply using infrared LEDs and photodiodes, relying on the principle of complete or partial blocking of the beams by hand geometry. Since a few points suffice for identification, the procedure is also fast enough and uncomplicated for the user, who only needs to place the hand in the instrument which performs the measurement.

Fingerprinting

The best-known physiological biometric procedure must surely be identification by fingerprinting. In the electronic version, this no longer requires pressing fingers coated with black ink on to paper. The thumb or other fingertip is placed on a transparent plate, and a camera mounted beneath it records the skin surface without any contact. As a rule, comparison with the reference pattern employs the basic features of Henry's classification system: arches, loops and whorls. The type, position and orientation of 20 or so of these features are stored, and the data is used to establish the reference pattern.

This procedure is often disliked by certain groups of users, since it is known to have been used for many years in the fight against crime. Problems with perfect identification may also arise from small injuries to the fingers. In addition to the optical recording unit, many systems possess sensors for recording finger temperature or pulse. This is designed to ensure that amputated fingers are not used for identification purposes.

Despite all this, fingerprinting systems are widely used, since they are relatively unproblematic in terms of technical complexity and user acceptance. In addition, the time needed for recording and subsequent processing remains within reasonable limits.

Figure 8.3 summarizes the biometric procedures described, and their essential properties.

Method	Length of examination	Size of reference pattern (bytes)	Probability of false rejection	Probability of false identification
Face	—	—	≈ 10%	≈ 1%
Skin surface	1.5 –7 s	40 –80	0.5×10^{-2}%	10^{-9}%
Hand geometry	1 –2 s	10 – 30	0.8%	0.8%
Fingerprint	1.5 –9 s	300 –800	1.4×10^{-2}%	10^{-6}%

Figure 8.3 Biometric procedures and their essential properties

8.1.2.3 Behavioural features

Biometric features based on behaviour may vary. Consider signatures, which change considerably throughout life. However, it is rare for these changes to occur suddenly – usually they are quite gradual and slow. That is why many systems use adaptive procedures which detect changes in the biometric feature, and in the case of correct identification, adopt them as a new reference pattern and store them in the Smart Card.

Writing rhythm

It has been determined that large differences exist between the way in which individuals type characters on a keyboard. These are mainly expressed in the pauses between typing separate letters. This phenomenon may therefore be used as an identifying feature. The procedure is as follows: the user types a specified character string on a keyboard, and the computer to which the keyboard is linked simultaneously evaluates the writing rhythm. The great advantage of this method is that one needs practically no additional hardware, since in most cases a keyboard and computer are already available.

Voice characteristics

Like the face, voices are unique to individuals and can be used for identification purposes. The user speaks one or more sentences into a microphone. The text must be different at each session, otherwise sentences recorded on tape during a previous identification could merely be played back. A Fourier analysis is performed of the waveforms of the spoken sentences, which provides a spectrum characteristic of the person. This is then compared with a reference value and the speaker's identity determined. The entire arsenal of modern IT processes is implemented here, such as fuzzy logic, neural networks and the like.

Of course, this method is also not without its shortcomings. The human voice depends greatly on the speaker's current state of health. Furthermore, all background noises must be filtered out to allow a reliable and unambiguous spectral analysis. As mentioned, a different sentence must be spoken each time to prevent recordings being played back, which complicates the procedure and makes recognition expensive. These technical difficulties, however, are offset by good user acceptance, which makes this biometric identification method of great interest.

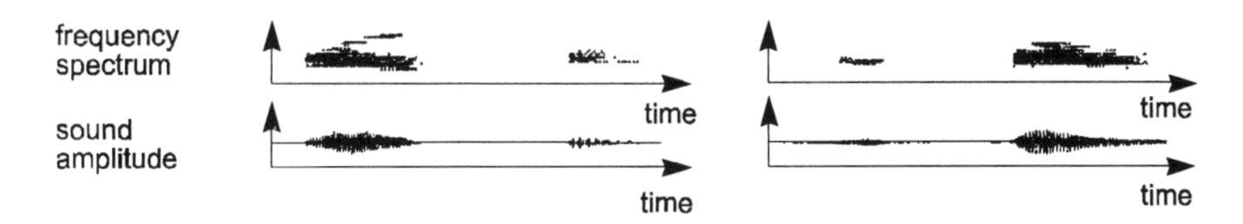

Figure 8.4 Volume and frequency spectrum of the name "Wolfgang" spoken by two different people

Dynamic signature

The only identification method common in everyday life is the provision of a signature. Due to its very individual character, a signature can also be used as a biometric feature. In the static version, a signature already supplied is evaluated. In the dynamic version, however, measurements are made during the signing process. The static signature is of a rather theoretical nature, since it cannot distinguish photocopied signatures from genuine ones.

The parameters measured in the dynamic signature may be the shape of the signature, speed, acceleration, pencil pressure on the template and the time taken to sign. Special pens or a surface sensitive to the parameters to be measured may be used. Figure 8.5 is an example of a possible configuration, in which an ordinary pencil is used on a special template and the measured signals used for identification. Pressure sensors are located at the intersections of the two series of wires, and the signal amplitude is transmitted to the computer via logic circuitry. The computer processes the measured data through various algorithms into a standardized format, and compares the result with the stored reference pattern.

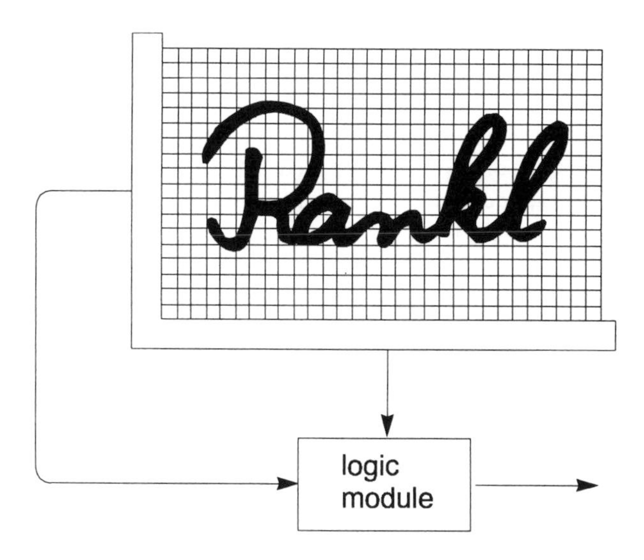

Figure 8.5 Example of a template used to measure a dynamic signature

The dynamic signature has the highest acceptance of all personal identification methods, since it is used daily by everyone in an almost identical fashion. But here too, the technical solutions are by no means simple since signatures change over time and are never quite identical. It is enough to consider the differences created if writing while sitting as opposed to while standing.

Method	Length of examination	Size of reference pattern (bytes)	Probability of false rejection	Probability of false identification
Voice	5 s	100 – 1000	1%	1%
Dynamic signature	2 – 4 s	40 – 1000	1%	0.5%

Figure 8.6 Comparison of biometric procedures based on behavioural features

8.2 AUTHENTICATION

The purpose of authentication is to check the identity and authenticity of one party in a communicating pair. When applied to the Smart Card, it means that the card or terminal decides whether the other party is genuine.

The two participants must possess some shared secret knowledge which is examined via an authentication procedure. This is significantly more secure than a straightforward identification process such as the one involved in a PIN check. There, one merely sends a secret code (the PIN) to the card, which confirms its genuine status or otherwise. The drawback is that the secret is sent to the card in uncoded text, so that anyone tapping in could obtain the secret, i.e. the PIN, very easily.

In an authentication procedure, however, it has to be impossible to discover the common secret through monitoring the phone line, so it cannot be sent publicly across the interface. A distinction must also be made between static and dynamic authentication. In the static procedure, the same static data are always used. However, dynamic authentication is so formatted that it is protected against playback of recorded data obtained in a previous session, because it employs a different set of data each time.

There is a further fundamental difference, namely that between a uni-directional and a bidirectional procedure. The former, if successful, leads to the authenticity of one of the two parties being established. The latter ensures that by the end of the process, both parties have been authenticated.

Authentications based on cryptographic algorithms are also divided into symmetric and asymmetric routines. Currently, Smart Card systems use symmetric procedures almost exclusively. The asymmetric ones, that is those based on RSA or similar algorithms, do not currently find significant practical application in this field due to their slowness, though this may well change in future. The principle of operation, however, is similar to that of symmetric methods. Device authentication is covered by several standards of which the most important is ISO/IEC 9798, part 2 of which describes the symmetric procedure and part 3 the asymmetric one.

Authentication involving Smart Cards always obeys the challenge–response procedure, where one party provides the other with a randomly generated query or challenge. The latter uses an algorithm to compute the response and sends it back to the former. Naturally, the preferred algorithm uses encryption with a confidential key which represents the two parties' shared secret.

8.2.1 Unidirectional symmetric authentication

Unidirectional authentication is used to ascertain the identity of a second party. It is necessary for both to possess a common secret, knowledge of which is being tested. The secret is a key to an encryption algorithm, on which the entire security of the procedure depends. If this key is revealed, an interloper would be able to authenticate himself just as if he were actually one of the communicating parties.

The principle is illustrated in diagrammatic form in Figure 8.7. For simplicity, the assumption is that the terminal is authenticating a card, i.e., it ascertains whether the card is bona fide.

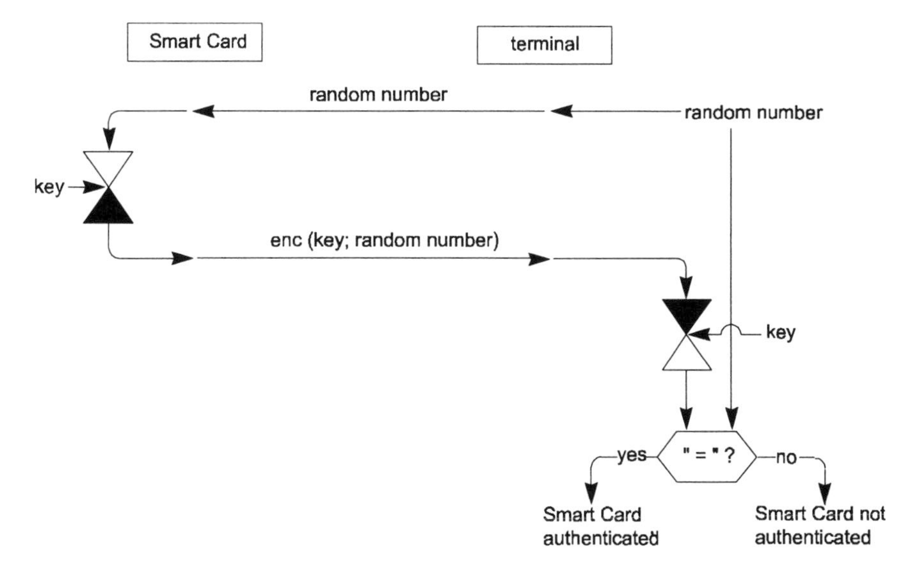

Figure 8.7 The principle of unidirectional symmetric authentication (authentication of Smart Card through the terminal)

The terminal generates a random number and sends it to the card, this being designated a "challenge". The card then encrypts it. The key used in this process is known only to the terminal and to the card. The security of the procedure depends on this key, since only those in possession of it can generate the correct response. The card sends the result of the encryption to the terminal. This is the response to the challenge. The terminal now decrypts it with the secret key and compares the result with the number originally sent to the card. If the two agree, the terminal knows that the card has the secret key and concludes that it is authentic.

An attack based on playing back a challenge or response previously obtained by bugging is not possible, since a different random number is generated at each session. The only attack with any chance of success would consist in a systematic search for the secret key. Since challenge and response are nothing but plaintext–ciphertext pairs, the key could be calculated through a 'brute force' attack.

If all cards in an application possessed the same key, and if the key were revealed, then the entire system would be compromised. It is precisely to prevent this eventuality that practical systems only use keys unique to the card. Each card then has an individual key which can be derived from a non-confidential card feature. This may be the chip's serial number, written to it during production, or some other number which varies between chips.

To be able to calculate the unique key, the terminal requests the chip's number from the card. The number generated becomes individual and unique to the whole system, i.e. no second identical card exists.

The secret authentication key, also individual to the card, is a function of the card number and the master key which is only known to the terminal. In practice, part of the card number is encrypted with the master key and the result is an individual authentication key for the card. Either a DES or triple-DES algorithm can be used. However, it is vital to realize that if the master key, which is only known to the terminal, were to be compromised, this would breach the entire security system, since it could be used to calculate any key assigned to an individual card. Therefore, the master key must be highly protected within the terminal (e.g. stored in a security module), and, if possible, actively erased in the event of interference.

Once the card's authentication key has been calculated by the terminal, the usual challenge–response routine continues. The card receives a random number from the terminal which it encrypts with its individual key and returns to the terminal. The latter carries out the same calculation in reverse and compares the two results. If they agree, then the Smart Card and the terminal share a common secret, namely the confidential card-specific key, and the card has been authenticated by the terminal.

The authentication is somewhat time-consuming, as the DES algorithm has to be called up and data transmitted to and from the card. In some applications this can be a problem. The time needed for unidirectional authentication can be roughly calculated by making the following assumptions: a card with a 3.5 MHz clock, T=1 transmission protocol, a 372 divider and a DES algorithm which requires 17 ms per block. All the card's internal routines are assumed to require 9 ms in total, which simplifies the calculation and has only an insignificant effect on the end-result.

The calculation demonstrates that a single authentication requires about 65 ms. Normally this should not cause major difficulties to most applications.

Instruction	Time required for transmission	Time required for calculation	Total time
INTERNAL AUTHENTICATE	38.75 ms +	26 ms	64.75 ms

Figure 8.8 Calculation of time required in Smart Card for unidirectional authentication, taking into consideration transmission time requirements

8.2.2 Mutual symmetric authentication

The principle of mutual authentication relies on double unidirectional checking. It is also possible to perform two unidirectional authentications in turn, one for each communicating party. In principle, this would then become a mutual authentication. However, since communications overheads need to be kept as low as possible to save time, a procedure exists which integrates the two unidirectional authentications into one. This also achieves much greater security than two consecutive authentications, since it is far more difficult for an attacker to interfere with the process.

For the terminal to calculate the card's individual authentication key from the card number by using the master key, the card number needs first to be obtained. Once this is done, the unique key for this card is computed. Then the terminal requests a random number from the card and also generates a random number of its own. The terminal now places the two numbers one after the other, encrypts them with the secret authentication key and sends the resulting cipher to the card.

The card decrypts the received block and checks to see whether the random number previously sent to the terminal agrees with the one sent back. If this is the case, the card knows that the terminal is in possession of the secret key. This authenticates the terminal vis-à-vis the card. Then the card swaps the two random numbers around, encrypts them and sends the result to the terminal. Swapping is designed to create a distinction between the challenge and the response.

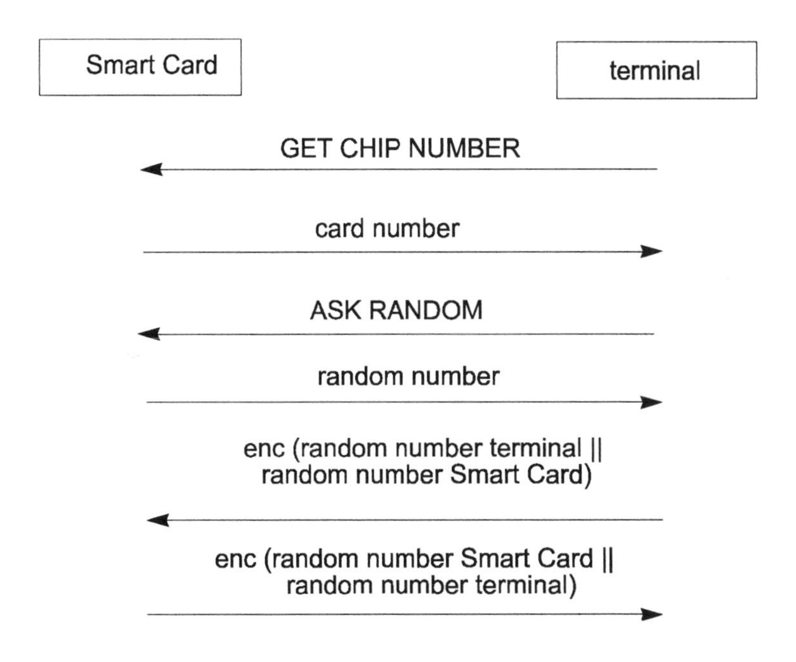

Figure 8.9 Mutual authentication using card individual key and symmetrical crypto-algorithm (mutual authentication of Smart Card and terminal)

The terminal decrypts the received block and compares the random number previously sent to the card with the one received. If they agree, the card has also been authenticated by the terminal. This concludes the mutual authentication, and both card and terminal know that the other party is genuine.

The card can send the random number together with the card number in order to minimize communication time. This is of interest when mutual authentication takes place between a Smart Card and a background system. The card is then addressed directly by this system, transparently to the terminal. Data transmission speed is often very slow in these situations, and the communication process must therefore be simplified as far as possible.

The following calculation illustrates how time-consuming this procedure can be, particularly when compared with unidirectional authentication. The basic assumptions are similar to those in the previous example. It is evident that mutual authentication takes nearly three times as long as using a uni-directional procedure.

Instruction	Time required for transmission	Time required for calculation	Total time
ASK RANDOM	28.75 ms	26 ms	
MUTUAL AUTHENTICATE	68.75 ms	95 ms	
Total	97.50 ms +	121 ms	218.50 ms

Figure 8.10 Calculation of time required in Smart Card for mutual authentication, taking into consideration transmission time requirements. It was assumed that no additional keys were used (GET CHIP NUMBER therefore not necessary)

8.2.3 Static asymmetric authentication

Only very few Smart Card microprocessors contain an arithmetic unit which can perform RSA calculations. This results mainly from the fact that it would require additional space on the chip, thus increasing its price.

However, since an additional asymmetric authentication procedure would mean increased protection, since an attacker would need to break not one cryptographic algorithm but two, it is often desirable to add this level of security. One way to get around the problem of the card's lack of an arithmetic unit is a static authentication of the card by the terminal. This only requires a signature routine in the terminal. An additional arithmetic unit in the terminal increases the cost only insignificantly compared with its overall price, so this solution is often chosen as it is considerably more cost-effective than the use of special Smart Card microprocessors. Furthermore, the procedure is significantly faster, as only one asymmetric encryption is necessary, compared with two in dynamic authentication.

Nevertheless, this compromise is bought with a reduced security of the authentication procedure. The static method does not, of course, offer protection against playback. Hence it is only used as an additional check on card authenticity, once it has first been checked with a dynamic symmetric procedure.

Basically the method works as follows. When personalized, each Smart Card stores its own individual data. This may be, for example, a card number and the holder's name and address. A digital signature is calculated during personalization using a secret key for this data which does not change during the card's lifetime. The key is used throughout the system. When the card is used at a terminal, the latter reads the signature and the signed data from a file in the card. The terminal possesses the public key valid for all cards, and can decrypt the signature and compare the result with the data it reads. If the two agree, the card is authenticated by the terminal.

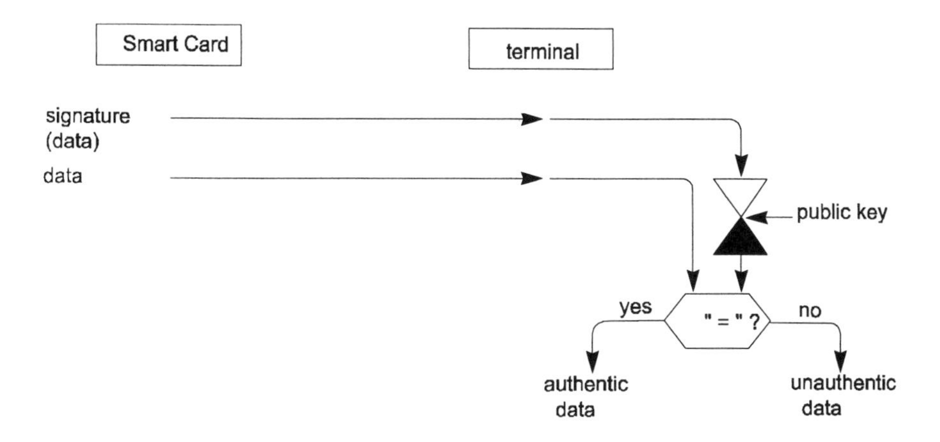

Figure 8.11 The principle of unidirectional, static and asymmetric authentication of a Smart Card through the terminal using a global key

Besides the lack of protection against playback, the above procedure suffers from another disadvantage. Global keys are used for producing and checking the signatures. The key held in the terminal does not provide any protection, of course, since it is public. In principle, large systems should not use uniform keys for all cards. If such a key is damaged or becomes known in any way, then this authentication becomes worthless across the whole system. This would make it necessary to introduce individual key-pairs for each card.

This, in turn, would create a problem concerning the terminal's storage capacity, since each terminal would now need to know all the public keys. Even in small systems, such as those with one million cards and assuming 512-bit RSA keys, this would require a 64 MB memory in each terminal for the keys alone. This would drive the price of terminals into regions which system operators would no longer find acceptable.

In symmetric methods it is possible to derive the individual key from a master key in a fairly simple fashion[1]. In asymmetric ones this does not work, because of the way in which the keys are generated. Hence the approach is different where individual keys are needed. The public key is stored in the card together with the signature. Although 64 MB is still required in a system with one million cards for key storage, the memory is distributed across one million cards in 64-byte packets. The terminal reads the public key from a file

[1] See also 8.4.1 Derived keys.

held in the card, and uses it to check the signature. This circumvents the problem of having to store all of the system's public keys in every terminal.

However, an interloper could generate a key-pair and use it to sign the data on a forged card. The terminal would read the public key and would eventually conclude that the card was genuine. Hence the procedure still needs to be refined to some extent. The public and individual key stored in each card would have to be signed with a global secret key, and this signature would then be stored in the card.

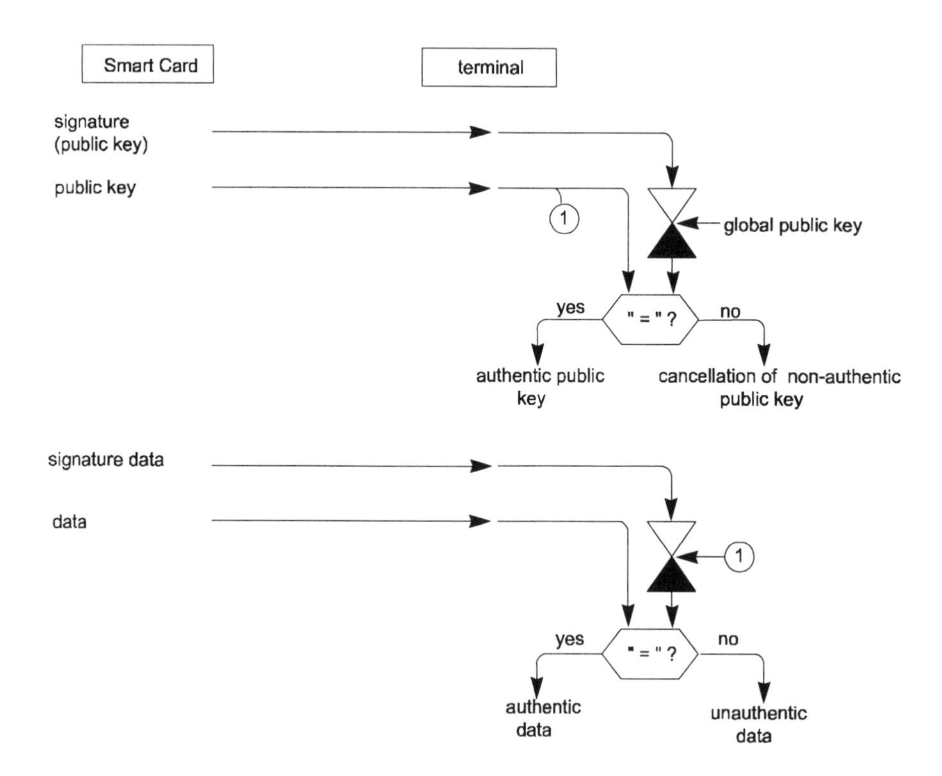

Figure 8.12 The principle of unidirectinal, static and asymmetric authentication of a Smart Card through the terminal using a derived key

The terminal would now proceed as follows. First it would read the public and individual key from the card, and use the global public key to check the authenticity of the individual one. If it is genuine, only then does it read the actual data and checks them through the public key stored in the card.

Both the above procedures are already implemented in many systems, and will no doubt become even more widespread over the coming years. They will quickly lose their attraction, though, as soon as the arithmetic units for asymmetric crypto-algorithms no longer lead to a significant price increase in microprocessors. The most serious drawback is the absence of protection against playback, which although it can be somewhat countered through a number of tricks (such as reuse of the signed data in subsequent symmetric crypto-algorithms), this can never quite offset the difference between this method and the dynamic procedures.

8.2.4 Dynamic asymmetric authentication

The static asymmetric authentication procedures described above suffer from several disadvantages, which can be overcome by the use of dynamic procedures. These offer protection against the playback of previously recorded data into the system. The usual approach uses a random number which serves as the starting value for a cryptographic algorithm. However, this type of authentication requires the Smart Card to contain an arithmetic unit which can perform it.

Figure 8.13 illustrates unidirectional authentication using a global public key. If card-specific keys are required, then the procedure described in Section 8.2.3 for the storage and authentication of such individual public keys is required.

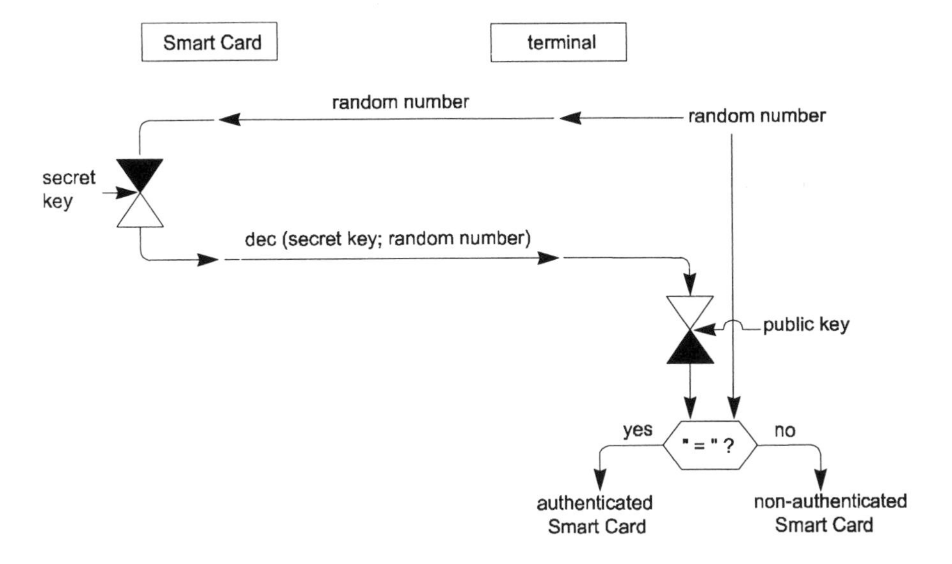

Figure 8.13 The principle of unidirectinal, static and asymmetric authentication of a Smart Card through the terminal

Similarly to the symmetric method, the terminal generates a random number and sends it to the card. The latter decrypts[2] it using the secret key and returns the result to the terminal. The terminal holds the global public key used for the encryption of the received random number. If the result of this calculation agrees with the number previously sent to the card, then the card will be authenticated by the terminal.

---

[2] The reason for a decryption operation when producing a signature lies in the convention that in asymmetric crypto-algorithms, decryption always takes place using the secret key and encryption always using the public key.

Mutual authentication of card and terminal is basically analogous to the uni-directional authentication described above. However, it is relatively time-consuming due to the high data transmission overheads and the complex asymmetric encryption procedure, so it is very rarely used at the time of writing.

8.3 DIGITAL SIGNATURE

Digital signatures, often also known as electronic signatures, are used for authenticating electronically transmitted messages or electronic documents. Examination of the digital signature makes it possible to determine whether these messages or documents have been altered.

It is a property of a signature that it can only be generated correctly by one person, but can be checked by all recipients, or at least by those who have already seen the genuine signature or have it available for comparison. This should also be an essential feature of digital signatures. Only one person or Smart Card can sign a document, but anyone can check whether the signature is genuine.

Due to this requirement, asymmetric cryptographic procedures represent the ideal starting point. Seen from an IT perspective, a digital signature is a type of cryptographic checksum similar to a MAC (Message Authentication Code) over a given data string.

As a rule, these strings are several thousand bytes long. In order to keep the computation time needed to create the checksum within acceptable limits, it is not calculated over the entire string but rather a hash value is first obtained from it. Hash functions[3] are, in simple terms, one-way data-conversion functions. The conversion is irreversible, in other words the original cannot be obtained from the converted data. Since the calculation of hash values is very fast, it is the optimum solution for the calculation of a digital signature.

The term "digital signature" is normally only used in connection with asymmetric crypto-algorithms, since they are very well suited to this process due to the separation into public and secret keys. In practice, though, "signatures" based on symmetric algorithms are also used. However, in this latter case, it is only possible to check the authenticity of a document if one is in possession of the secret key which was employed in generating the signature. The signature is no longer a signature in the strict sense, but in practice it is frequently designated as such though the word "digital" is omitted to identify the procedure used.

The routine used to generate a digital signature can be visualized in quite simple terms. The message, i.e. a file produced by any one of the range of word-processing programs, is used to form a hash value via a hash algorithm. This value is decrypted using an asymmetric algorithm, which in the example given here is RSA. The result of this calculation is the actual signature, which is added to the message.

---

[3] See also 4.3 Hash functions.

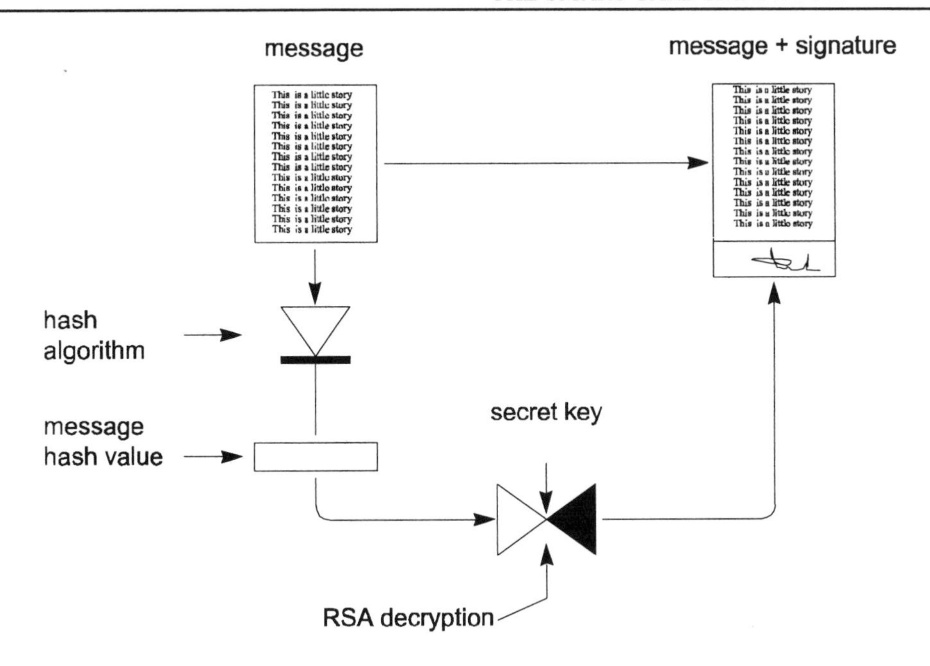

Figure 8.14 Signing a message using the RSA algorithm

The message containing the signature can now be sent to the recipient over an unprotected channel and separated again into message and signature at the destination. The message is converted using the same hash algorithm as at the sender. The digital signature is decrypted using the RSA public key, and compared with the result of the hash calculation. If the two values are the same, the message has not been altered during transmission, otherwise either message or signature has been altered. Authenticity has been compromised, and the message contents can no longer be assumed to be the same as they were originally.

The card's task in this scenario can be stated very simply. It stores at least the secret RSA key and decrypts the hash value of the message, thus producing the signature. Everything else, such as generating the hash value or subsequently checking the signature, can in principle also be performed by a PC.

The optimum situation, nevertheless, would obtain if the card received the message across the interface, calculated the hash value and sent it back to the terminal with the signature. Checking the signature could also be performed within the card. Although this procedure is no more secure than simply computing the signature, it is considerably more user-friendly, since hash algorithms and RSA keys could then be changed simply by replacing the card, without changing PC programs or data.

Both examples use global keys, keys which are identical for all the cards in the system, to generate and check the digital signature. If for reasons of security this approach needs to be modified so that each card carries its own digital signature key, a procedure such as the one described in Section 8.2.3 must be employed.

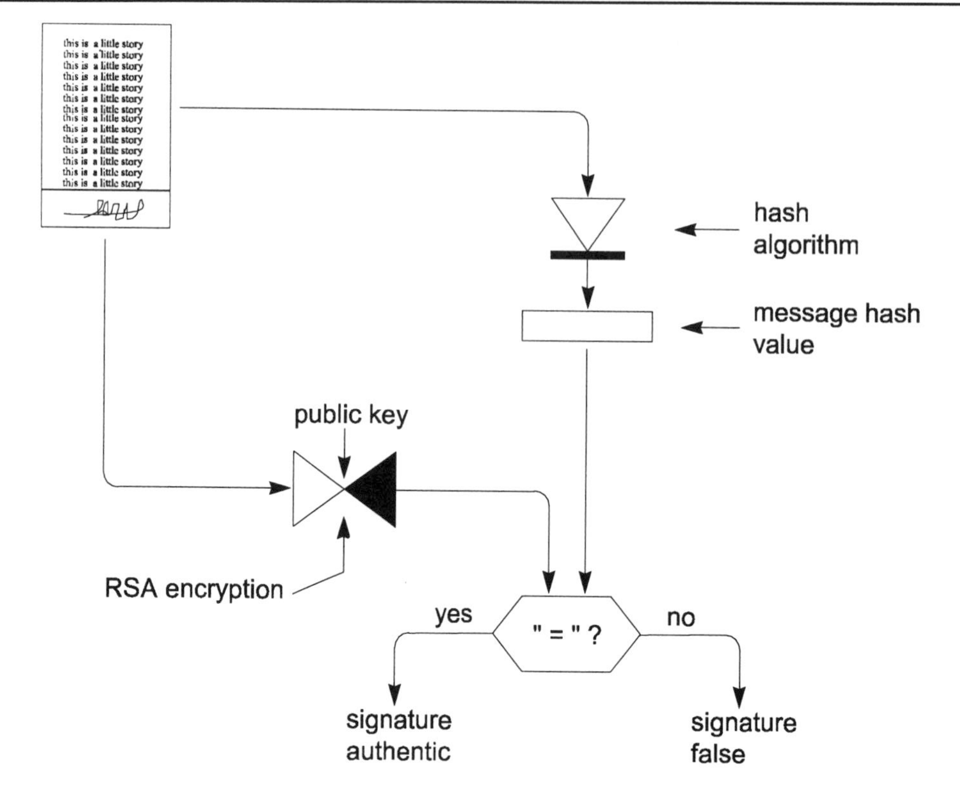

Figure 8.15 Checking a message signed with an RSA algorithm

It is not only the RSA algorithm that can be utilized to produce digital signatures. There is also a cryptographic procedure which has been especially developed for such use. This is the DSA (Digital Signature Algorithm) proposed in 1991 by NIST (US National Institute of Standards and Technology), with which signatures can be produced as well as checked. Unlike RSA, however, it cannot serve for encryption and decryption (although an encryption method has since been found). In view of the strict export limitations on RSA, this is of great interest to the international market.

8.4 KEY MANAGEMENT

All the principles governing keys for cryptographic algorithms have only one purpose, namely to limit damage to the system and the Smart Card application should one or more of these keys ever become public knowledge. If one could be sure that the keys always remained secret then a single key per card would suffice. Such an assurance, of course, is something which can never be given.

One result of the security principles applying to the use of keys for cryptographic algorithms is the steep increase in the number of keys. If all the procedures and principles listed here were implemented in a single card, the keys would take up over half of the user data storage space.

Not all of these procedures and principles need to be implemented – that depends on the field of application. For example, it is not really necessary to support several key generations if the card will only remain valid for a short period of time, as the extra overheads in terms of management and memory would not be justified.

8.4.1 Derived keys

Since, unlike terminals, Smart Cards can be taken home by anyone and analysed (albeit with a considerable investment of time and effort), they are subjected to the most severe tampering. Where the card does not contain a master key, the repercussions from an unauthorized reading of the data can be minimized. Hence only keys derived from the master key are present in Smart Cards.

The derived key is generated by a cryptographic algorithm. The input values are a master key and a feature individual to the card. The algorithm is normally DEA or triple-DES. For the sake of simplicity, let us assume that the card's number serves as the individual feature. This is a number which is written to the card during manufacture. It is unique throughout the system and can be used throughout it to identify the card.

A derived key is thus one of a kind, and can be generated, for example, with the following function:

- derived key = enc (master key; card number)

8.4.2 Key diversification

In order to minimize the consequences arising from a key being compromised, a dedicated key is often used for each cryptographic function. Thus, for example, it is possible to distinguish between keys for signatures, secure messaging, authentication and encryption. It follows that each key must have its own master key from which each individual one is derived.

8.4.3 Key versions

As a rule, it is not enough to supply a Smart Card for its entire lifetime with a single key generation. Consider the case in which a master key is computed by an external agent tampering with the system. All the application providers would then have to immobilize their systems, and card issuers would have to recall all the cards. The resulting losses would be immense. All the latest systems are therefore capable of switching to new key generations.

In the worst scenario, conversion is dictated by a key being compromised or it may be performed routinely according to a fixed or variable schedule. The end-result is an exchange of all the keys in the system for new ones, without card recall becoming necessary. Since the master keys are located in the terminals and in the system components, only secure data exchange is needed to supply them with new, as yet unknown, master keys.

8.4.4 Dynamic keys

In many applications, particularly in the field of secure data transmission, dynamic keys[4] represent state-of-the-art technology. One of the two communicating parties first generates a random number, or some other value to be used in the current session, and notifies it to the other party. The Smart Card and terminal then encrypt it using a derived key. The result is a key which is only valid for this session.

- dynamic key = enc (derived key; random number)

The great benefit of dynamic keys lies in their transient nature, which makes an attack considerably more difficult. Care must be taken, though, when they are used to generate signatures, since the temporary key will be needed to check the signature and can only be generated using the same random number which was used for the signature. It follows that the use of a temporary key requires the random number to be retained for checking, and therefore stored.

ANSI standard X9.17 provides for a somewhat more complex procedure for the generation of dynamic and derived keys, which nevertheless finds wide application in the field of electronic payment. What is needed is a time- or session-dependent value T_i, as well as a key KeyGen which is reserved for generating the new key. Subsequently one can calculate any necessary new keys with the start key Key_i. This key-generating procedure enjoys the additional advantage that it is a one-way function, i.e. it cannot be computed backwards.

- Key_{i+1} = enc (KeyGen ; enc (KeyGen ; (T_i XOR Key_i)))

8.4.5 Key data

In order to be able to address the keys held in the Smart Card from outside, an extremely simple mechanism is required. Furthermore, the card's operating system must ensure in every case that the keys can only be used for their designated purpose. For example, the system must be able to prevent an authentication key from being used for data encryption. In addition, key addressing also requires a key number, which is the actual key reference. Finally, the version number is needed in order to be able to address the specific key.

Some Smart Card operating systems provide for the incrementation of a key error counter when a prohibited activity, such as authentication, is performed. This is quite a reliable way of preventing a key from being revealed merely by trial and error, although due to the long computation times in the card this kind of risk need not be taken too seriously. If the error counter reaches its maximum value, the key is barred and can no longer be used. The error-counter is reset to zero when the procedure has performed successfully. Such a mechanism must be implemented extremely carefully, though, since a wrong master key in a terminal can very easily lead to mass failures of cards. As a rule, error-counter resets are only performed at special terminals after checking the holder's identity.

Some systems prohibit the reuse of previous key versions. A warning signal is attached to the key and is activated as soon as a new key with the same key number is addressed.

---

[4] Other names: temporary keys or session keys.

key number
version number
function
prohibited activities
error counter
maximum value of error counter
length of key
key

Figure 8.16 Necessary information about Smart Card keys

8.4.6 Example: key manager

The following is an example of a possible key manager for a large Smart Card-based system. Its purpose is to demonstrate a complete situation and to illustrate in outline the previously explained principles. Real-life large systems, in comparison, are often designed with far more stages and are more complex than this one. In small systems there is frequently no key hierarchy at all, since a global secret key is used for all the cards. The system used as an illustration represents a compromise between very simple systems and large ones, and is thus very instructive.

In this system, the keys could be used to credit an electronic wallet and use it for payment. In any case, these keys are important to the system and are relatively well protected by the hierarchy described. The individual derivation functions are not described in detail, but DEA or Triple-DES could be used for all of them. Key length is another variable feature which is not specified completely. For security reasons, the keys further up the hierarchy are normally derived using more powerful cryptographic functions than the ones further down the scale.

The highest key in the hierarchy is the general master key. It is unique to any key generation. One generation may be valid for one year, and the system may switch over to a new generation during the following year, i.e. to a new general master key. This key is the most sensitive in security terms. Should it become known, then all keys in this generation could be calculated and the system is breached for the duration of this generation. The master key may be generated from a random number. It is also possible to construct the master key from dice thrown by several independent persons, each of whom only knows part of the key. It should never be known to any one individual, and its derivation must under no circumstances be repeatable.

The actual master keys are derived from this general one depending on function. The function of a key may be crediting or payment with an electronic purse. In our example, a one-way function (e.g. modified DEA) is used to derive the master keys and to prevent calculation of the general key backwards from a main one. If, despite all safety precautions, a master key were to become known and no one-way function had been used to derive it, then by knowing the derivation data it would be possible to compute the general key. The reason for a one-way function at this point is the assumption that in our putative payment system, the master keys are stored in the local terminals' security modules. Thus, from the system security perspective they are exposed to considerably greater risk than the general master keys which never leave the background system.

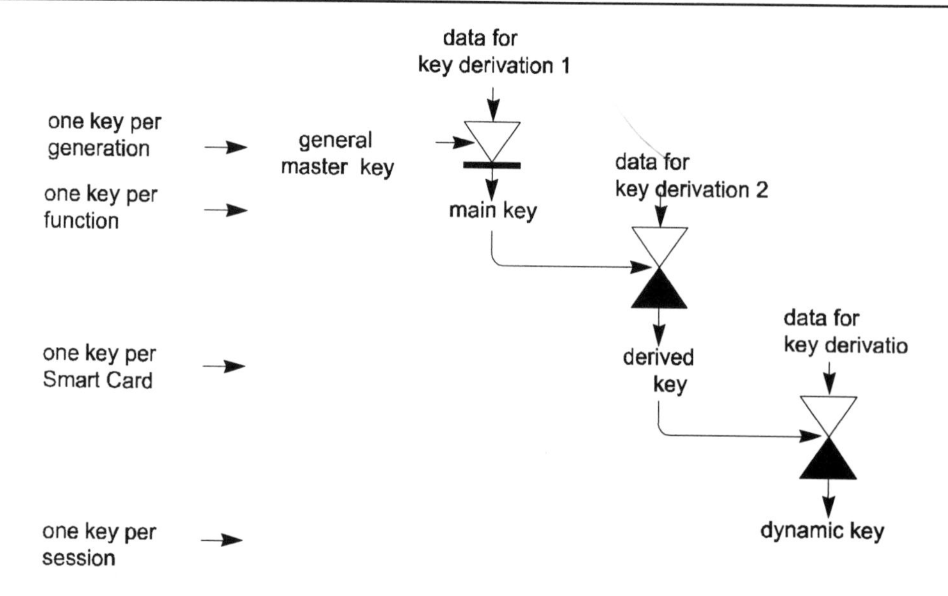

Figure 8.17 Example of Smart Card key hierarchy

The next step in the key hierarchy is represented by the derived keys. These are the keys located within the cards. Each card contains a set of derived keys, subdivided by generation and function. When such a card is used at a terminal, the latter can calculate the relevant derived keys from the derivation data. Naturally, the key derivation data are first read by the terminal from the card. The dynamic key, unique to the session and only valid for its duration, can now be calculated in the next step with the derived key. In typical Smart Card applications, the dynamic key would be used for between a minimum of a few hundred milliseconds and a maximum of a few seconds, and then scrapped.

The system just described may appear at first sight to be expensive and complicated, but in comparison with real systems this is not the case. The purpose of such a system is to specify a precise method for generating all of the keys. It also indicates implicitly what measures must be taken if a key should ever become known. If this is the general master key, a new generation must be introduced in order to allow the system to continue operating without security restrictions. In contrast, if a master key becomes known then only the branch below it needs to be barred or switched to a new generation. If it is a derived key which is no longer secret, then the only step which needs to be taken is to block the relevant card. Any additional key management measures would be quite superfluous. All these steps assume, of course, that the reason for the key being revealed could be discovered and could be avoided in future.

These key hierarchies naturally require a great many keys to be generated and stored in Smart Cards. However, several functions can be combined in the same key to save storage space. It is also quite possible to design the hierarchy in a different way. This depends to a large extent on the actual system for which the key manager was developed.

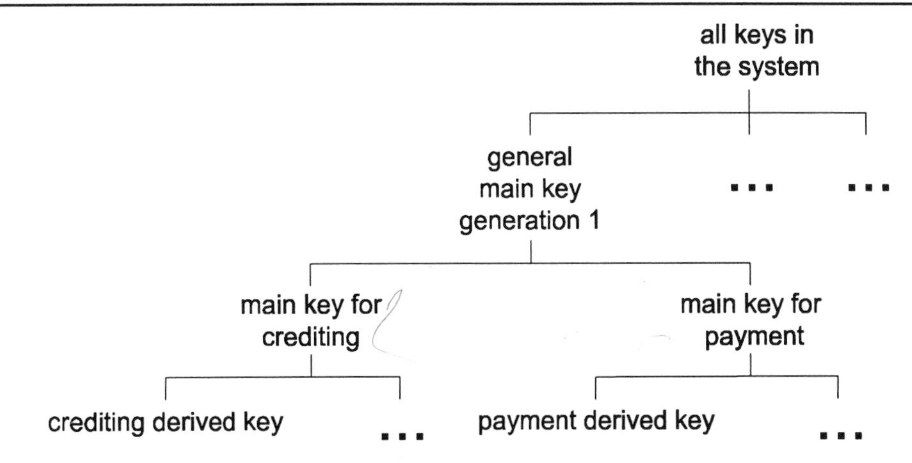

Figure 8.18 Example shows key in electronic purse with the functions of loading and payment. (Only stored keys are shown, i.e. dynamically created keys have been omitted in order to simplify matters)

8.5 SMART CARD SECURITY

The following four components guarantee the security of a Smart Card. The first security component is the body of the card, into which the microprocessor is incorporated. Many of the security features exploited here are not only machine-readable, but can additionally be checked visually. The technology is not specific to Smart Cards, and is used elsewhere in the card industry. The other components – semiconductor, system software and application – protect the data in the card's microprocessor.

It is only when all these components are present and their defence mechanisms are operational that a Smart Card's security is guaranteed. Where the card only finds its application in the machine-readable domain, the card-body component is unnecessary. The three other components, those which are distinct from the body, are vital, however, for the logical and electronic safety of the card in order to protect it from tampering. If even one of these three components fails or does not meet its specifications, the card is no longer secure. These components are, therefore, linked by a logical AND function.

8.5.1 Technical options for chip hardware

The technical options for protecting a Smart Card microprocessor against tampering are manifold. These options should be fully exploited. The special hardware solutions for microprocessors arose from technical security requirements. The Smart Card industry no longer uses even one single standard module; only specialized modules are employed.

Hardware protection may be divided into passive and active elements. The former are based directly on the semiconductor manufacturing technology. They cover all the possibilities and procedures which can be utilized to protect the memory and the rest of the microprocessor's functional elements from various types of analysis.

The semiconductor (passive) solutions are complemented by a whole spectrum of active chip components. Active protection means the additional integration of various sensors on the silicon crystal. These sensors are interrogated and evaluated by the card's software as and when necessary. Of course, this is only possible if the chip is connected to all of its power supply lines and these are fully activated. A chip which does not carry an electric current cannot sense any signals, let alone evaluate them. That is why for sensors, more than any other component, the gap between necessary defence elements and technical gamesmanship is often quite small. A light sensor on the chip which is designed to prevent optical analysis of the memory will not react when the chip is placed on a microscope-carrier if its power and clock supply lines have been disconnected. Such a sensor would also be easy to spot, and could well be obscured by a drop of black ink so that its protective function is neutralized without much difficulty even under operational conditions.

At the end of the day, only long-term functional security is meaningful. For example, a temperature sensor which detects transient and trivial overheating and causes the card's software to clear the entire EEPROM on security grounds does not contribute to increased functional security or protection against attack. For this reason, only a few sensors are implemented in the card as a rule.

8.5.1.1 Passive protective mechanisms

The various options available for passive mechanisms in Smart Card microprocessors are extremely diverse. The most important ones, and those most commonly used in practice, are explained below.

All microprocessors possess a so-called test mode, which is used to test the chip during manufacture and to perform internal test routines. These are performed either on the semiconductor while it is still part of the wafer, or in the module on the manufacturer's premises. However, this mode allows access to memory in ways which are strictly prohibited at a later stage. For technical reasons relating to production methods, it is absolutely necessary to be able to read data from EEPROM whilst in this mode.

Conversion from test mode to user mode must be irreversible. This can be achieved by adding polysilicon fuses to the chip. A voltage is then applied to a predetermined test point on the silicon crystal which causes the relevant fuse to burn through. The chip is now converted to hardware user mode. Generally speaking, this process is irreversible, but since it is at least theoretically feasible to bridge the relatively large and flat structures of a burnt fuse on the chip, additional protection is provided in an EEPROM area which cannot be erased. If this contains a particular unalterable value, then the chip switches irreversibly into user mode.

Chip structures (width of conductor tracks, transistor size, etc.) are at the limit of the technically feasible. Conventional widths range from 0.8 µm to 1.2 µm, which in itself no longer presents a particular technical challenge. Transistor density on the semiconductor is about as high as is currently possible using standard lithographic manufacturing processes. It is only these high-precision structures which make it almost impossible to obtain data from the chip through analytical processes.

The chip's internal buses which link the processor to the three memory types – ROM, EEPROM and RAM – are not connected to the outside world, and cannot be accessed, even using very complex methods. An interloper is therefore unable to tap into the microprocessor's address, data or control bus, or to tamper with them and thus read out data from memory.

Many Smart Card microprocessors scramble the internal buses used to control memory. This means that the individual conducting tracks do not lie next to each other in an ascending or descending sequence, but are swapped around and interchanged several times. This is an additional hurdle for the potential hacker to overcome, since at first, he will be unaware of which bus performs which function. But since this scrambling of the tracks is absolutely static, i.e. it is identical in every chip, it constitutes no huge problem in the medium term to identify the buses and to take this into account when tapping in.

The processor's design is also a security factor. It must draw an approximately equal amount of power when performing any machine instruction. If this is not the case, then the current consumed when the most recent instruction is executed may reveal confidential information.

Most programs are stored in ROM. The contents of ROMs commonly used in industry can be read bit-by-bit with an optical microscope. Combining these bits into bytes, and the bytes into a complete ROM code, is not a real problem. It is precisely in order to counter such analysis that the ROM is not located in the top silicon layers, the ones most susceptible to inspection, but in the lower ones. This prevents optical analysis. Nevertheless, sanding down the chip from the back would make it possible to read the contents of the ROM.

Another risk lies in the analysis of electric potentials in the chip while it is in operation. With sufficiently high readout frequency, it is possible to measure charge potentials, i.e. voltages, across very small crystal areas and thus obtain information about the data in the RAM during operation. This can be very reliably avoided by additional metallization over the relevant memory cells. If these metallic layers are removed chemically then the chip will no longer be functional, since it needs to use them as electrical conductors for its operation.

8.5.1.2 Active protective mechanisms

After microprocessor manufacture of the silicon chip, a passivizing layer is added to prevent oxidization or other chemical processes from attacking the chip's surface. The first step in any chip manipulation requires the removal of this layer. A sensory circuit can determine whether it is still present by making resistor or capacitor measurements. If it is absent or damaged, an interrupt can be triggered via the chip software or else the whole chip can be disconnected from the hardware, thus completely preventing any dynamic analysis from taking place.

Every Smart Card microprocessor contains a voltage regulator. It is responsible for a defined disconnection of the module when the upper or lower limits of operational voltage are exceeded in either direction. This guarantees to the software that operation is impossible in those regions in which the chip is not fully functional. If such a regulator were not installed, then in those areas the processor's program counter, for instance, might no longer operate in a stable manner, which would cause uncontrolled jumps within the program.

The power-on detector is a sensor present in all chips which is partially reliant on the voltage regulator. It is independent of the reset signal, and ensures that the chip always works in a defined domain when booted. The reasons for this are the same as for the voltage regulator.

The Smart Card's clock is always supplied from outside, so that computational speed is determined entirely externally. This theoretically permits the microprocessor to be run from outside in single step mode, which provides an excellent opportunity for analysis, particularly in the measurement of power consumption and electric potentials in the chip. A functional low frequency detection module is integrated into the chip, in order to combat tampering of this kind by disconnecting unauthorized applied clocks with such frequencies. Most chip specifications allow clock frequencies down to 1 MHz. Since detection of low chip frequencies necessarily introduces a fair amount of measurement scatter, the chip is usually switched off only if the frequency drops to below 600 kHz. This ensures that the chip can always be operated at the minimum value of 1 MHz. Sometimes high frequency detection is also implemented but modern hardware is so constructed that the chip cannot be operated if the frequency is too high.

Processor features can also offer increased security in Smart Card operation. They consist of additional functional units which can be tested by the terminal in parallel with chip software. Both analogue and digital modules are used. The security of these features is based on camouflage and varies with the application, resulting in application-specific chips.

One sensor variant which is used for temperature supervision is implemented in some chip types but its purpose is controversial. A transient increase in temperature beyond the specifications does not cause permanent damage to the chip nor does it necessarily mean that an attempt at tampering is taking place. Switching the chip off in such borderline situations could lead to artificially increased failure rates, rather than offering any additional security to the suppliers of Smart Card systems.

8.5.2 Software protection mechanisms

The mechanisms which protect system software must build on those used for hardware. No gap in this ringfencing must be overlooked, since the protection mechanism's three components (hardware, operating system, application) are linked through a logical AND function. If one mechanism fails, the whole security setup falls apart. The operating system itself supports the actual application, whose data and routines must be protected.

During operating system initialization, the most important sections and those of the hardware need to be tested. A RAM test is imperative, since it stores all access conditions during run-time and the failure of a particular bit may cause the whole security system to crash. The calculation and comparison of checksums over the EEPROM's most important sections is equally vital.

Completion of the operating system, i.e. the loading of tables, program code and configuration data into the EEPROM, represents not merely increased flexibility but is also an important security factor. The result is that the chip manufacturer, who receives the complete and assembled ROM program code for mask fabrication, has incomplete knowledge of the operating system. Those system parts which are located in EEPROM remain unknown to him, so that analysis of the ROM code does not reveal everything there is to know about security mechanisms and system functions.

Layer separation, with clearly defined interfacing parameters between the individual layers, is always a feature of stable and robust Smart Card operating systems. The adverse effects resulting from programming or design faults which may exist within the operating system are minimized by clean separation of the layers within it. Although this does not mean that errors cannot occur, their repercussions are not as serious as in a system programmed by using very compressed and concise code, since the separation means that errors cannot easily be propagated from one layer to the next.

Another highly important security element, namely I/O control, protects the memory from unauthorized access. The entire communication to and from the card takes place across the I/O interface controlled by the operating system. No other access is possible. This represents the strongest memory protection in the card, since this is the only way in which it is possible to ensure that the system retains control over memory access in all situations.

The transmission protocol controlled by the transport manager must intercept all possible false inputs. There must be no way of influencing data transmission by means of manipulation of the transmission blocks resulting in false data being sent without authorization from the memory to the terminal.

File organization, and in particular the headers (i.e. the file descriptors), should be secured via checksums. If some data is altered inadvertently, at least this must be detectable by the operating system. In view of the fact that the respective object-oriented access conditions are located in this part of the file, this requirement is evidently very important.

Some systems encapsulate and insulate the individual DFs containing the applications with their files from each other. These concepts are based only on pure software protection and are unsupported in the chip hardware. This makes protection less effective than it might otherwise be. However, in the event of a fault this is of great benefit, since it makes it impossible for the file manager to exceed the boundaries of a DF without prior explicit selection. Hence the results of a memory error in a file are, at least, restricted to that file.

All the memory sections in the EEPROM which are of vital importance for the card's operating system, must be protected by checksums (EDC). When these sections are accessed or interrogated, the consistency of their contents must first be established through the checksums, so that the system's stability is not endangered by EEPROM memory errors.

The writing of data to the EEPROM requires the chip's charge pump to be switched on. This increases the card's power consumption considerably, and this can be measured using a simple device. This means that system design must take into account that writing to the EEPROM can always be detected outside the card. The software must prevent an interloper from making use of this knowledge. It is therefore very important for current measurement not to allow any useful information to be deduced about internal routines and decisions made within the machine program. For example, if a current measurement made it possible to make a reliable deduction about PIN comparison before the completion of an instruction and receipt of the return code, this could be used to good effect to analyse the PIN.

Early Smart Card applications were always based on a centrally administered access mechanism. These centralized access automata suffered from the disadvantage that software or memory errors affected the entire security of the card. Modern, object-oriented file management systems with access rights localized to individual files have the advantage that in the event of memory errors only one file is affected and the other files' security remains intact. This is a fundamental property of distributed systems. Their programming is somewhat more costly, but they have considerably more resistance to tampering or failure, thanks to the security offered by this local authorization.

The operating system must be capable of completely deactivating the card. This is very important during the last phase of its life. Statistical methods, together with the collection of expired but fully functional cards, make it possible to perform very accurate analyses of the chip's software. In order to prevent this, the operating system must have mechanisms at its disposal which enable it to deactivate itself completely together with all its programs, and thus render electrical or run-time analyses impossible.

8.5.3 The application's protective mechanisms

The application's protective mechanisms are based on those which are present in the hardware and the operating system. The application has to rely on the two lower layers performing their protective tasks with the utmost rigour, since it cannot intercept a hardware or system error. For example, if the EEPROM can be read by an analytical procedure, then the most complex and secure encryption methods become useless, since the keys can be obtained by the attacker directly from the EEPROM. However, an application must be so constructed that if it compromises the card, the whole system is not compromised as a result.

It is of great importance to have a unique card number which is not duplicated in any other card. This allows the card to be identified unambiguously across the system. In addition, this number can be used for key derivation and offers the possibility of setting up warning lists so that suspect cards can be removed from circulation.

A hacker's task is made considerably harder if the files on the card are protected not only by the access conditions attached to the object, but also by instructions and corresponding parameters being fixed and predetermined by a state automaton. This stops trial-and-error experimentation with instructions, or combinations of instructions, from uncovering system-specific characteristics which may eventually be exploited. If the instruction sequences are supervised by a state automaton, only the instructions defined in the application can be performed, all the others being blocked by the state automata before they can be implemented. This largely reduces the options open to a hacker through the manipulation of instructions.

Data transmission in potentially insecure environments contains several risks. It is possible, through relatively straightforward technical manipulation of the card/terminal interface, to add almost any desired data during a session or delete it from the normal run. If this happens during the transmission of sensitive data, an attacker may, under certain circumstances, obtain some benefit from it. The secure messaging procedure can be implemented so as to interfere with such tampering which is neither excessively expensive nor difficult. Complete data encryption should nevertheless be avoided as far as possible, and should only be used to transmit such data as secret keys. The reason for this is the double encryption required for the transmitted data, which reduces effective transmission speed rather heavily. Another argument against complete encryption derives from data protection legislation. Almost all the data written to the card's memory is in the public domain. If it is encrypted, then nobody can check what was actually written to the card or read from it. In order to counter any legitimate objections, the data should not be encrypted during transmission if possible.

If a session is interrupted by an undefined breakdown, or the details of a previous session are fundamentally unclear, it is very useful to have application-specific protocol files available in the card. During the session, these are updated by the operating system using the current application states and any signatures or other terminal data received. This data is then stored in a cyclical file, the oldest record always being overwritten and thus lost. If this file contains, say, 20 records, then the data concerning the most recent 20 sessions can be stored for analysing the session. This allows many ambiguities to be removed, and disputed transactions and events clarified. Another argument for maintaining detailed transaction records is that error recovery functions are then made possible. This would allow the card's previous state to be automatically restored after an undefined breakdown, without the need for human involvement in the analysis of the precise run and sequence of events during which this had occurred.

At the end of the day, one-sided authentication (as practised on magnetic cards) means that only the terminal can be used to check whether the card is genuine. Due to its passive nature, this type of card cannot check to see whether the terminal itself is authentic. The introduction of Smart Cards has changed this situation fundamentally, and the card is also capable of checking to see whether it has been inserted into a genuine terminal. This has far-reaching consequences in security terms, since it renders the card actively able to take steps against unauthorized access. The options made available by this ability to carry out mutual authorization are enormous, and are nowhere near exhausted. At the very least, a Smart Card should be able to lock itself against further unauthorized access, in whatever form, as long as the terminal is unable to authenticate itself correctly. This makes it impossible to undertake backroom analysis of Smart Card operating systems, even if this is only done so as to examine all the available instructions.

Terminals with security modules can run Smart Card applications wholly independently. Naturally, periodic uploads and downloads to and from a background system are necessary, but in the normal course of events this only happens rarely. In large applications with a large number of cards in circulation, the terminals must be capable, if necessary, of establishing communication with the background system without delay, so that the latter can communicate directly end-to-end with the card. This becomes increasingly important with the size of the system, since the larger the system the more benefits are available to a hacker. With direct connection between the Smart Card and the background system, the latter can access its current database and lock the card if appropriate. Furthermore, the keys are much more securely stored in a background system than in the many terminals out in the field, even if they are fitted with a security module. In addition, these sporadic end-to-end communications allow the background system to make very efficient statistical analyses about diverse data contained in the cards. Of course, all these arguments are mainly of interest to the electronic wallet application. The on-line status can be forced on the card by random variables or by time limits stored within it. Counters in the card which can demand an on-line connection with mutual authentication when they reach a particular value are just as effective. The background system can reset the counter once connection is established.

8.6 TYPICAL ATTACK AND DEFENCE MECHANISMS

This section presents and explains several examples of types of interference which have almost become classics in their field. Protective measures against these attempts are also described. They, in turn, can be circumvented by slightly modified attack scenarios resulting in the well-known cat-and-mouse game of measures and countermeasures, attack and defence.

The scenarios described are not meant to be an encouragement to breach Smart Card security, as they have long since entered the public domain. They do not represent serious threats to the security of today's modern cards, these types of attack having already been taken into account. However, a few years ago they might still have enjoyed some success.

The attacks are divided into those which take place at the chip hardware level, and those in which attempts are made to crack the Smart Card system at the logical level. Physical attacks or analysis methods can be further subdivided into static and dynamic types. During static analysis the chip is not in operation but can be switched on electrically. In the incomparably more difficult dynamic chip analysis, the chip is running and is fully functional during the analysis.

8.6.1 Attacks at the physical level

Manipulation of the semiconductor requires considerable investment in high-tech equipment. Depending on the exact scenario, this may include a microscope, laser cutter, micromanipulator, chemical separation facilities and a very fast computer for the analysis, recording and evaluation of the chip's electrical behaviour. Very few people possess these devices and know how to use them, and this sharply reduces the likelihood of physical manipulation. Nevertheless, in principle any card or semiconductor manufacturer must assume that a potential attacker may be able to put together all the necessary instrumentation and facilities required for such an attack, and build appropriate safety features into the hardware.

The first example of an attack at the physical level is really a lesson in the failure of a defence mechanism. The basic idea is as follows: if it were possible to read a Smart Card EEPROM, then at least a hacker should be prevented from discovering the PIN in this way. Naturally, this type of attack can be guarded against. It is only necessary to encrypt the PIN with a one-way function, and the result is stored in EEPROM as a reference for the PIN comparison. One could employ card-specific keys for the one-way function, so that the reference values for two identical PINs in two different cards would vary.

If one could now read the reference values from the EEPROM, then at first sight it would appear to be impossible to derive the PIN from the data. But a clever attacker would not even set out along this path. If the conventional, four-figure decimal PINs are used, then their range lies between a lower limit of '0000' and an upper one of '9999'. Thus the number of possible PINs is exactly 10000. If an interloper could read the entire memory of the Smart Card, he or she would also know the one-way function and the corresponding card-specific key. Using this information, a hacker would now embark upon the one-way encryption of all possible PINs. After 5000 attempts, on average, he or she would have generated a result which agrees with the reference value stored in the card. The attacker now knows the PIN. One can see that in this case, a one-way function for PIN storage does not confer any great benefits.

Conversion of the microprocessor from test mode to user mode proceeds in various different ways, as determined by the semiconductor manufacturer. Often it is a polycrystalline silicon fuse, which after the completion of tests during the chip's manufacturing process is burned through. Due to the fuse's manner of operation, it forms a relatively large structure on the chip's surface. It is conceivable that after partially removing the passivizing layer from above the fuse, it could be bridged mechanically. This would return the microprocessor to its test mode, and memory contents could be read by using the more extensive access options available in this mode. Once the contents of memory were completely known, the Smart Card could be easily cloned.

In order to guard against this type of attack, most semiconductor manufacturers now reserve an area in the EEPROM for a conversion mechanism, in addition to the silicon fuse. The chip is no longer in test mode even after it is bridged, since this is prevented by a further logical switch in the EEPROM.

When judged overall, this type of attack is one of the most dangerous for the system generally, since if successful it would allow the attacker access to the whole memory. No attack in which the fuse on the chip was successfully bridged has come to light so far. Furthermore, the EEPROM switch would also need to be altered, and this is more difficult to achieve.

A variation on the above attack would also make it possible to read data throughout the memory. After removing the passivizing layer from the chip's surface it is at least theoretically possible to contact the ROM's and EEPROM's address, data and control buses. If an electrical connection can be established with all three conductors, it is very simple to address individual memory cells and read out any area desired. The chip need not be switched on, and may be connected in a variety of different circuits. The consequences of such access would be just as catastrophic as in the previously described approach.

As already indicated, it is not merely very difficult but almost impossible to establish an electrical connection with individual conducting tracks on the chip. The number of connections needed for this attempt is 16 for the address bus, 8 for the data bus and between 1 and 4 for the control bus. Altogether, at least 25 simultaneous connections would have to be set up between an external computer and the chip's conducting strips. The very small structures in the semiconductor make this impossible at the moment, even with modern micromanipulation techniques.

If, despite this, an attempt should succeed, then the buses would also have to be unscrambled before the data could be read, since the individual tracks are not located on the chip next to each other in their correct sequence.

When in a few years' time and using vastly improved technology, it becomes possible to contact the buses of today's microprocessors, this would still have little effect on security, since by then semiconductor structures will have become even smaller than those in use today. Micromechanical methods will probably always lag behind optical semiconductor manufacturing technology. It appears probable that this type of tampering will never become feasible for compromising Smart Card security.

The random numbers generated by the card are used during authentication in order to individualize a session, i.e. to render it unique and different from any other previous and subsequent sessions. The point of this procedure is to make successful insertion of earlier, tapped sessions impossible. One variation of this attack is to let the card generate so many random numbers that they become predictable. Another possibility is to request random numbers from the card for such a long period that the random number generator's EEPROM memory no longer functions and the same numbers are always produced.

All such attacks, if successful, could circumvent terminal authentication by the card. However, without exception, they only work with first generation Smart Cards and would fail if applied to modern operating systems. The period of random number generators has become so large that the same numbers never turn up within the card's lifetime. It would also be pointless to generate so many random numbers that EEPROM problems occur. Should this be the case, then their generation would simply be blocked and any further authentication prevented.

8.6.2 Attacks at the logical level

The main precondition for attacks on Smart Card security at the logical level is an understanding of communications and of the flow of information between terminal and Smart Card. It is less necessary to understand hardware processes than those associated with software routines. From the perspective of information technology, the sample scenarios illustrated here are attacks at the level above that in which the main targets are hardware features.

The easiest type of attack to visualize is one involving a self-programmed Smart Card, extended through a variety of analytical and protocol functions. Up until a few years ago it was almost impossible to tamper with the card in this way, since only a few companies were in a position to purchase Smart Cards or their associated microprocessors. In the meantime, it has become feasible to buy both Smart Cards and a number of configuration programs from several sources. This opens up the options available to an attacker. A small investment in a plastic plate, a standard microprocessor in an SMD housing and a little skill will result in the construction of a functional Smart Card, or at least a card which behaves like a genuine one, electrically and during data transmission.

By using a dummy card it is at least possible to record part of the communication with the terminal, and evaluate it later. After several attempts some of the communication routines could probably be performed in a manner indistinguishable from that of a genuine Smart Card.

Whether one could obtain any benefit from this is doubtful, since all reasonably professional applications include cryptographic protection for important actions. As long as the secret key remains unknown, tampering is only possible up to the authentication stage. It would only be successful if the key were known, or if the entire application were running without cryptographic protection. Even if such an application existed, it is highly unlikely that the advantage gained by tampering would be cost-effective enough to justify the huge expenditure involved.

The instructions and their classes supported by a given Smart Card are rarely published, but they can be discovered very easily. This would be useful not so much for subverting the card's security, but rather for investigating its instruction set in detail. It is conceivable, though, that it could form the basis for card manipulation.

The method of investigation is as follows. First an APDU instruction is generated and sent to the card from a freely programmable terminal. The class byte is changed from '00' to 'FF'. If a return code other than "invalid class byte" is received, then the first valid class byte has been discovered. Usually there are two or three instruction classes which could be used to try out all the instruction bytes in the next round. APDU instructions are sent to the card using an altered instruction byte until a return code other than "unknown instruction" is received. If this method is used, and assuming appropriate terminal software is available, it will be clear within a couple of minutes which instructions are supported by the card. To some extent it is also possible to discover some of the instruction parameters in a similar way.

The reason why this simple search algorithm for instructions, instruction classes and parameters might be successful lies in the fact that practically all instruction interpreters in card operating systems inspect the incoming instructions from the class byte onwards. Processing is interrupted when the first value is spotted as invalid. An appropriate return code is generated and sent back to the terminal.

However, the approach just described only works if the card does not have a global state automaton governing the sequence of instructions. Even if it did, the sequence of instructions could still be discovered step-by-step.

The benefit to an attacker may not appear to be particularly great, since the instruction set is not normally secret. But all instructions could be investigated very simply and quickly. In any case, this method would be very useful for discovering whether a system manufacturer had incorporated undocumented instructions into the card.

The security of a Smart Card application is based on the cryptographic algorithm's secret keys. The terminal must always authenticate itself before it can be permitted to perform certain access or other actions, and this is done with the aid of a secret key. It is understandable that terminal authentication by the card is an important target for the attacker. On the other hand, card authentication by the terminal during an attack on the card is of no interest, since it can be manipulated at will by a (dummy) terminal.

When the terminal is authenticated by the card it receives a random number, encrypts it and sends it back. The card then performs the same encryption and compares the result with that sent by the terminal. In the event of agreement, the terminal is authenticated and receives a return code accordingly. The card sends a different code where authentication fails. The point on which an attacker concentrates is an analysis of the processing time between the instruction and the response. Some early crypto-algorithms exhibited considerable variations in execution time, which depended on the key and on the plaintext. Using this reduced key space as a lever, the attacker could search for the secret key using a "brute force" approach. The length of the search depended sharply on the noise of the algorithm, but the greater the time difference, the smaller the key space and the simpler and quicker the search for the key.

In principle, this attack could be very dangerous for card security, but since it has been known for quite some time, all current Smart Cards use noiseless algorithms, i.e. encryption and decryption time is always independent of input values, thus blocking this type of attack. For additional security, the authentication keys in some applications also possess error counters, so that only a particular number of unsuccessful authentications can be performed. Once the counter reaches its maximum value, the card locks itself against all further attempts.

A slightly modified card can be used to hack into data during a session, and if necessary manipulate it. An electrically isolated dummy contact is glued to the I/O contact field. The original I/O interface is no longer connected to it. The new dummy contact and the original I/O contact are connected to a fast computer. Depending on how it is programmed, it can delete or insert data selectively during terminal/Smart Card communication. If the computer is sufficiently fast, this manipulated communication is not discovered by either the terminal or the card.

Obviously, this method can fundamentally affect the way a session proceeds. Whether an attacker derives any benefit from it depends above all on the application in the card. A recognized design criterion states that neither the hacking, omission or insertion of data during communication may adversely affect security. If this criterion is not observed, then certainly an attacker may gain an advantage thereby.

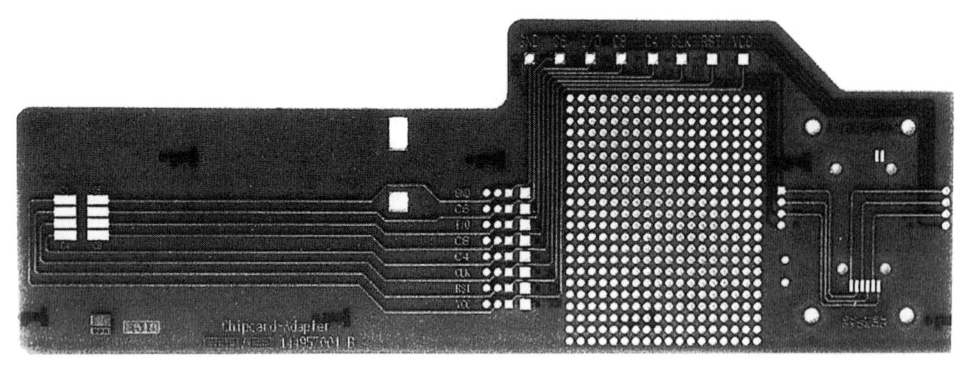

Figure 8.19 Adaptor with which a Smart Card, in conjunction with the body of a card reader, can be used as measuring device

In order to offer additional security against this attack, some terminals have so-called shutters which isolate any wires connected to the card. This is also a perfect arena for the implementation of secure messaging, with which the manipulation of data transmission can reliably be detected.

Since many terminals can only be operated under supervision, it is not quite so simple to use fake cards hard-wired to a portable computer. With hindsight, this line of attack could be classified as quite interesting and promising in theory, but rather unlikely in practice.

A technically very interesting attack on comparison features, such as the PIN, could be launched through a combination of the physical measurement of a parameter and the modification of a logical value. This type of attack would affect any mechanisms by which data is sent to the card and compared to a value stored therein and an error counter incremented as a result. The principle used involves current measurement of the card, which may be made, for example, via the voltage drop across a resistor connected to the V_{CC} wire. If the relevant instruction is sent to the card with the comparison data, it is possible to determine through the current measured before the return code is received whether or not the error counter has been incremented. If the return code is sent out before the error counter is incremented, this could be used to discover the comparison value. All the possible variants are then sent to the card, and in the event of a failed comparison it is always switched off before the counter is incremented. A successful comparison can be unambiguously recognized through the appropriate return code, which is sent out before incrementation.

Two fundamental methods exist to counter this attack, both of which happen to be patented. The simplest consists in incrementing the error counter before each comparison, and decrementing it again if necessary. Regardless of when the power supply is interrupted, the attacker can never benefit from it since the counter has already been incremented. The second variant is somewhat more complex, but performs the same protective function. After comparison, the counter is incremented in the case of failure and otherwise written to an unused EEPROM cell. Both these steps take place at the same point in time, so that no conclusions can be drawn about the comparison. It is only through the return code that the result of the comparison is notified. By then it is too late to prevent incrementation of the error counter by interrupting the power supply.

Quality assurance and testing

Quality assurance and its related test procedures and methods are of particular importance in the case of Smart Cards. The card manufacturer needs to issue a product in very large numbers, of a high quality and at a low cost. In contrast to other branches of this industry, though, it contains in addition a relatively expensive and sensitive microprocessor, as well as software which cannot later be modified.

If, for example, one compares this starting point with standard software for PCs, the fundamental difference is very obvious. In the latter, the established convention has been dictating for some years that the first generation (mostly denoted by '0' at the end of the version number) is replaced within a few weeks, or at most one to two months, by revised and improved versions (ending in 'a', 'b', 'c' etc.). This would be impossible in the Smart Card domain. Due to the storage principle used, the mask-programmed software cannot be updated, and the large number of cards issued cannot be replaced through a recall, however organized. Even in the case of cards used other than in the particularly sensitive area of financial transactions, the issuer's reputation would suffer greatly for many years, and the resulting costs would be immense.

That is why quality assurance and testing are of fundamental importance in the production of Smart Cards. After their production and issue, it is simply impossible to "squeeze in" an improved version after a short period of time. Naturally, this results in increased investment to manufacture the product with as few faults as possible.

When discussing the various tests, a fundamental distinction must be made between qualification and production tests. Qualification tests mean that a judgement is being made on whether the relevant card can be introduced in the first place. Such tests usually take place before the launch of a new card body, chip, module or operating system. If the new or improved product meets the requirements set for it, then it is "qualified" for production and can be manufactured in large numbers. From this point onward, qualification tests are only carried out on random samples at large time intervals.

Production tests make use of other methods. These can all be performed quickly and straightforwardly, so that the indispensable demands of mass production can be met after a short run-time and high throughput. Mostly, these tests only measure simple mechanical and electrical boundary parameters, as well as employing appropriate test instructions to the card's microprocessor.

9.1 TESTING THE CARD'S BODY

Only one international standard currently exists for testing cards with and without chips, the ISO/IEC 10373. In Europe there is also standard EN 1292, but which is only concerned with Smart Cards, terminals and their electrical boundary conditions. In addition, however, one can also often find isolated tests and test procedures mentioned in standards relating to cards, which cover the testing of properties listed in the standard.

The following pages give an alphabetic list and outline of many of the usual tests. Standardized environmental conditions are vital for these tests, which means that lab temperature must be maintained at 23°C ± 3°C and the relative humidity kept between 40% and 60%. The cards being tested must be acclimatized to these conditions for at least 24 hours before the actual test.

Adhesion or blocking

(Based on ISO 7810; test regulation: ISO/IEC 10373)

This test is designed to check whether the card's behaviour changes after storage under certain environmental conditions. Five non-embossed cards are stacked together and subjected uniformly for 48 hours to a pressure of 2.5 kPa at 40°C and 90% relative humidity. The cards are then inspected for delamination, discoloration, surface changes and other visually discernible changes.

Bending stiffness

(Based on ISO 7810; test regulation: ISO/IEC 10373)

In order to establish the required bending stiffness, the card is first clamped 3 millimetres from its left side and held with the front facing down. Bending is first measured without loading, and then the outer edge is subjected to a force of 0.7 N. The difference in bending is measured, the result expressing the card's stiffness.

Card warping

(Based on ISO 7810; test regulation: ISO/IEC 10373)

This test measures the card's warping. The card is placed on a level surface, and the warping measured with a profile projector. It is designed above all for cards stamped out of foils supplied in a roll.

Delamination of multilayer cards

(Based on ISO 7810; test regulation: ISO/IEC 10373)

The following test is only employed in multilayer cards, which are put together from laminated plastic foils. The covering foil is separated from the core foil with a sharp knife at some point. Starting at the separation, the examiner attempts to pull the two laminated foils apart. The required force is measured, and compared with the reference value.

Dimensional stability and warping with temperature and humidity

(Based on ISO 7810; test regulation: ISO/IEC 10373)

Since certain types of plastic severely distort both in shape and size with air humidity, the card's stability in accordance with the standards must be measured under these conditions. A card is laid flat on a substrate, and the temperature and humidity are varied. The following four boundary conditions are employed: –35°C, +50°C, +25°C at 5% relative humidity, and +25°C at 95% relative humidity. Size and warping are measured 60 minutes after each of these tests, using the known procedure.

Dimensions – format and thickness

(Based on ISO 7810; test regulation: ISO/IEC 10373)

This test measures the height, width and thickness of a non-embossed card. It is subjected to 2.2 N and its height and width measured with a profile projector. For thickness measurement the card is divided into four equal rectangles, and measured at the centre of each with a micrometer at a force of between 3.5 N and 5.9 N. The measured maximum and minimum values are compared with the standard.

Dynamic bending stress

(Based on ISO 7816-1; test regulation: ISO/IEC 10373)

The dynamic bending tests are illustrated in Figure 9.1 The card is bent at a frequency of 30 times a minute (= 0.5 Hz) either 2 cm longitudinally or 1 cm transversely (f). The card must survive without damage at least 250 bending cycles in each of the four possible directions (i.e. a total of 1000 bending cycles).

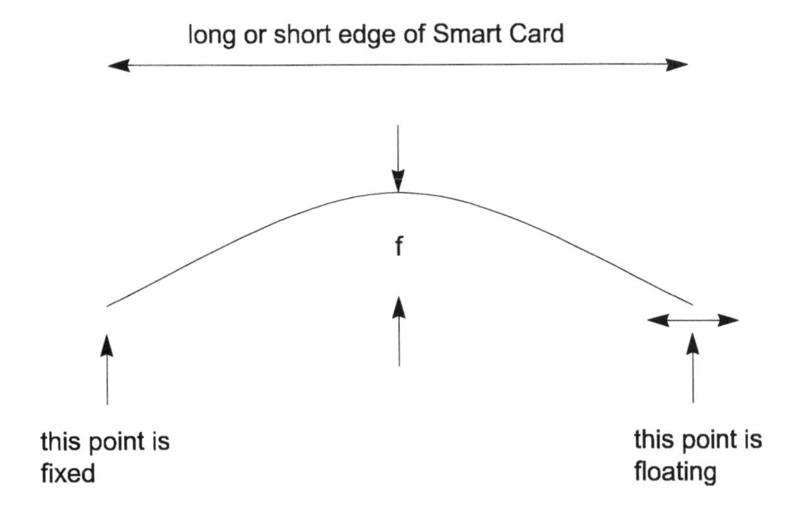

Figure 9.1 Dynamic bending test

Dynamic torsion stress

(Based on ISO 7816-1; test regulation: ISO/IEC 10373)

In the dynamic torsion test, the card is subjected to ±15° torsion around the longitudinal axis at a frequency of 30 torsions per minute (= 0.5 Hz). The standard requires 1000 torsion cycles without functional chip failure or visible mechanical damage to the card.

Electrical resistance and impedance of the contacts

(Based on ISO 7816-1/2; test regulation: ISO/IEC 10373)

The contacts' electrical resistance is an important criterion for the reliable power supply and data transmission to the card's microprocessor. Resistance is measured with two test contacts, which are pressed onto the two opposite corners of the smallest contact area with a force of 0.5 N ± 0.1 N. The resistance between the two gold-coated contacts, whose points are rounded to a radius of 0.4 mm, must be less than 0.5 Ω.

Electromagnetic fields

(Based on ISO 7816-1; test regulation: ISO/IEC 10373)

In this test, the card is pushed into a static electromagnetic field of 1000 Oe (= 79***580 H) at a maximum speed of 1 cm/s. The card's memory contents must not change.

Embossing relief height of character

(Based on ISO 7811-1; test regulation: ISO/IEC 10373)

In this test, the card's height is measured at the embossed points with a micrometer, with a force of between 3.5 N and 5.9 N.

Flammability

(Based on ISO 7813; test regulation: ISO/IEC 10373)

A card's flammability is measured by holding it by an edge at an angle of 45° for 30 seconds in the defined flame (diameter 8.5 mm, height 25 mm) of a Bunsen burner.

Flux transition spacing variation

(Based on ISO 7811-2; test regulation: ISO/IEC 10373)

This test determines whether the magnetic flux transition in the encoding of individual bits in the magnetic stripe is uniform and sufficient. A reading head is traversed across the stripe, and the field fluctuation recorded. The measured value is compared with the one specified in ISO 7811-2.

Height and surface profile of the magnetic stripe

(Based on ISO 7811-2/4/5; test regulation: ISO/IEC 10373)

This test serves to measure the stripe's height as well as uniformity. It uses a special measuring device described in detail in the standard, with which a height profile can be produced.

Light transmittance

(Based on ISO 7810; test regulation: ISO/IEC 10373)

Some cards contain an optical barcode on an inlet foil. The test determines the covering foils' and the rest of the body's optical transmittance. The card is illuminated, and light transmission measured on the other side with a detector sensitive to 900 nm.

Location of contacts

(Based on ISO 7816-2; test regulation: ISO/IEC 10373)

This test is used to measure the location of the contacts, by placing the card on a flat surface and subjecting it to a force of 2.2 N ± 0.2 N. The contacts are then measured by any method whose inaccuracy does not exceed 0.05 mm, relative to the card's edges.

Resistance to chemicals

(Based on ISO 7810, ISO 7811-2; test regulation: ISO/IEC 10373)

The card's body and magnetic stripe are tested as follows. Individual cards are placed in precisely defined liquids at a temperature of between 20°C and 25°C:

- 5% aqueous solution of sodium chloride
- 5% aqueous solution of acetic acid
- 5% aqueous solution of sodium carbonate
- 60% aqueous solution of ethyl alcohol
- 10% aqueous solution of sugar
- benzine (according to ISO 1817)
- 50% aqueous solution of ethylene glycol

The card is removed from the solution after one minute and assessed visually, and the stripe's data contents checked in a reader.

Signal amplitude in the magnetic stripe

(Based on ISO 7811-2; test regulation: ISO/IEC 10373)
This measurement tests signal amplitude and resolution in the magnetic stripe. This is done with a standardized read/write head, driven over the stripe at a precisely defined speed.

Static electricity

(Based on ISO 7816-1; test regulation: ISO/IEC 10373)

This test, which is only meaningful in Smart Cards, checks the chip's resistance to electrostatic discharges (ESD). A 100 pF capacitor is charged to both ±1500 V, and discharged through a 1500Ω current-limiting resistor and the chip's various contacts. There must be no damage either to the chip's functional units or to memory contents.

Surface profile of contacts

(Based on ISO 7816-1/2; test regulation: ISO/IEC 10373)

This test relates the individual contacts' surface profile to the rest of the card's surface. It is designed to ensure that the contacts lie approximately in the same plane as the overall surface.

Surface roughness of the magnetic stripe

(Based on ISO 7811-2; test regulation: ISO/IEC 10373)

The same device is used to measure the magnetic stripe's surface roughness as is used for the height and surface profile. It is, however, supplemented by a special reading head which measures the stripe's roughness. This test is important since surface roughness is one of the essential factors in the wear of read/write heads in magnetic stripe readers.

UV radiation

(Based on ISO 7816-1; test regulation: ISO/IEC 10373)

Since (E)EPROM cells lose their contents when irradiated by UV, a special test is used to ensure their resistance to them. The Smart Card is irradiated for between 10 and 30 minutes with UV at a wavelength of 254 nm and incident energy of 15 W***s/cm^2. The (E)EPROM data contents must not change as a result.

Vibrations

(Based on test regulation: ISO/IEC 10373)

Cards are often subjected to severe oscillations during transport and use (e.g. mobile phones in cars), so this test is designed to conform to the above regulation. It requires that the card be tested on a vibration table in all three axes, with an amplitude not exceeding 1.5 mm and across the frequency range 10 Hz to 500 Hz. Neither chip functions nor memory contents must be adversely affected.

Wear test for magnetic stripe

(Based on ISO 7811-2; test regulation: ISO/IEC 10373)
In order to establish the stripe's magnetic behaviour after wear, test data is first written to it. Then a dummy read/write head, with a hardness of 110 HV to 130 HV, a radius of curvature of 10 mm and subjected to a load of 1.5 N, is passed 1000 times (each way)

across the magnetic stripe. The data is then read. Signal amplitude must remain within the limits specified in ISO 7811-2.

X-rays

(Based on ISO 7816-1; test regulation: ISO/IEC 10373)

In analogy with UV radiation, the contents of (E)EPROM memory cells are changed by X-ray irradiation. In order to test their resistance, the chip is irradiated with X-rays with an energy of 70 kV. Memory contents are then examined for any changes, and to see whether it is still possible to write to it.

Naturally there are many other tests, such as the number of insertion cycles, resistance to colour wear, softener stability, resistance to perspiration and saliva. The appropriate tests are selected and performed, depending on the card's planned area of use.

9.2 MICROPROCESSOR HARDWARE TESTS

Besides the card's body, quality assurance must also ensure the functionality of the microprocessor, which is the most important and most susceptible component in modern Smart Cards.

The CPU and memory are extensively tested as early as the semiconductor's manufacturing stages. In order to be able to carry out these tests, each microprocessor possesses a so-called test ROM, which contains various programs for accessing the CPU and memory externally. In addition, special connections are sometimes present which permit arbitrary access to the processor's central bus. During production, pin adapters are used to contact the chip's appropriate pads and execute the necessary test programs. These connections are cut off when the chips are separated from the board, so they cannot be used later to manipulate the processor. It is then impossible to connect the internal bus for fraudulent purposes.

Once the dies have been packaged in modules, another test is, of course, carried out via the module contacts. Often this is restricted to a booting sequence, where the test involves a check on whether an ATR can be received. If this is possible, one assumes that during packaging the semiconductor has not suffered serious damage, and all contact wires are correctly connected. A similar ATR test is also used directly after module implantation in the body of the card. Its purpose is to determine whether the short-term temperature increase during this process has damaged the module.

The microprocessor is then tested comprehensively before the card is initialized, using instructions specific to this manufacturing step[1]. After successful running of the test program, which lasts between 10 and 100 seconds, these instructions are irreversibly barred from further use. These time-consuming tests can be carried out during this stage without seriously affecting throughput, since the machines employed run massively in parallel and

---

[1] See also 7.12 Hardware testing instructions.

the test duration is not reflected individually in each card. For example, one tests here whether all EEPROM bytes can be written to and then cleared, and whether the RAM is fully functional. A scratch on the chip suffered during bonding could result in some EEPROM cells being incapable of being written to, or certain ROM regions containing faulty data.

After card initialization and personalization, various final tests are performed depending on the manufacturer. Usually this is done with fully automatic and self-calibrating testing tools. They read out from the card all the data relevant to the test, and use them as necessary.

Besides the relatively simple and fast tests undergone by all the cards, however, there are additional tests carried out only on a sample batch. These cards, taken from current production series, can of course also be subjected to destructive testing if necessary.

The standard EN 1292 was published to cover qualification tests and sample tests, and it defines many different testing methods for the microprocessor. Typical microprocessor tests are:

- Rise and fall times at the I/O contact (EN 1292)
- Number of possible write/clear EEPROM cycles
- EEPROM data retention
- Detection of CLK over- and under-frequency
- Detection of V_{cc} over- and under-voltage
- Voltages at the I/O contact (EN 1292)
- Current consumption at the CLK (EN 1292)
- Current consumption at the reset contact (EN 1292)
- Current consumption at V_{cc} (EN 1292)
- Current consumption at V_{pp} (EN 1292)

Naturally, every card manufacturer also employs special in-house tests for his own relevant built-in microprocessor. Thus, for example, there are special tests for the various sensors on the chip.

9.3 TESTING MICROPROCESSOR SOFTWARE

The testing of physical components, represented by the card itself and its modules, can largely be carried out using traditional methods. Electrical behaviour can also be investigated very effectively by automatic test rigs. However, the situation with regard to the processor's software is somewhat different. The debugging of software has, naturally, been constantly refined over the past 40 years since the first programs, and many recognized procedures exist for the generation of reasonably error-free programs. Nevertheless, software errors occur relatively frequently in daily practice.

In most applications this does not present a problem, since issuing a revised version can quickly neutralize the error. This, however, is not quite so straightforward with Smart

Cards, as the software is largely located in the processor's ROM. A new version necessitates a complete production run by the semiconductor manufacturer, which ranges between 8 and 12 weeks. Once the cards are issued to users, it is practically impossible to modify the current software. It follows from these very strict boundary conditions that Smart Card software must be extremely error-deficient. The term "error free" would have been better, but this cannot even be approached with state-of-the-art software engineering.

As is well known, the topic of software testing is extremely wide-ranging, and is described in many books in all its variations and approaches. We can only present a short sketch of this specialized field, which has by now become almost a free-standing branch of information technology. Therefore, the following sections will only discuss the testing of Smart Card software which present some special features. A good introduction to this subject is provided in G. J. Myers' book *The Art of Software Testing* (Wiley, New York, 1995). We will only add that the military field, in particular, contains many good and long-proven procedures for the generation and testing of software, in the form of standards.

9.3.1 Security tests

Due to their property of being able to store data securely, Smart Cards are mainly employed in sensitive domains. However, their benefits are not restricted to the safe storage of data: they are just as useful in the area of secured execution of cryptographic algorithms.

The field of electronic financial transactions is one of the foremost future expanding market for Smart Cards. Because their wide dissemination leads to the movement of enormous amounts of money, the relevant application provider or card issuer must have very great confidence in the semiconductor manufacturer, operating system manufacturer and also the card's personalizer. The application provider must be absolutely certain that the software executes the required transactions without errors, and that no security gaps, or even trapdoors deliberately introduced into the software, are present.

For example, a secret instruction to the card might be used to read out the PIN and all secret keys. In the case of GSM or Eurocheque cards, the attacker would be able to clone any number of fully functional cards and then sell them.

The security requirements, though, relate not only to the card's manufacturer, but just as much to the initialization and personalization steps, since it is there that the secret keys and PIN are loaded into the card. The issuer must put a high degree of trust in the card's supplier, in anything regarding security.

The same also applies to the fundamental security of the card's software. There must not have been a deliberate trapdoor installed for the purpose of obtaining data. However, it may well be the case that a malfunction makes it possible to read data from the card or write them to it with a combination of instructions not employed in normal use. Although the likelihood of such a coincidence is extremely low, nevertheless it is impossible to guarantee that a program will always be error-free, even with current state-of-the-art software engineering. One may be certain that in the future, Smart Card software firms will not be able to shrug off all responsibility with such legalistic formulas.

Only two ways remain to the application provider to test a product's trustworthiness. Either he tests the card and its software under all circumstances himself, or he lets it be tested by a trustworthy party. The first approach, card testing by the application provider, is often only possible to a very limited degree, since he may not possess all the required technical expertise and capabilities. The second method, that of submitting the card to be tested by some other organization, is currently regarded as an acceptable solution for all concerned.

This difficulty is hardly new: it has existed for many years in the field of software and system development for military use. In order to make the trustworthiness of software products objectively measurable, NCSC (US National Computer Security Center) issued in 1983 a catalogue of criteria for evaluating the trustworthiness of information-based systems. NCSC was established in 1981 by the American Department of Defense. The Trusted Computer System Evaluation Criteria, or TCSEC for short, was published in 1985. The book, published in an orange binding, is commonly known as the Orange Book. These criteria serve as guidelines to the NCSC for the certification of information-based systems.

TCSEC has become a worldwide model for practically all IT criteria catalogues. Independent European criteria have been defined, but are still based on it. They were first published in 1990 as ITSEC (Information Technique System Evaluation Criteria), and the current revised version dates from 1991.

Since ITSEC is supposed to apply to all possible IT systems and the document is only around 150 pages long, the security criteria must be described in very abstract form. Consequently it is very difficult to read, and also, like Acts of Parliament, must be interpreted from time to time by those implementing it.

The 1991 version of ITSEC presumes the existence of three fundamental threats to any system, namely unauthorized access to data (confidentiality), unauthorized alteration of data (integrity) and unauthorized impairment of functionality (availability). The security criteria are based on these three threats, which can also be reduced outside the system through various appropriate measures. For example, it is possible to vastly improve the security of an insecure computer system by limiting physical access to it. All these are overall criteria which must be taken into account in any evaluation. The remaining threat can then be assessed in line with the ITSEC evaluation catalogue.

The degrees of protection against threats are divided in ITSEC into several functional classes. These summarize the security functions in their respective special combinations, and predict which level of threat can still be safely withstood by the system. There are six functional classes (El to E6). E1 is the lowest level, and E6 the highest.

In all these classes, the formal demands made on the development process and the environment in which it takes place are laid down in very abstract fashion. The functional classes also contain details and instructions concerning documentation as well as the eventual operating environment. This data is produced in a form that is applicable to all the technical variants of software development.

The fundamental process in the certification of a system in accordance with ITSEC is the evaluation of the mechanisms which maintain security with regard to the three defined basic threats. The evaluation takes place in dimensions similar to academic grades. The mechanisms are assessed according to the following elementary security criteria:

- Process documentation
- Guarantees and functionality
- Error correction
- Identification and authentication

- Rights management
- Authorization checks
- Transmission security
- Reprocessing

In a real-life Smart Card project, the steps are as follows: the card issuer together with the operating system manufacturer determine the operational requirements and the threats to be taken into consideration. Agreement is then reached as to the necessary functional class for certification. The operating system is produced in accordance with the functional class and the corresponding procedure, and the necessary documentation attached. In the final step, the operating system with all its components can be certificated by the appropriate independent organization, in line with the previously agreed functional class.

ITSEC certification offers a number of benefits to the card issuer and application provider, since they can be sure, with regard to many security aspects, that the essential mechanisms are clearly defined and functional in their conception and execution. This certainty, however, presents drawbacks in the fast-moving Smart Card market. Development time increases substantially where certification is needed, even in the lower functional classes. Also, the additional expense due to documentation results in increased development costs, which at the end of the day the system manufacturer has to recoup through product price. However, the greatest disadvantage is quite different in nature. Even in the case of a certificated system it is impossible to guarantee that all the routines and mechanisms function as described in the covering documentation. Certification does not mean that the certifying organization tests the product completely, only that it examines the papers received together with the product. These three reservations with regard to certification must be borne in mind in all cases and taken into account.

9.3.2 Software testing methods

As already emphasized repeatedly, Smart Card microprocessor programs are not very large compared with other software. Nevertheless, they have their own peculiarities. In order to make this point more concrete, consider a few of the figures for a typical operating system. In a current average chip with a 16 kbyte ROM and 8 kbyte EEPROM, the software requires around 20 kbytes of memory which leaves 4 kbytes of EEPROM available for any applications. At the moment, the software would be completely programmed in Assembler and in our example would run to about 30***000 lines of program code. If this were printed out at 60 lines per page, one would need 500 sheets of paper. Given that the number of steps are around 2000, even an experienced programmer would need around 9 months to produce the 20 kbytes of Assembler code.

This numerical example demonstrates quite clearly that in a Smart Card operating system we are dealing with a rather complex piece of software. On top of that we have software which is implemented almost exclusively in sensitive fields where security is one of the main factors. Hence the required freedom from errors cannot simply be checked during or after programming with a few tests. One needs a corresponding testing strategy.

9.3.2.1 Fundamentals of Smart Card software testing

In order to be able to establish a testing strategy in the first place, one needs to be aware of the life-cycle of Smart Card software. One may utilize the waterfall model proposed by W. W. Royce, known since 1970 and published in many forms. However, since it is designed for very extensive software projects for PCs or mainframes, in the following we will use a simplified version specially adapted for Smart Cards.

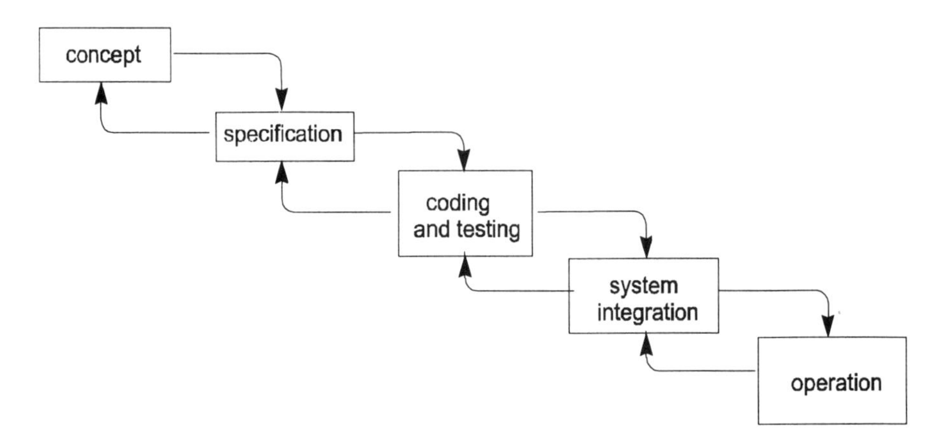

Figure 9.2 Waterfall model depicting manufacture of software for Smart Cards. The individual steps can be seen as activities or periods of time

The five specified steps are normally performed sequentially. It may well happen during a particular step that problems crop up, making it necessary to go back one or even several steps. This should be avoided as far as possible, though, as such iterations cost time as well as money.

In order to meet economic demands such as time-to-market and high software development turnaround, it is often necessary to perform the steps not strictly one after the other but to overlap them to some extent. In this method, known as simultaneous engineering, parts of the software are split into modules as early as possible, which are then sent down the waterfall together. Thus, it may be possible for a card only provided with a transmission protocol to be at the system integration stage, whilst a cryptographic algorithm also relating to the application is still being specified.

Concept

This step deals with the establishment of a fundamental objective definition, as well as the collation of the requirements in the form of a catalogue. All the requirements are specified which must later be met by the software. The concept stage also includes the drafting of initial solutions. In simplified form, this step determines "what" the finished software has to perform.

Specification

The next step establishes "how to do it". A precise specification must be produced which is not subject to interpretation, and completely determines one of the possible solutions for the requirements made by the concept. Formal specifications are best, since they can be used to determine unambiguously and indisputably the software's properties, functions and performance. The optimal approach uses specification in pseudocode, which can be tested by computer programs for consistency and freedom from error. The specification can then be directly used in CASE (computer aided software engineering) for source code and test program generation. In this context one often also speaks of "executable specifications", i.e. specifications produced in a form which can be interpreted and further processed by computer.

Coding and testing

Once the specification is finalized and accepted, flowcharts for Assembler programming may be produced. This is followed by programming and the relevant tests. The outcome, on completion of this step, is a fully programmed and tested Smart Card operating system.

System integration

Since Smart Cards can only work in the context of a system, this step deals with the integration of its components. Its result is full and error-free cooperation between all the parts, as well as the final documentation of the whole system.

Operation

This last step in the software development process can only be used to modify any global parameters which may be present in the issued cards. Large-scale software adjustments or modifications are no longer possible at this stage.

It has long been known, in connection with all kinds of projects and particularly in the field of software development, that error correction becomes more expensive the further advanced the project is. As a result of this state of affairs, time and expenditure are correspondingly heavily invested during the early steps in the waterfall model, since if a concept is incomplete, or a specification faulty, the cost of altering these problems rises exponentially during the later stages.

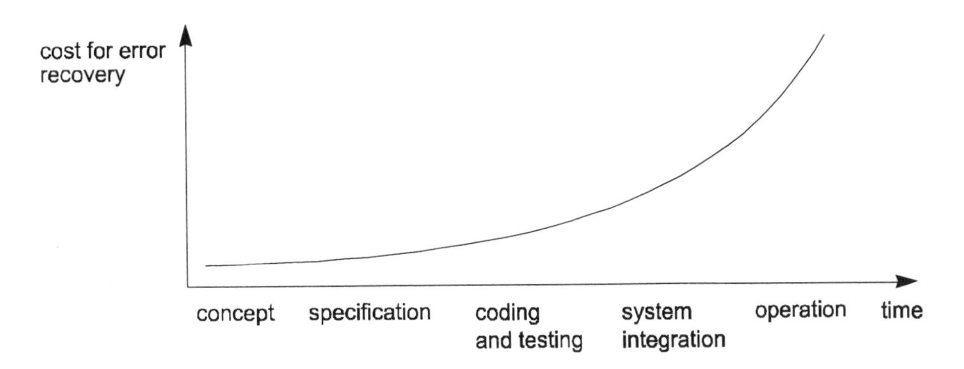

Figure 9.3 Cost of error correction, independent of error detection time

9.3.2.2 Test procedures and testing strategies

It is impossible to be unaware of the large number of diverse software testing methodologies and procedures currently in use. However, one only needs a few proven methods to test Smart Card programs. This is one of the few benefits of the, as yet, unavoidable programming in Assembler. In testing programs written in this language, one can fall back on decades of experience and a large number of publications. Incidentally, software testing always means attempting to discover program errors, not to demonstrate their correctness.

All testing procedures can be divided into static and dynamic ones. In the static procedures, the program code is analysed and evaluated either manually or automatically, using various methods. The two most commonly used static testing methods are described and explained briefly as follows:

- Static program assessment using software tools (various program code features are analysed by static procedures, e.g. ratio of comments to program, breakdown of program code, number of functions, dead code, nesting depth etc.)
- Review (formal analysis and evaluation of program modules by a team of assessors, in special cases also known as Code-walk-through or Code Inspection)

In contrast to the static procedure, dynamic program analysis tests the program during operation, manually or with the help of computers. There are two fundamentally different approaches, plus a hybrid one.

Blackbox test

In the Blackbox test one assumes that the tester knows nothing about the program's internal routines, functions and mechanisms. As a result, one only considers the input and output data in relation to each other, in line with the specification.

Blackbox tests are the standard in Smart Card operating systems. They are also used for terminal and computer system security modules. It is, however, often falsely assumed that in addition to errors, these tests can also discover any Trojan horses or suchlike. This is then supposed to make a relatively complex and expensive program code analysis unnecessary. This test may make it possible to recognize simple and unsophisticated, or inadvertently programmed, system trapdoors, but an experienced programmer can easily create access possibilities which can never be detected by a Blackbox test. This may be visualized with a small example. It is not meant to serve as a model for the famous Trojan horses, but to draw attention to the need for code inspections in security analyses.

Almost all Smart Card operating systems contain an instruction for the generation and output of random numbers (GET CHALLENGE). It is conceivable to modify this instruction so that only the first 8-byte number is generated by the pseudorandom number generator. The subsequent "random numbers" are each 8-byte EEPROM data, XOR-ed with the first one. A simple program can then externally read out the entire memory, including all the keys.

A Blackbox test is incapable of determining that a Trojan horse is concealed behind this instruction. Even a statistical analysis of the random numbers obtained would not detect any significant deviation from conventional pseudorandom numbers. The only way to recognize this manipulated program is to inspect the entire operating system's code. This example illustrates only one of many possible options for altering a normal instruction in order to obtain data from memory. Since the necessary program code only runs to a few lines, the only effective method to exclude such trapdoors is a complete search through the source code.

Whitebox test

The Whitebox test is frequently also known as the Glassbox test, which clearly illustrates the philosophy behind it. All internal data structures and routines are known and can be completely reconstructed. The relevant program documentation serves for the test's conception and generation, but the specification is always the sole authority referred to. The documentation format conventionally used in all Assembler dialects is the program flowchart, which in the Whitebox test usually also serves as the basis for evaluating the software's internal functions.

Since this enables the tester to know the exact program execution, naturally he would want to test all the possible paths through it. There are several ways to achieve this. In statement coverage, every program instruction is run at least once. This type of analysis makes it very easy to discover if the program contains dead code which is never implemented. This test is too weak to ensure the desired functionality. It is better to use decision coverage, since then all the decisions in the program code must be run at least once in all the possible variations.

An extension of the decision coverage criterion requires all program decisions to be run once in every possible combination. This covers all the paths through the program. However, due to the restriction on testing duration this is only possible in very small programs with a few hundred bytes of code. Even in programs of the order of magnitude of 1000 bytes, testing all the possible combinations cannot be carried out in a sensible time.

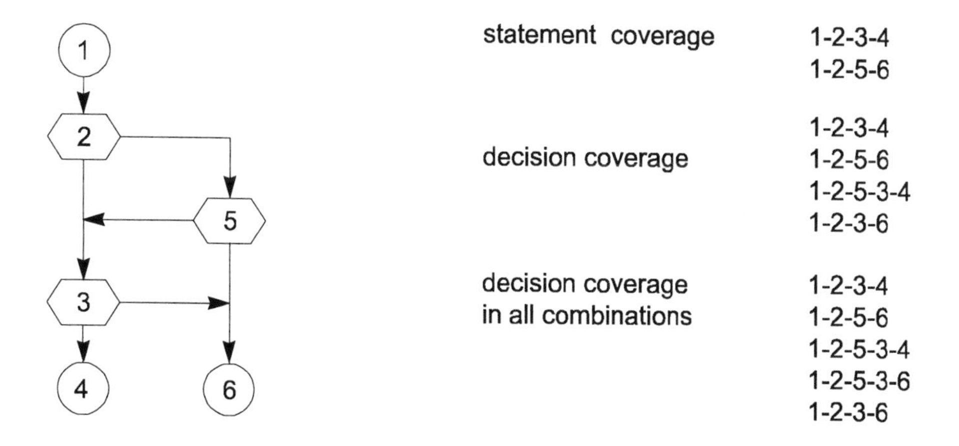

Figure 9.4 Example of decision coverage

Figure 9.5 summarizes the situation, using a typical instruction interpreter for Smart Cards. The function of this program module is to recognize an instruction placed in the card's input buffer in terms of class and instruction byte, and subsequently to perform a test of P1, P2, Lc and Le. This program runs to around 200 bytes and contains 18 branchings in its code. The possible output values are five return codes as well as calls to 26 different instructions.

Two further criteria for path coverage are used particularly in the testing of Smart Card operating systems: input and output coverage. The objective is to generate all the possible input and output values. Often the output values are restricted to the available return codes, otherwise the number of variations would be too large.

Since the number of possible input values rapidly reaches orders of magnitude which make them unusable in tests due to their volume and the time needed to process them, so-called equivalence classes are usually employed. This reduces the number of possible input values to a small number which can be tested in a reasonable amount of time. Equivalence classes are formed by checking for borderline cases on either side of the tested decision, and in addition using a typical value in the centre of the possible domain.

For example, if the card's command interpreter permits the range from 20 (incl.) to 50 (incl.) for the P1 byte, the following values would be used as equivalence classes: 19, 20, 50, 51, with, for example, 35 as central value. This would then have tested the essential interrogation conditions in the program. After this test one could assume, with a high degree of probability, that interrogation of the value range has been correctly implemented.

Unfortunately, more than anywhere else, it is necessary in Assembler programming to take the target hardware's properties into account when specifying the equivalence classes. For instance, all arithmetic operations which due to computer architecture (8 bit, 16 bit) cause an over- or underflow in the processor, must be considered when forming the equivalence classes. It is only thus that one can ensure that carry operations are correctly taken into account.

Decision coverage	Number of possible test cases	Duration of tests
1 million random input values	1000000	≈ 8 h
Instruction coverage	10	≈ 0.3 s
Decision coverage	50	≈ 1.5 s
Decision coverage in all its possible variations	50000000	≈ 17 days
Input coverage in all possible variations (5 byte header)	$\approx 1.1 \times 10^{11}$	≈ 1000 years
Input coverage with equivalent classes	15	≈ 0.5 s
Output codes with return coverage	6	≈ 0.2 s

Figure 9.5 A number of different test cases dependent on different methods of coverage in the Whitebox Test. The example used was a 200 byte command interpreter for Smart Cards. The test length was calculated on an average processing time including data transmission of 30 ms

Greybox test

The Greybox test represents a hybrid form between the Blackbox and Whitebox tests. Parts of the software are known here, such as internal program routines. In the Smart Card field it is mainly implemented during the integration phase, since it allows errors in the meshing together of the individual components to be very quickly and effectively detected and rectified. Naturally, on the key manager's part this requires appropriate public test keys. Once this part of the integration test is successfully concluded, one can check the results with the life (real) key.

9.3.3 Dynamic testing of operating systems and applications

It is important to be clear from the start that program testing can merely demonstrate the presence of errors, not their absence. Assuming that only roughly half of all errors are found anyway, one can say with safety that an average program still contains a number of weak points. In practice, in Assembler programming one assumes an error rate of 0.7 per 100 lines of code after all testing. Using the above figure of 30***000 lines of program code for a Smart Card operating system as a representative one and regarding two-thirds of all lines as comments, it is possible to calculate that a fully tested and issued operating system still contains about 70 errors. Most of these will never become apparent, but in certain circumstances one of these will suffice to circumvent all of the system's security measures. It is very useful always to remember this fact as a motivation for careful and considered testing.

Of course, even tests are subject to real-life constraints. In contrast to research projects, in customer-oriented projects both the available time and maximum cost are very limiting criteria. In addition, testing becomes ever more difficult and demanding the fewer the remaining errors in the program. Thus, at some point the search for the last errors must come to an end, since the time and resources one can spend on it are fundamentally finite.

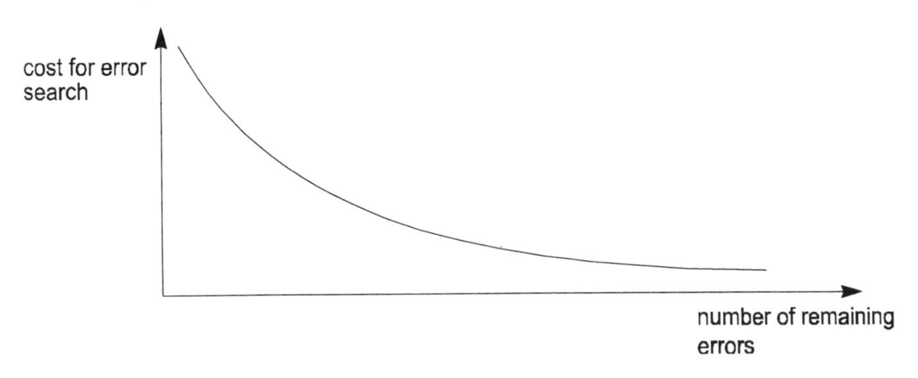

Figure 9.6 Error detection costs depending on existing errors

When issuing new versions one might assume that they always contain fewer errors, since it is possible to analyse faults discovered during operation and eliminate them. Interestingly, however, the reduction in the number of errors does not continue without limit; after a minimum usually in the second version, one generally notes a renewed increase. This comes about simply because the necessary error corrections are based on the original specification and program code. From a particular point onward (and this point can vary greatly), it is likely that the correction of one error gives rise to one or even several new ones. The result is the curve shown as Figure 9.7. Hence, after a particular number of versions it is substantially better to make a completely fresh start than to build on outdated concepts and repeatedly improved code. Incidentally, this is true in almost all technological fields.

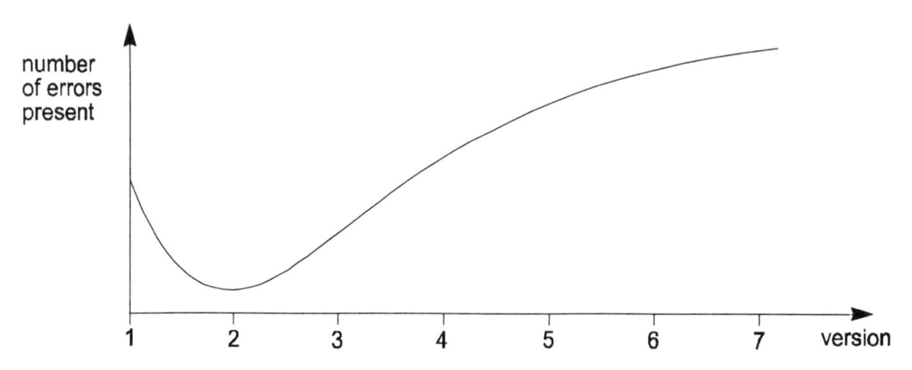

Figure 9.7 Empirically arrived at dependence of errors still in a program depending on the version

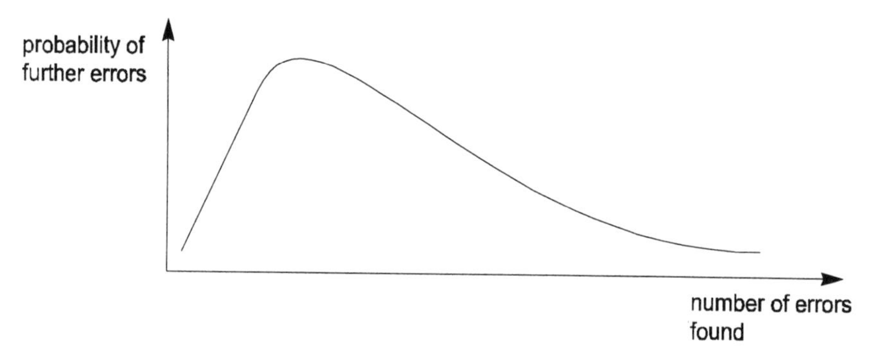

Figure 9.8 Empirically arrived at dependence of errors still contained in a program dependent on errors already detected

Testing procedure

There is a great difference between testing a new operating system and a new application. When testing a Smart Card operating system, the entire program code must be checked for the various application cases. This requires a large number of different tests. In a new application, which consists only of a DF and several EFs, the tests are reduced in line with the data used and the relevant defined identification and authentication procedures.

If a new operating system needs to be tested, one normally produces a number of test applications similar to a few typical real ones. Equivalence classes are formed, so to speak, for the usual applications. The individual tests can then be performed on these.

The following approach has become established for some years and across the most diverse projects for testing new Smart Card operating systems: data transmission is always tested first, since it forms the basis for all further actions. Then all the available instructions are tested. In the case of applications, this is followed by testing the files. If all those tests are successful, one can begin to test the fixed routines.

Currently there are no international standards which regulate the construction and execution of tests for Smart Card operating systems or applications, only a European standard (EN 1292) which defines a few tests for the ATR and the transmission protocol T=1. In order to gain an overall view, the following sections present a selection of possible tests in the conventional order. This list does not pretend to be comprehensive, and is only meant as a detailed illustration. The purpose of the quoted tests is to check a new operating system, including one or several applications, with regard to its essential overall parameters.

Data transmission tests

- ATR (detection of parity errors, character repetition if T=0 present, ATR structure and contents)
- PTS (PTS structure and contents)
- Data transmission test at OSI layer 2 (start-, data-, stop-bits, clock rate conversion factor, transmission convention)
- Transmission protocol T=0 (detection of parity errors and character repetition, various routines)
- Transmission protocol T=1 (CWT, BWT, BGT, reject, resynch, error mechanisms, various routines)
- Secure messaging

Instruction tests

- Testing all the possible class bytes
- Testing all the possible instruction bytes
- Testing all available instructions with equivalence classes of the supported functions

File testing

- Testing whether all files are present at the correct position (MF, DF)
- Testing for correct file size
- Testing for correct file structure
- Testing for correct file attributes
- Testing for correct file contents
- Testing the defined access conditions (read, write, block, unblock etc.)

Routine testing

- Testing the specified micro and macro state automata (instruction sequence)

As it is not difficult to imagine, despite the formation of equivalence classes and a few other optimizations, a relatively large number of individual tests is required. We may assume that for a 20 kbyte Smart Card operating system, it is necessary to define between 4000 and 8000 different tests to cover the essential cases. Tests which for example send several hundred different values to the card in a loop, count here as one test. Execution of all the tests takes in the region of one to two days. This large number of tests can only be carried out at reasonable cost with the help of a suitable database, which at the same time can also store the test results.

Formal description of the tests can be done, for example, in TTCN (the Tree and Tabular Combined Notation, standardized in ISO/IEC 9646-3). This allows the definition of any test cases in a general and standardized form, from which an interpreter can then automatically generate the instruction APDUs for the card being tested. Largely automated test runs can then be defined from this.

The structure of an idealized test tool is shown in Figure 9.10. The specification for the card's software is written completely in pseudocode, and is located in an appropriate database. If the specification changes, this leads automatically to the necessary modification of the tests. Another database contains all the tests in a high-level language which can also be directly read by a computer. The two databases feed a test-pattern generator, which generates the instructions (i.e. TPDUs or APDUs as relevant) for the card being tested. In parallel, a simulation of the real card is being run, largely defined by the specification. Since there are incompletely predictable processes in the real card (e.g. generation of a random number), additional data must be sent to the simulated one. The real and simulated cards send the responses to the instructions to a comparator. If they are identical, then the real card has provided the correct outcome, insofar as the simulation is the recognized reference. All the data generated during a test run is stored in a protocol database, where it can be evaluated manually at a later time.

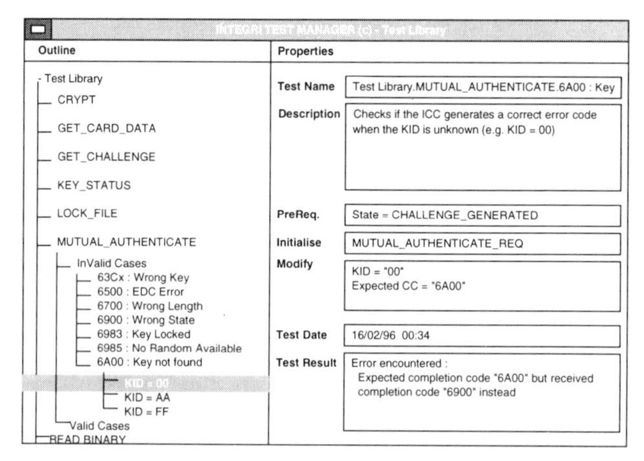

Figure 9.9 On-screen representation of object-oriented and databank-supported test tools for testing of Smart Card operating systems and applications.The figure defines an instruction (MUTUAL AUTHENTICATE), as well as the condition necessary for processing and the possible return codes.
(Copyright by Integri Acceptance)

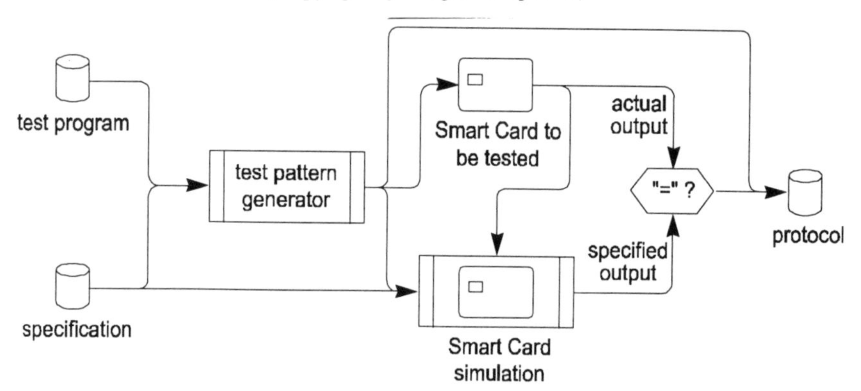

Figure 9.10 Ideal test tools for checking Smart Card operating systems and applications

Smart Card manufacturing

10

A Smart Card consists of two entirely different components. The first is the card's body with printed and embossed text, security marks and possibly a magnetic stripe. The second component, which turns it into a complete Smart Card, is the module which contains the microprocessor.

Figure 10.1 illustrates in rough schematic form the production of a Smart Card from the manufacturing of the two basic components, body and module, through use and finally recycling. At the beginning, there are two separate processes in which the card body and the module are created separately. At a certain point in the production process, the two are joined together to create a Smart Card.

The Smart Card manufacturing process consists of many different production stages, which are numbered sequentially here. The stages, which can be grouped together, represent the various production phases which may be implemented separately by different companies. The manufacturing steps themselves are divided into three branches, denoted by CM (chip and module), CB (card body) and SC (Smart Card). This is followed by actual utilization, and finally by disposal or recycling.

The technical manufacturing process for a Smart Card containing a single application is described below. In the example illustrated, no additional functions can be incorporated into the card during its lifetime. This simplifies manufacturing to some extent, since neither memory nor keys for future files need to be provided. The card's body is manufactured in the conventional way, by the lamination of several foils. In the example, the user receives the card by registered post, already personalized with the necessary data.

Naturally, all the manufacturing stages are accompanied by appropriate quality assurance. Since Smart Cards are products which are mainly used in security-related fields, production reproducibility within the meaning of ISO 9000 ff is guaranteed as far as possible. This means that all the manufacturing stages have to be logged, together with the batch and chip numbers. It should be possible, at any point after manufacturing, to reproduce each stage for each individual card. If any defects emerge subsequently, this makes it easier to analyse the possible causes. Since all microprocessors are uniquely identifiable after the semiconductor production stage, it is relatively easy to locate a problem chip, using the chip number. However, in order to keep track of the card's body it is still necessary to maintain a careful log. If the body is simply a carrier for the module and does not exhibit any special identifying features, it is sufficient to log the cards by batch, as is normal in the case of bulk products.

CM1 – software engineering

Due to the small memory available in the processor, operating system software and any applications based thereon must be implemented in Assembler programming language[1]. Ideally, programs should be written in a high-level language such as C or Pascal, but even highly optimized compilers generate program code which is between 20% and 40% longer than optimized Assembler code. This is the main reason why Smart Card software was developed in Assembler. Naturally, this is strongly reflected in the software's overall development time, and thus its cost.

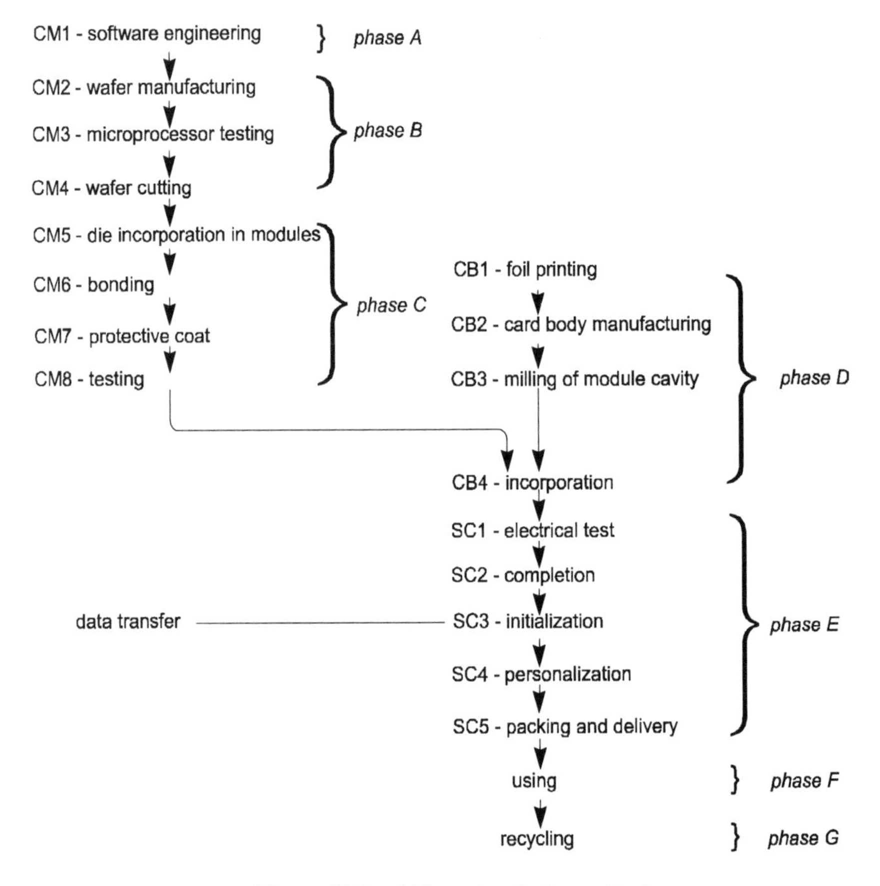

Figure 10.1 Life-cycle of a Smart Card

[1] See also 9.3 Testing microprocessor software.

Testing the software, which is largely located in the processor's ROM, is a complex and lengthy process, since software errors can hardly ever be corrected after chip manufacturing. This always involves the development of a so-called ROM mask, i.e. software which is later located in the processor's ROM and to which any modifications are no longer possible. Software defects which are only discovered in one of the subsequent production steps lead to a repetition of all the previous ones.

In order to use the available memory as efficiently as possible, the program code must be individually optimized for each chip type. Therefore transferability to other chips will involve additional expense. As a result, development time for a complete ROM mask is in the range of nine months. Nevertheless, it may be significantly shorter if it is possible to rely on existing program code (software libraries). Once the ROM mask is completed, it is officially handed over to a semiconductor manufacturer.

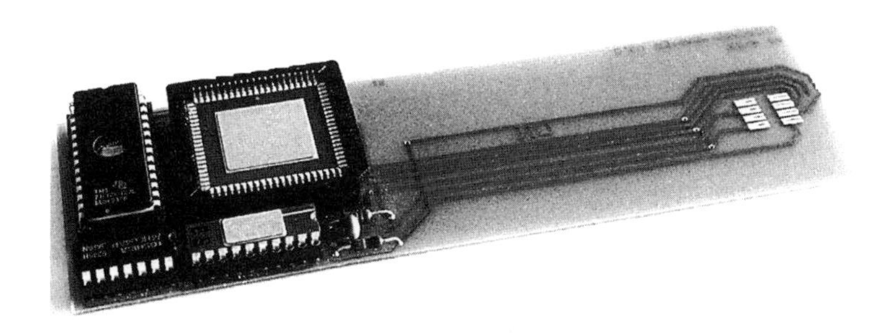

Figure 10.2 Smart Card mini-emulator, in which the ROM mask is replaced in the DIL housing by a replaceable EPROM. The large chip is a microprocessor for Smart Cards whose buses are freely accessible (Bond-Out-Chip)

CM2 – Wafer manufacturing

The semiconductor manufacturer uses the software received on an EEPROM or disk to produce an exposure mask for the processor's ROM. The mask containing the program code is referred to by operating system designers as a ROM mask. However, it is only one photographic mask out of the 20 or so needed for microprocessor manufacturing. Construction of the eventual processor takes place on an appropriately prepared, extremely pure silicon plate, known as a wafer. Wafer diameter in the Smart Card industry is usually 6 to 8 inches (15.2 – 20.4 cm). A 6-inch wafer carries about 700 microprocessors, assuming 0.8 μm technology.

Photographic masks for the entire wafer were in use up until a few years ago, so that the 700 processors could be exposed all at once. This is no longer possible, however, with the reduced dimensions of chip structures, as the resulting yield is too low to be acceptable. In all modern production processes, the masks carry only the image of a single chip and not that of the whole wafer. After exposure, the wafer is advanced by one chip width and the next one exposed.

Semiconductor manufacturing follows the usual procedures of doping, exposure, etching, washing, etc. The 400 or so production steps, depending on the manufacturer and the particular production method used, take between six and 12 weeks. However, the net processing time is less than a week. These extremely long processing times result mainly from the queuing technique used in semiconductor mass production.

For reasons of economy, many wafers must always be sent together in batches through the various production stages. A typical batch consists of 12 wafers, and this is often also the standard minimum production run. This represents 6000 or so chips. Quantities of less than a whole batch generate considerable extra cost during production, so that manufacturing costs become just as high as for a whole batch.

The normal total yield in semiconductor manufacturing is around 80% in an optimized process. Thus, of the original 700 processors on a single wafer only about 560 remain for the subsequent steps. Where innovative production methods are being used, however, yields may well be around the 60% mark when averaged over time, severely affecting throughput and profitability.

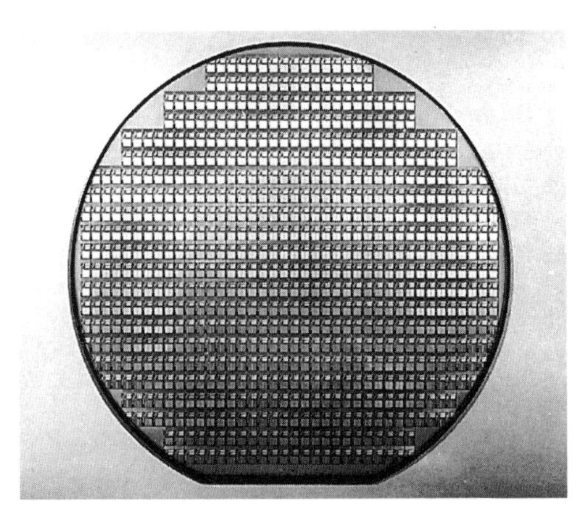

Figure 10.3 Figure of a 6-inch wafer, carrying about 700 microprocessors, SLE 44C80. The primary flat is used to mark out the wafer in the manufacture of semiconductors. (Siemens)

CM3 – Microprocessor testing

In the next manufacturing step, the microprocessors on the silicon plate are connected to metal pins and tested individually. This involves addressing the 700 processors on the wafer either individually or in groups of up to eight, and performing an electrical function test. Since no additional contact fields exist for the processor even during this step, only the five contacts available later on the card can be used.

The functional module test at this stage is substantially more intensive and wide-ranging than those performed later on, since the processor is still in its so-called test mode. All the memories (RAM, ROM, EEPROM) can be read and written to without restriction. Processors which need to be discarded as a result of the test are dabbed with a small dot of paint placed on top. Thus the failed modules can be detected visually in subsequent production steps, and discarded after separation of the wafer into its individual modules.

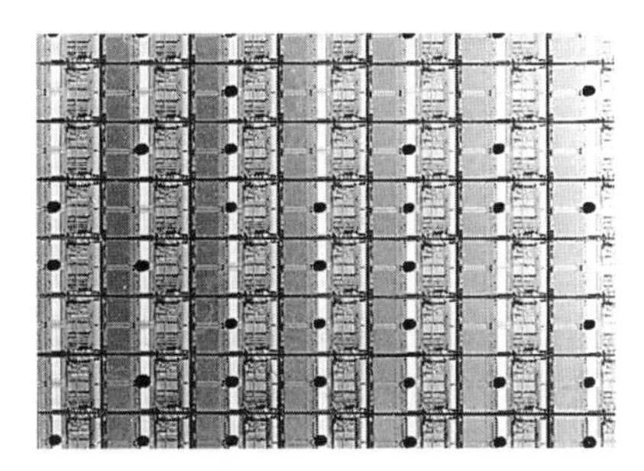

Figure 10.4 Wafer with semiconductor microprocessor. The points on individual chips identify defective steps

The write access available during the test mode is used for writing individual data into each chip's EEPROM. This might be an individual or a consecutive serial number. This is what makes each chip, and therefore each future Smart Card, unique. An important advantage, besides the security aspect, is that this always guarantees that the manufacturing of each individual chip can be logged and tracked as required by ISO 9000 ff.

CM4 – Wafer cutting

After testing the modules on the wafer, they are separated in the next step. The silicon plate is cut so that each section carries a single processor. The 25 sq. mm or so of silicon crystals carrying the microprocessors for the future cards are called dice (die in the singular). After separation, defective dice identified by a spot of paint can be discarded and destroyed.

Prior to this manufacturing stage, it is impossible to determine whether the incorporation of the ROM software in the semiconductor has been performed without errors. At this point, about ten dice are selected from each batch and installed in a ceramic DIL unit.

The software manufacturer receives this first sample and can test it to establish whether the processor's ROM software functions properly. The whole chip can also be tested. If an error is discovered now in the software or hardware, any further processing is stopped and the entire batch is only worth its scrap value. Then the problem needs to be corrected, and production restarted at the "CM2 – Wafer manufacturing" step. The time lost as a result cannot be made up even through an acceleration of all the production phases.

Figure 10.5 Example comparing a single die to a match. The diameter of the die measures 0.12 mm

CM5 – Die incorporation in modules

The next manufacturing step after wafer-cutting involves incorporation of the dice into the modules. This increases the stability of the very fragile crystals, and the contact fields located at their surfaces represent subsequent electric connections to the terminal.

Figure 10.6 Example of a 35 mm wide strip of film with modules mounted in pairs

Under normal circumstances, two modules are supplied which are mounted next to each other on a 35 mm-wide plastic strip with edge perforations. This is the standard film format for photography and cinematography, which originated in the early days of Smart Card manufacturing technology. In order to be able to make use of economical transport and packaging methods which require the minimum of R&D, the 35 mm film format was used as a carrier. This allowed utilization of the many film-winding devices available on the market. Since conversion to other formats was no longer practical once this approach had become widespread, this carrier material is still used today.

The dice are irreversibly attached to an adhesive with the silicon layer, i.e. the module's underside, facing upwards. They can thus be electrically connected to the corresponding contact surfaces on the underside during later production stages.

CM6 – Bonding

Once the dice are attached to the modules, electrical connections are made to the underside of the contact fields. This is achieved with the aid of very thin gold wire, soldered to the fields provided on the dice as well as to the corresponding contact points on the module's underside.

The wire is shaped into an arch, to ensure it does not break in the event of severe temperature fluctuations. However, this arch must not be too large otherwise the wire will not be fully covered by the protective layer, increasing the risk of corrosion.

CM7 – Protective coat

After bonding, the reverse side of the module and the mounted dice is coated with a darkened layer of epoxy resin. The resin protects the sensitive crystal against environmental effects such as moisture, torsion and bending. The resin is opaque since most semiconductor modules are very sensitive to light and electromagnetic waves in the near spectrum.

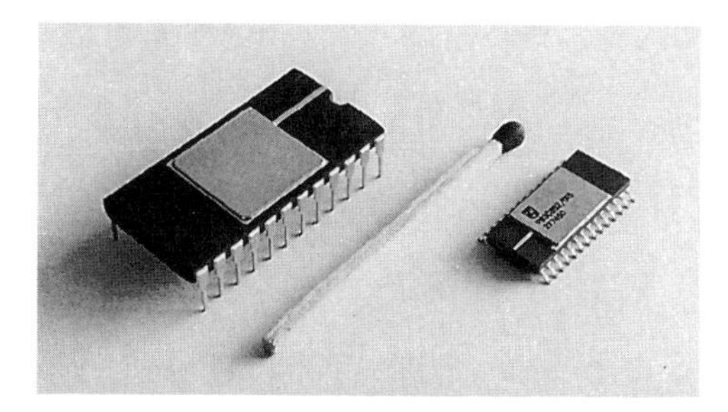

Figure 10.7 Microprocessors in various ceramic housings to test software

The carrier strip containing the modules is wound on to large film spools and packed in cartons. In the case of short production runs, it is also possible to pack the modules individually in plastic boxes, but this is avoided in long series since it makes automated module incorporation more difficult.

Where new processors are being launched or modified and chip hardware is being tested, the protective layer is often the last production stage. This is followed by the relevant testing and certification steps, and only if no defects appear is a new batch initiated, the software being adjusted for mass production. Completely new operating systems are treated similarly. Hardware production also stops at this point and the necessary certification is performed over weeks or sometimes months. Finally – if necessary – the correction loop is performed once again, using an improved system.

CM8 – Module testing

The previous manufacturing stages involving cutting, fixing, bonding and coating result in about 2% of the dice becoming unusable. A further testing stage is normally performed before packing and delivery. The contact fields on the module's upper face need to be electrically connected to a test rig. This device first switches the processor from test mode to user mode by burning through a polysilicon fuse. This is preceded by writing a special byte to a particular address in EEPROM. From now on, it is impossible to access the memory externally for reading or writing without satisfying the predefined security conditions.

After an ISO booting sequence, the test computer attempts to detect a valid ATR. If this is achieved, it tests the chip's instructions as integrated into the mask software. If all these tests are successfully performed, then the previous production steps have not caused a defect in this module and it may be installed in a card.

CB1 – Foil printing

Two different mass production techniques are used for printing the card's body. The first is sheet printing in large plastic foils, from which the cards are then pressed out. A standard sheet size can incorporate 21 to 48 blanks. The sheet passes through the individual colour-printing stages, which involve one or more runs through silk-screen or offset printing presses. The printing of one sheet is necessary for the card's front and back, since both sides consist of a single foil.

The second method involves individual printing of cards already pressed out. This process takes place after manufacturing of the card's body (CB2) and before milling of the module cavity (CB3). The throughput of presses dealing with individual cards is of the order of 12***000 units per hour.

A fundamental distinction is made between the two printing processes, silk-screen and offset. Offset printing can provide finer detail on the card than can silk-screen printing. The pigments harden under UV radiation. This irradiation takes place immediately after printing, which allows the cards to be stacked at once. However, UV-hardened lacquer cannot accept magnetic stripes or holograms through the hot foil-stamping process[2], as it has no thermoplastic properties. The situation is similar in inlet foils for laminated card bodies, since lamination requires the foils to be thermoplastic. In this case, a silk-screen printing process is employed in which the pigments harden by evaporation of one of the components and behave like thermoplastics. The remaining card elements can then be added easily to this printed surface. In practice, these two printing processes are often combined. The large plain areas and the background for the magnetic stripe and hologram are silk-screen printed. The finer detail which cannot be achieved by silk-screen printing is added by offset printing in a second run.

Thermotransfer is a third printing method which is sometimes used. Coloured foils are applied to white cards under heat. In principle, this can be used for all the colours and processes. However, this method is slow and expensive and is only employed for short series.

CB2 – Card body manufacturing

The classic process involves manufacturing the body from several individual layers of various plastic foils. Hot glueing of these foils under pressure is called lamination. The printed foils are additionally protected on both sides against scratching and rubbing by transparent laminated coating foils. Depending on the client's requirements, signature and magnetic stripes and security features are incorporated in the multilayer body during these production steps. Once all the foils have been bonded together and the necessary additional features integrated, the large sheets need to be cut. The cards are pressed out in the required format. The burr visible or felt along the edges of some Smart Cards is the result of worn presses.

After the card body is correctly dimensioned, a hologram can be embossed on it. Holograms are supplied in rolls, and are irreversibly attached to the body via the hot foil-stamping technique. Removal is only possible through destruction of this security feature. The magnetic stripe can also be attached to the body by hot stamping.

CB3 – Milling of module cavity

After the card's body is manufactured, a cavity must be milled out to hold the module. In some processes, the foils are pre-pressed to leave a space which is used after lamination for incorporating the module. However, as a rule, this space is milled out later. Since the

---

[2] A hot adhesion technique which requires a thermoplastic base.

module is thicker on its underside where the protected die is located, the cavity must be appropriately formed. A single, non-stepped cavity would be unsuitable, since due to the very small contact area between module and card body the attachment would not be permanent.

The first hollow is milled out the same size and depth as the contact field. An additional indentation is then milled out for the die. This results in a stepped cavity[3].

Milling must be very precise, since the card's thickness at the deepest indentation is only 0.15 mm. This production stage is carried out in fully automated machines, the cards being supplied and removed in magazines.

CB4 – Module incorporation

Regardless of which manufacturing method is used to make the body and to create a cavity for the module, the subsequent step needs to incorporate the module into the card.

For example, a double-sided adhesive tape with a hot-adhesive component can be used to attach the module to the card. However, adhesion is only required where the module overlaps the card; the die in the centre remains exposed. The adhesive tape needs first to be so pressed out that only the module's edge is covered. The module and the tape are then attached to the card. Adhesion depends on three factors, namely pressure, temperature and time. A problem arises from the fact that the module experiences a transient temperature rise to about 180°C. If the process, which normally takes about one second and requires some technical expertise, lasts too long, the module is destroyed by heat.

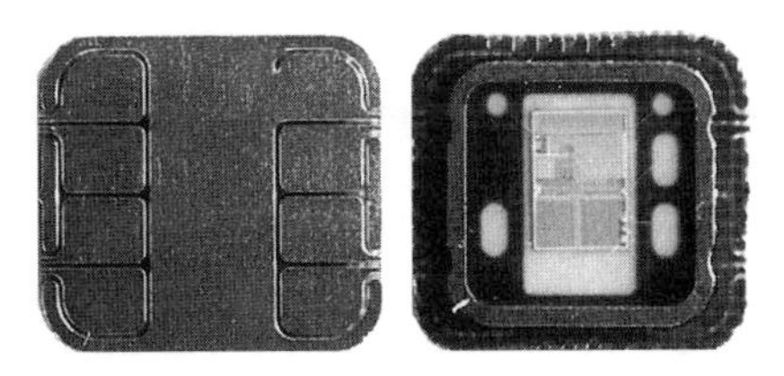

Figure 10.9 Geometry of module's front and back

[3] See also 3.2.2 Production methods.

Alternative attachment techniques, such as the use of liquid adhesive, are also employed, but the hot adhesive method is always regarded as the most reliable. The problem with liquid adhesive is that when it is poured into the milled indentation it tends to harden too quickly.

Once the module is implanted in the card body and the latter provided with all its general features and wording, the Smart Card is mechanically complete.

Data loading

The application provider needs to supply the card issuer with all the data relating to the application to be used for personalizing the card. This will include such items as the application name, file tree format, necessary files and file structures. These data are loaded when the card is initialized. All the customer- and system-specific data such as secret card-specific keys and the card holder's name and address are also required. All this data may be input via disks, magnetic tape or data transmission.

Since this data is almost always sensitive, the channel and method must be secured accordingly. It is therefore usually encrypted. Of course, the key needed for decryption is delivered to the personalizer separately from the data. If the data is intercepted or lost it is useless, as decryption is impossible without the key.

SC1 – Electrical test

The first step in this phase is an electrical test of the Smart Card. The basic test performed is the ISO booting sequence of the card, which must send a valid ATR in response. If the ATR is received and matches the expected response, this ensures that the processor's core functions are working properly. Special tests are then performed on the hardware components such as the ROM, EEPROM and RAM. In order to achieve high throughput in these tests which may take several seconds, special machines are used which can process a number of cards simultaneously. Typical machines have a circular production line and a throughput of up to 3500 cards per hour.

SC2 – Completion

Most operating systems are only partially present in the card's ROM mask. Look-up tables and some program codes are only loaded into the EEPROM after authentication using a secret key. This loading of EEPROM sections is known as completion of the operating system.

This approach makes it possible to create minor adaptations of the ROM code for special applications without the need for a new ROM mask. Only after writing the system data into EEPROM is the card's operating system complete, at which stage all application instructions such as SELECT and READ RECORD can be executed.

Completion, for which the data is identical for all Smart Cards running a given application, is also performed simultaneously on machines with very high throughput, just like the functional tests.

SC3 – Initialization

Completion is the technical precondition for loading all the uniform data into the card, which constitutes the following step. This is all the application data which does not vary from card to card and all the data, not related to a specific individual, which is also the same in each card. This step is known as initialization. All the data associated with an application which is identical in every card is loaded into the card at this point.

Initialization generates all the necessary files (MF, DFs and EFs) and writes as much application data as possible to them. Modern operating systems use the CREATE and UPDATE BINARY or UPDATE RECORD instructions to do this. This manufacturing step is the last one in which all the cards can be treated identically. That is why initialization can also be performed on fast machines running simultaneously. Card-specific and personalized data is only loaded at the next stage.

Figure 10.10 Initialization machine running 40 times its usual speed (Mühlbauer)

The reason for this technical distinction between general and specific (personalized) data is related to the optimization of processing costs. Personalization machines capable of writing individual data to every card under the required security conditions are technically very complex and have a low throughput of about 700 cards per hour. They are usually equipped with slow text-writing units for the card's body. Naturally, this results in high unit costs for loading of data. That is why all the global data which are not card-specific are loaded where possible using the simpler and faster test machines (throughput of about 3500 cards per hour).

The initialization and personalization bottleneck is created by the transmission and subsequent writing of data into EEPROM. Technical limitations mean that write access to EEPROM cannot currently be altered. However, transmission time for initialization and personalization data can be sharply reduced by increasing the clock frequency and reducing the data transmission divider. The result is that this process can be speeded up to twice its current normal rate.

SC4 – Personalization

The next step towards a finished Smart Card is personalization. In its widest sense, this term means that all the data which are related to a single individual or a unique card are entered into or onto the card. For example, these may include the name and address as well as card-specific keys. The only important criterion is that these data are individual to the card.

Figure 10.11 Personalization machine with lasergravure ability (Giesecke und Devrient)

A fundamental distinction is made between physical and logical personalization. Embossing of characters and lasergravure of text and images in the body of the card are the physical part of personalization. The logical part involves the loading of personal data into the processor and the writing of data to the magnetic stripe.

Embossing of names or similar alphanumeric, card-specific data is achieved through a machine which presses metal characters at high speed and with great force on to the back of the card. Since this is a relatively simple but very loud and vibration-generating procedure, these machines are usually located at some distance from the rest of the manufacturing process. Lasergravure, in which a laser darkens the area directly below the uppermost layer of the card, is very often used instead of embossing. This procedure is particularly secure when a photograph needs to be applied to the card.

Chip data is written to memory in the same way as initialization. To the extent that this may involve secret keys, however, cryptographically secure data transmission is often used[4], so that tapping into the transmission link to the card would not help a hacker. An even more complex procedure is sometimes implemented for cards used for financial transactions. The personalization data encrypted by the card issuer are re-encrypted in a special security module in the personalization machine and loaded directly into the card. The advantage of this approach is that the personalizer does not know the card's confidential data nor can they be revealed by bugging any of the transmission channels.

SC5 – Packing and delivery

The final production step is the packing and delivery of the cards. This is not necessary in all card types, since the cards are often supplied en masse to the retailer, as in the case of telephone cards. For the more advanced and valuable cards, however, the rule is that the cardholder receives a personal item in the post. Depending on the type of card, the personal data such as name and address are read from the card or from the production database and printed by laser on a carrier, i.e. a pre-printed letter. The letter contains two pre-cut slots,

[4] See also 6.6 Secure messaging.

into which the card is pushed and is thus secured. Alternatively, a residue-free and easily peeled adhesive tape is used. The carrier is then folded and placed in an envelope. After appropriate franking, the Smart Card with the personalized letter is ready to be posted to the end-user.

In order to keep postage as low as possible, it is usual to pre-sort the letters by postcode before delivery to the post office. This is simplest when the card production sequence follows the sorting criterion required by the post office (usually postal district and postcode). Similarly, a letter containing the PIN is created where necessary, which is posted separately and on a different day. This letter is so designed that even during preparation it is impossible to read the PIN without opening the letter.

Part/final step	Smart Card (low end) ≈ 6 kbyte ROM ≈ 1 kbyte EEPROM ≈ 128 byte RAM	Smart Card (high end) ≈ 16 kbyte ROM ≈ 8 kbyte EEPROM ≈ 256 byte RAM
Die	50.0%	65.0%
Module	25.0%	15.0%
Card body	12.5%	10.0%
Initialization/personalization	12.5%	10.0%

Figure 10.12 Costing for two kinds of Smart Card with microprocessors of differing capabilities

The manufacturing steps and phases described above illustrate the sequence during mass production, such as might apply to a GSM card. Other applications and other card issuers might require different production methods. Thus, some GSM cards are personalized on the spot in the shop and handed directly to the customer. From the latter's viewpoint, this receipt of a personalized card directly after registration and payment demonstrates competence and efficiency, but ultimately, this situation is also influenced by the issuer's own marketing and security preferences. In contrast to the above, manufacturing of the card's body and module are, in principle, independent of the eventual issuer or his marketing strategy, and thus largely the same for all applications.

Recycling

The recycling of Smart Cards is not yet particularly well developed, since most cards do not end up as waste but are kept by collectors for an indefinite length of time. With the steady proliferation of cards this is likely to change very soon. Nevertheless, Smart Card recycling is particularly problematic. Laminated out of many individual layers and of different types of plastics, they are very heterogeneous. Separation into homogeneous components is currently practically impossible. The only way to deal with multi-layer cards, therefore, is to burn them at high temperature. In the long term this solution is unacceptable, since 1 million cards with an individual weight of 6 g, for example, would weigh 6 tonnes!

In addition to environmental protection, there is another important reason to hang on to redundant or expired Smart Cards, namely, that some of them contain valid and secret keys. Statistical hardware and software analysis of a very large number of units is substantially more likely to lead to valid conclusions than is the case with individual cards. Hence, besides the environmental concern it is also necessary to collect expired cards and dispose of them carefully for security reasons.

If the card's body is fabricated by extrusion rather than by lamination, steps CB1 – Foil printing, CB2 – Fabrication of the card body, and depending on the extrusion process, CB3 – Milling of the module space, may all be omitted. If the body is only an extruded part it can be printed individually as required. It is then possible to proceed directly to CB4 – Module incorporation and the rest of the production sequence.

When the body consists of a single plastic layer, only the lamination is omitted during CB2 – Fabrication of the card body. The rest of the sequence is identical to that for the multi-layer cards already described.

The manufacturing process illustrated here does not, of course, take place within one single company; the production stages are shared among many separate companies. These are referred to here as "phases", and marked as such in Figure 10.1.

The reasons why card manufacturers are also developing the associated microprocessor software result from the (albeit very recent) history of Smart Card production. The same firm performs both the first phase (A) and the last one (E). Phases B and C follow at a semiconductor manufacturer's premises. Mounting the dice in modules and bonding can be performed either by the semiconductor manufacturer or by the card manufacturer. The whole of phase E, i.e. initialization, personalization, etc., is always performed by the card manufacturer. Normally, this is also the site from which Smart Card projects are managed; the other firms are usually merely subcontractors.

The process described above contains very little information about the time required for each manufacturing step. However, this must not be underestimated since many different companies need to cooperate, and some of the stages last for several weeks. Figure 10.13 illustrates some typical values for the duration of the more important production steps.

The following boundary parameters apply to this example. The assumption has been made that 50***000 Smart Cards need to be manufactured, the operating system can be constructed on the basis of available software libraries, all participating companies have medium capacity, and the relevant production processes can start immediately upon receipt of the necessary parts.

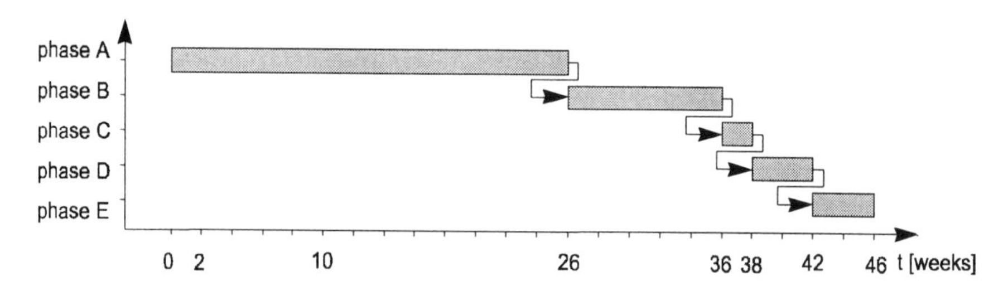

Figure 10.13 Gantt-diagram for progress of Smart Card project
phase A 6 months, phase B 10 weeks, phase C 2 weeks, phase D 4 weeks, phase E 4 weeks

Smart Card terminals

The only connection between a Smart Card and the outside world is via the serial interface. Hence an additional device is needed which sets up electrical connection to the card. This book always refers to this device as a "terminal". However, other terms are used, such as IFD (interface device), CAD (chip-accepting device), CCR (chip-card reader) and Smart Card Adapter. The basic functions, namely to supply the card with power and to establish a data-carrying connection, are identical in all these devices.

In contrast to Smart Cards, which are all very similar in their technical construction, terminals are implemented in many different ways. Fundamentally, a distinction is made between mobile and stationary terminals. Mobile terminals are battery-powered and stationary terminals are normally powered from the mains or via the data interface. Terminals can also be classified by their user interface. Portable devices may possess a display screen and a very simple keyboard so that the most important functions can be accessed on site.

Stationary terminals are often also equipped with a display and a keyboard, as well as a permanent connection to a host computer system. If no man–machine interface is available at the terminal (i.e. no display and no keyboard), then it follows that direct connection to a computer must be present to enable the Smart Card and the user to communicate.

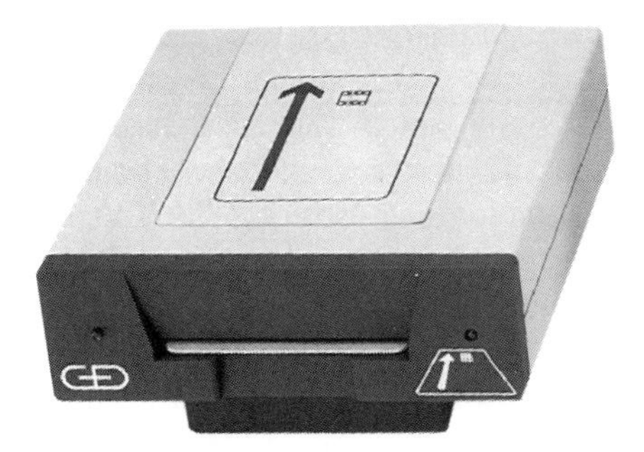

Figure 11.1 Typical Smart Card terminal for connection to a computer via a serial interface (Giesecke und Devrient – CCR2)

The division into portable and stationary terminals leads to a further distinguishing feature, namely the operating method. An on-line terminal has uninterrupted connection to a remote computer during operation, which takes over part of the control function. The typical representatives of this class are terminals for physical access control, which are completely controlled by a permanently connected background system.

The counterpart to this type, the off-line terminals, run completely independently of the supervisory system. There are many on-line terminals, but hardly any dedicated off-line terminals. All terminals need to exchange data with a background system via a communications channel, if only to request a new barring list or an update of terminal software.

In typical applications within a building, the physical channel between a terminal and a computer is an electrical conductor or an optical waveguide. However, it can consist of a telephone connection to the nearest computer centre, as in the case of financial transactions at a cash dispenser. Whether the terminal dials this computer on demand or whether there is a permanent connection depends on the particular application. Since the provision of permanent lines is expensive, however, the increasing tendency is to reduce costs by using the telephone only as required, so the terminal needs to be equipped with a dialling modem.

Figure 11.2 Smart Card terminal with off-line capability, integrated keyboard and display (Giesecke und Devrient)

Smart Card terminals in the form of a PCMCIA card cannot really be included in the above classification. They can be run both on-line and off-line, and can be used in conjunction with both desktop and laptop computers. In principle, they are nothing more than a simple and economical hardware interface between the Smart Card and the computer. The only precondition for their use is a PCMCIA socket which, depending on the manufacturer, must be either a type I (3.3 mm in height) or type II (5 mm in height). Some PCMCIA Smart Card terminals possess, in addition to the card interface, an internal supplementary memory for the card and arithmetic units for mass data encryption and decryption. These terminals, which are only a few millimetres in height, are probably the most flexible of all, since to some extent they open up totally new Smart Card applications. They make it possible for the first time for Smart Cards and standard PCs including standard software to work together without additional cables, power supply and external hardware. The spectrum of possible applications is very wide, ranging from access protection for certain PC functions through software copying protection to securing and signing e-mail during data transmission.

In order to make data exchange between Smart Cards and PCs possible, diskette terminals have recently been proposed. A 3.5-inch disk can be converted to accept a Smart Card. There is sufficient room in the 3.3 mm-thick diskette to hold a very thin card reader, a microprocessor, a battery and a spool for data transmission to and from the PC. All that is needed is a suitable software driver for the PC to handle the data exchange. This would be an ideal solution for integrating Smart Cards into existing systems in an uncomplicated and economical way.

Figure 11.3 Typical Smart Card terminal in the form of a PCMCIA card (Gemplus – GPR400)

Many years of R&D activity have led from the earliest Smart Cards using two-chip technology to today's cards with their very powerful microprocessors. The terminals have undergone a similar technical evolution during this time. Their original design was often very primitive in mechanical and electrical terms, resulting in frequent damage to the cards' microprocessors and their premature failure. These "teething troubles" have been overcome by most terminal manufacturers, and a development stage has been reached in which it is not so much the similar technical features offered by all manufacturers which influence a buyer's purchasing decisions as the external design of the equipment.

In functional terms, a Smart Card terminal consists of two components, the card reader and the computer. The reader, into which the card is inserted and then electrically contacted, has essentially only a mechanical function. The computer is needed to control the reader electrically, administer the user interface and establish a connection with the higher system. In the simplest case, it can consist of a single microprocessor, or in technically more advanced solutions, a single-chip computer.

11.1 MECHANICAL FEATURES

When a Smart Card is inserted into a terminal, the card's contact fields must be connected via a conductor to the terminal's computer, a function performed by the addressing unit. The card's insertion must also be detected by the terminal, and this is done by means of a microswitch or a photoelectric cell. The latter suffers from the drawback that dirt or a transparent card may affect the reliability operation. A mechanical switch is the most effective device and this has always been the case.

The greatest differences between card readers are exhibited in the area of contacts and contacting units. As for the shape of the contacts as well as the largest permissible forces, the GSM 11.11 specification offers restrictions on which almost all readers are based. According to this specification, the reader's contact elements should be rounded and not sharp. The radius of curvature at the tips should be at least 0.8 mm. This helps to prevent scratching of the card's contacts. In addition, the initial application of pressure to the contact using an element whose contact point is smooth and rounded is considerably more effective than if the contact point is sharpened.

In the case of GSM, the maximum force exerted by a single contact element must not exceed 0.5 N under any circumstances (0.6 N according to EMV-1). This is designed to protect the chip located beneath the contact fields which is made of a single silicon crystal since it might break under greater stress.

Although the position of the contacts on the card is regulated internationally by the ISO and should really be identical worldwide as a result, there is a French national standard, AFNOR, which provides for a chip position nearer the card's upper edge.

For this reason, card readers have been constructed which are equipped with two contact heads. This makes it possible to support both the ISO and AFNOR contact positions. This technically demanding solution is of interest in systems where Smart Cards with ISO and AFNOR contact positions are used together. This is only to apply during a transitional period, since according to the ISO the AFNOR position should no longer be used. Thus, several French banking applications employ terminals with double contact heads. During the conversion period, they allow both the old AFNOR position and the contacts in more recent Smart Cards based on the ISO standard to be accepted.

It is mainly when using portable or mobile in-car terminals that difficulties are encountered in creating a contact between the reader and the card. This type of terminal is often exposed to high vehicle acceleration, which can lead to short-term contact separation. For instance, a vehicle travelling over cobblestones at speed will cause sprung contact elements to start oscillating at their resonant frequency. If the card is electrically energized at the time, it is simply impossible to predict what might happen.

In an extreme case, all the contacts might separate simultaneously and then reconnect, which could well cause the card to execute a power-on sequence and then send out an ATR. However, the electrical booting sequence is unlikely to accord with the ISO standard, which if repeated several times might lead to card failure. Quite independently of this, the entire status reached in the card during the session would be lost through the brief power failure. This means that depending on the application, the PIN might have to be entered again or the user authenticated.

When only one contact is separated, the result depends on which contact it is. If I/O contact is lost, then only the communications channel will be temporarily down and this can be rectified by the error recovery mechanisms. If another contact is separated, a card reset occurs which causes a complete re-establishment of communications.

The connecting force can be increased to prevent separation through acceleration forces, but 0.5 N per contact field is the upper limit. There is no simple and satisfactory technical solution to this problem, but the probability of contact separation can be minimized by sensible design of the card reader. For example, one option is to position the contacts at right angles to the main direction of acceleration.

Even so, the terminal software has to be capable of re-establishing communications when contact is temporarily lost. The millions of GSM sets in use demonstrate that the use of Smart Cards in portable devices is possible and not fraught with difficulty.

The lifetime of contacts and the technical construction of the readers vary immensely. They are also strongly affected by environmental conditions such as temperature, atmospheric humidity, and so on. An MTBF (mean time between failures) of 150***000 insertion cycles is an acceptable figure for terminals.

Contacting unit with sliding contacts

The simplest and thus also the least expensive card readers have sliding contacts only, in the form of leaf or plate springs. No other mechanical contacts are present in these simple readers. However, a spring-based unit drags the contact elements across part of the card and the contact field during card insertion and removal, leaving scratch marks. These are undesirable not only on aesthetic grounds, but also for technical reasons.

Scratching off the card's gold-plated contact fields gradually wears away the chemical protection offered by the gold surface, causing the exposed metal surface to oxidize. This adversely affects the electrical contact, and the user sometimes needs to insert the card and remove it a number of times until the oxide layer is rubbed off and a satisfactory connection is re-established.

Mechanically driven contacting unit

The next category of card reader does not have fixed sliding contacts, but rather a mechanism whereby insertion of the card causes the contact unit to press against the contact fields. The force exerted through card insertion is converted via a lever arrangement into one which moves the unit perpendicularly. A relatively small movement is created parallel to the card when contact is established. This ensures secure electrical contact and removes small dirt particles from the contact surface. The elements themselves are also individually sprung in order to set up a defined force against each field.

Electrically driven contact unit

The most complex solution, but the one which is technically and mechanically superior, consists of card readers with an electrically driven contact unit. The parallel contact elements are driven by a motor or electromagnet and contact is established vertically from above using a relatively small sideways movement. As a result of this complex electromechanical arrangement, the reader's dimensions have to be relatively large.

These devices, however, are the ones most suitable for professional solutions in which many millions of contact cycles must be achieved without maintenance. Typical applications are ATMs, cash dispensers and the personalization machines used in Smart Card manufacturing.

Ejection

The Smart Card is normally inserted manually, that is without any assistance from the card reader. Devices which draw the card in only exist in cash dispensers, and they pull the card across the reading head within the machine with the aid of a special mechanism. However, this is not the case in an ordinary card reader. Variations only exist between different methods of ejection. Simple readers do not do this automatically, so that the card must be manually removed from the device. This process, in turn, is governed by two different principles – push–push and push–pull. In a push–push card reader, the card is inserted by hand as usual and must be pressed again and then pulled to remove it. This is not ergonomically efficient, since it is inconsistent with natural human movement. As a result, it often happens that people simply pull the card brusquely out of the reader because the contacts have not been released by the mechanism. This causes the body of the card to be dragged across the contact fields. Push–pull reader implementation accords much better with natural human behaviour, as the card is pushed into the device and simply pulled to withdraw it.

In card readers with automatic ejection, a spring is tensioned during insertion which the terminal's computer can release by means of a magnet. This causes only partial ejection of the card so that the user can grasp it and pull it out completely.

Ejecting devices offer great advantages compared with other designs. Ejection of the card signals to the user very clearly that the session has terminated and at the same time it acts as a reminder not to leave the card behind, often also emphasized through beeping. This argument, based on practical experience, is the greatest benefit of ejecting card readers.

Cash dispensers are the devices which are most frequently fitted with the capability of retaining Smart Cards. Since they can draw in the cards towards the contact head as a standard feature, it is technically feasible to direct the card not to the exit slot but to a special retention bin in the machine. Thus, from a practical viewpoint, card retention presents no major problems, as long as the terminal is large enough for the additional mechanisms and retention bin. However, in certain circumstances this can lead to legal problems where the user is the card's legitimate owner.

Removal options

The reliability of a system based on Smart Cards can suffer severely when users are able to withdraw their cards from the reader at any time during a session. For one thing, this causes the card to be disconnected from the power supply without the specified shutdown sequence. For another, it could lead to an interruption of EEPROM read or write procedures so that the file's contents would remain undefined. This could cause the card to fail completely. That is why it is advantageous to use ejector readers in which manual removal of the card is rendered impossible through appropriate machine design.

A hidden mechanical emergency ejector is available in the event of a power failure, so the card can be removed from the reader. Normally, however, the terminal can determine when the card is returned to the user, without the latter being able to interfere in the execution of the various procedures.

11.2 ELECTRICAL PROPERTIES

With the exception of the card reader, a terminal consists mainly of electronic components. These are used to provide interfaces to the user and the background system and to control the card reader. The card reader's electromechanical parts and the card itself need to be supplied with electrical signals. The only information provided directly by the reader is whether the card has been inserted or not. On the other hand, the only signal sent to the reader is the activation of the electrical ejector, where such a device is present in the first place.

The card interface consists of five contacts: earth, power supply, clock, reset and data transmission. It is very important for the card, and has a major impact on its life expectancy, that after the electrical connection is established by the contact elements the booting sequence is followed exactly as per ISO/IEC 7816-3. If this does not happen, then excessive electrical demands on the chip's semiconductor can cause increased failure rates. Similarly, it is important to observe the shutdown sequence otherwise the same problems will occur as with a faulty power-on sequence.

There is a particular problem in the case of simple card readers in which it is possible to remove the card manually. When the reader detects a card being withdrawn, the terminal's electronic circuitry must immediately execute a shutdown sequence. Only thus can a situation be avoided in which sliding contacts which may still be energized rub across the card's contact fields. This has little in common with a standardized shutdown sequence, but the consequences of such unauthorized card removal may be even worse, since worn or slightly bent contacts may cause short-circuits between two wires. The mild sparking resulting from the discharge of capacitors damages the reader's contact elements and the contact fields on the card.

With regard to electric circuitry, almost all terminal manufacturers have discovered by now that resistance to short-circuits is an absolute imperative. If this point is ignored, than a single card with shorted contacts can cause the electrical demise of many terminals. Incidentally, shorted cards keep cropping up, partly due to vandalism and partly due to technical defects.

Short-circuit resistance should extend to being able to connect each contact to any other contact and connect all of them together without any repercussions. Ideally, there should also be complete isolation between the card's electrical control circuitry and the terminal's remaining electronics. This is standard practice in public telephones, since it also helps to prevent adverse effects even where voltages are applied externally to the terminal.

The voltage needed for writing and erasing the EEPROM's memory sections is generated by the processor via a charging pump on the chip. The card needs currents of up to 100 mA lasting for a few nanoseconds. The same effect, though not quite so powerful, can be produced through the switching processes of transistors in the CMOS circuits. Even very fast control circuits in the power supply are stretched by these short spikes. The end-result is that the card's supply voltage collapses due to the high current consumption, and the EEPROM's write or erase process fails. In extreme cases, the voltage drop can be so severe that the processor is no longer maintained in its stable functional region, and a system crash occurs.

The solution is to place a capacitor as close as possible to the contact elements. Ceramic capacitors of about 100 nF are suitable as they can release their charge very fast. The wires to the card should be as short as possible, so that the increased current demand can be met at the required speed without great ohmic resistance and inductance. Should a transient power surge occur, it can be discharged from the capacitor until the voltage regulator can re-establish normal power. This is a simple and economical way to avoid current supply problems.

State-of-the-art technology, particularly in the area of electronic payment, dictates that terminals possess built-in real-time clocks for reliable transaction logging and for user protection. According to the EMV specification for credit card terminals (EMV-ICCT), the clock may not deviate by more than 1 minute per month. This does not represent a technical challenge, since suitable and precise clock modules are obtainable even as single-chip solutions. Additionally, clock adjustment can take place during every on-line connection to the supervisory systems.

Radio receivers in the terminal for the reception of broadcast clock signals have not so far attained practical importance, since signal reception is too strongly affected by the site's screening behaviour. For example, most conventional timing signals cannot be received from within reinforced concrete structures.

11.3 Security methods

Terminals may contain a very wide range of security mechanisms. The spectrum extends from mechanically protected housings to security modules and sensors for the various card features. In purely on-line terminals, which only perform electrical conversion of the signals emanating from a background computer system and from the card, there is normally no need for additional built-in security facilities. This is entirely the responsibility of the computer controlling the terminal.

However, as soon as data needs to be input to the terminal or where it is able to operate independently of the monitoring system, it is necessary to incorporate appropriate mechanisms for additional system security. The existing options are almost unlimited but depend very strongly on the relevant Smart Card and its security features.

In typical Smart Cards, in which the body is very basic and only serves as a carrier for the microprocessor, there are usually no security features on the body. This makes the checking of such features by the terminal unnecessary. In contrast, financial Smart Cards are usually hybrids, i.e. they incorporate both a magnetic stripe and a chip so as to retain compatibility with older systems. However, hybrid cards also possess the conventional features which enable the terminal to check their validity independently of the chip. The corresponding sensors must then also be available in the terminal.

Terminals which run completely or temporarily off-line must of necessity contain the main keys for the cryptographic algorithms being used. Card-specific keys cannot be created without them. These main keys are very sensitive, since the system's entire security is based on them. In order to guarantee their security and confidentiality at all times, they are not stored in the terminal's standard electronic unit but in a mechanically and electrically separate module within the terminal, one which is particularly well protected.

This security module might consist of a single-chip computer cast in epoxy resin, for example, which can only exchange data with the terminal's main computer across one interface. The secret main keys must never leave this security module, and are only used internally for calculations. In the typical application, the security module receives an individual card or chip number from the Smart Card via the terminal's computer, which it then employs to create a card-specific key. This is then used within the security module for signature computation or for authentication.

Modern designs of this module, which is normally the size of a matchbox, contain extensive sensors for detecting attacks, and are also largely independent electrically in order to be able actively to resist tampering, even when the power supply is cut off. Detection and response to any attack usually cause all the keys to be erased, so that an attacker ends up analysing only an electronic circuit in an epoxy resin and metal housing, devoid of any data content.

Due to the high cost of good security modules, the tendency in recent years is to switch over to Smart Cards instead. Although this causes certain restrictions in terms of memory size, sensors and independence, the level of security generally suffices even for electronic payment applications. In order to limit the physical size, it is increasingly rare to use ID-1 formatted cards and more common to employ plug-ins or DIL ceramic housings.

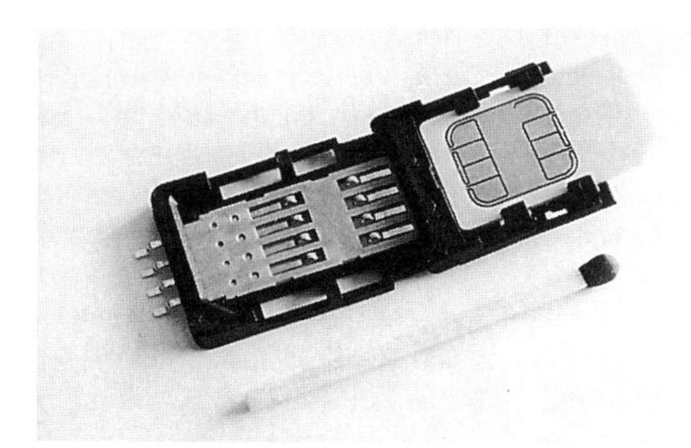

Figure 11.4 Example of plug-in security module in Smart Card terminal

Since security modules in Smart Card form are not built permanently into the terminal but can be replaced, they are eminently suitable for extending the terminal's hardware. The following example will illustrate this. Static unidirectional RSA authentication will become increasingly important in the next few years, since it is also required by the international EMV specification for chip-carrying credit cards. Since an RSA authentication is so computationally intensive that it can no longer be performed by normal terminal processors within an acceptable time frame, fixed security modules present a problem. On the other hand, terminals which use plug-in formatted Smart Cards as the security module can simply be replaced. Somewhat more expensive Smart Cards fitted with a supplementary arithmetic unit are employed, and once the terminal software has been modified appropriately, the terminal will be able to execute an RSA computation at high speed.

In the future, the various card issuers will be launching chip-carrying debit and credit cards. All these cards use various keys, key derivation and authentication procedures. Furthermore, the card issuers are unlikely to be willing to reveal the secret data and processes to the manufacturers of security modules. In all probability, the approach chosen will be that of one or several card issuers jointly issuing a so-called terminal card which will support all the processes relevant to a given system security and will be capable of executing them from within the terminal. The terminal card will be addressed via one of the two standardized transmission protocols T=0 and T=1, and, by and large, will behave just like a standard Smart Card. The only difference will be that the card will contain functions for the secret main keys, the creation procedure and the collection of security-related data (e.g. the cash balance). The terminal will only handle user interfacing and uploading and downloading of data in conjunction with the background system. All the technical security aspects will be taken over by the card. This will make it necessary for the terminal to work together not just with one terminal card but with several. The user will select a particular card, depending on the card issuer and the chosen function. The requirement for being able to handle several independent terminal cards has been taken into account in recently designed terminals. Some of them have up to four card readers for the plug-in format. They can thus run several different card issuers' terminal cards in parallel without interference between them.

Besides terminal protection using mechanically robust housing which can only be opened using a special tool containing a security module, further mechanical protection often exists against unauthorized tapping into the card's data transmission. This is a type of guillotine, which after card insertion cuts off any wires or cables which may lead from the card to the area outside the reader. The purpose of this device, also known as a shutter, is to prevent tapping and manipulation of messages sent between the card and the terminal. It can be operated electrically or merely by inserting the card. If cutting is impossible due to the thickness or nature of the material, the card-reading slot will not be closed completely. The terminal's circuitry then detects this fact and the card is not powered so no communication can take place.

Communication between the terminal and the Smart Card must be fundamentally so designed that tapping or tampering cannot affect the system's security. Accordingly, shutters are not really necessary. Nevertheless, security can be improved somewhat if a putative attacker's life is not made too easy. There is a great difference between data exchange being capable of capture and alteration without much effort and several major hurdles first having to be overcome. On the other hand, shutters render the card reader larger and more expensive, and very few designs are still capable of accurate closure after several thousand cycles. Hence, system design should not rely completely on mechanical protection such as this.

Smart Cards in electronic payment systems

12

The original application of Smart Cards with microprocessors was mainly for user identification in the field of telecommunications. However, in recent years they have established themselves in another market sector, namely electronic payment. Due to the large number of units used, the market potential of this sector is huge, a point underlined by the issue of over one hundred million credit and cheque guarantee cards. The future applications of electronic purses extend from the replacement of conventional means of payment such as banknotes and coins, through shopping over global networks to individual payment for television broadcasts.

The Smart Card's special properties make it particularly suitable for electronic payment applications. The data can be stored in the chip easily and secured from tampering. Handling the cards is also easy for everyone, due to their size and robustness. Because Smart Cards can also actively execute complicated calculations, independently of external influences, it is possible to implement totally new types of transactions. This can be seen very clearly in the case of electronic purses in the form of Smart Cards, which can only be realized in this medium.

Electronic payment and electronic purses exhibit distinct advantages for all participants. Banks and traders enjoy reduced outlay on the handling of cash. Off-line electronic purses largely obviate the costs of data transmission. The risk of robbery and vandalism is reduced, since cash can no longer be stolen by the criminal.

The faster payment process is also an important argument for traders, since it allows optimization of cash flow. Simpler and cheaper vending machines can be constructed, as the expensive coin and banknote checking unit is no longer needed. Since electronic funds can be transferred across any preferred telecommunication channel, the weekly or monthly collection of cash from the machine is made redundant. The customer, too, enjoys the benefits of the new payment method, though they are not quite as numerous. The problem of unavailability of the right coins is solved, and the result is faster payment at both cash tills and vending machines.

At the end of the day, however, it is the putative purse users who decide the success or failure of a payment system. When the advantages are too minor, they will choose other options instead. Electronic purses, after all, are not a new type of payment likely to replace existing methods such as credit cards and cash, but rather one of a range of options on offer. It is unlikely that those methods, reliably in use for many years, will be driven out by Smart Card purses.

12.1 CARD-BASED PAYMENT TRANSACTIONS

The simplest solution for card-based payment transactions is the use of magnetic stripe cards, whose data is stored for online authorization. After comparison with the warning list and a validation of liquidity, a transfer directly from the holder's bank account to that of the trader can be carried out. This scheme can be slightly modified in the case of Smart Cards, but the principle is identical. The card is logically connected to a bank account, and after unidirectional or mutual authentication of background system and card, the transfer of a previously input amount is executed. A PIN check is, of course, also performed in the card or in the background system during the course of the transaction.

Both the above scenarios are based on an all-controlling background system, and in no way exploit the capabilities of Smart Cards. However, below we describe various procedures and payment methods made possible through their effective and consistent implementation.

12.1.1 Electronic payment transactions with Smart Cards

There are two fundamental models for electronic payment transactions which use Smart Cards: credit cards, in which payment is only made after receiving a service, and debit cards, in which payment must be made before the service is received. Below we describe these two models, plus a variation on them.

Credit cards

The original idea of using a plastic card for payment against goods or services originated with credit cards. The principle is simple: a person pays with their card, and the corresponding amount is debited from their account at a later stage. The costs of this procedure are borne by the trader. Up to now, this function was largely fulfilled by cards without a chip. The drawback of low security against forgery causes the card issuers a not inconsiderable loss due to the trader being guaranteed payment. Evidently, thus far this loss has been lower than the cost of introducing cards with a chip. However, in the not too distant future, credit cards are likely to be supplemented with a chip, to reduce the increasing cost of fraud.

Electronic cheques

The simplest form of electronic funds is the electronic cheque. In principle this is a special form of debit card, since it is also prepaid. Electronic cheques are of a fixed denomination (i.e. they are issued for a fixed amount), and thus can only be used in their entirety. Functionally, therefore, they are very similar to the well-known travellers' cheques. Therefore they must be prepaid. The user of an electronic cheque thus exchanges real money against an electronic form of the same amount, which is loaded into the Smart Card. During payment, the card holder exchanges the cheque for goods or services. The disadvantage of the fixed denomination can be minimized by carrying cheques for various amounts in one card. This comes very close to conventional money, in which arbitrary amounts are also absent[1].

Electronic purses

In electronic purse cards, electronic funds need to be loaded onto the card before payment is possible. During the actual payment process, the balance in the user's card is reduced and in parallel that in the second participant's electronic purse (usually the trader) is increased. The electronic amount received is then passed on to the purse operator against the corresponding funds.

The card user suffers from three considerable disadvantages. When loading the card, real money is exchanged for electronic money. From a financial standpoint, the purchaser is granting the purse operator an interest-free loan, since he may not use the electronic money for several weeks but the real money is immediately transferred to the operator's account. The interest lost by a single user may be small, but seen overall it represents a substantial additional income for the operator. In many field trials conducted so far it was found that the average amount in an electronic purse in industrialized countries is around £50. This mean amount held in an electronic purse is termed "the float". Assuming an issue of 10 million cards and a 5% rate of interest, the interest earned annually comes to £25 million. This figure is not offset by additional outgoings. Since the interest lost by the card holder in this example is only £2.50, this is not a major disadvantage.

A real problem would arise if the purse operator went into liquidation. The reason for this is as follows: the card user has exchanged real money, whose value is guaranteed by the state within certain limits, for electronic currency on a Smart Card. In the event of the operator's liquidation the electronic funds could suddenly lose their value, and the user loses all his money. As a result, efforts are now being made in some countries to regulate this trade, so that this type of electronic purse may only be issued by banks or similar institutions. Setting such a payment system would require at the very least a form of security lodged with an official regulator, so that in the event of liquidation the amount credited to cards is fully covered.

There is, however, a third distinct disadvantage for the user. What does the holder of an electronic purse do when it no longer functions properly? If it is anonymous, then the purse operator cannot establish the last stored amount. For the holder it is also practically impossible to prove his latest balance beyond dispute. Thus if the chip is destroyed, the electronic funds are irrevocably lost. Unfortunately, though robust, Smart Cards can be damaged. In practice, the current solution to this problem tends towards compromise. Since the last online credited amount is known and hence also the balance at that transaction, the amount still on the card can be approximately calculated. This is then paid to the client. However, if the same client repeatedly makes claims against the operator due to faulty Smart Cards, the latter restricts his goodwill payments and the client receives no compensation.

---

[1] E.g. there are no banknotes worth £11.97.

Open and closed system architectures

One needs to distinguish between open and closed payment system architectures. An open system is fundamentally at the disposal of many application providers, and can be used for general payment transactions between various organizations. In contrast, a closed system can only be used for payments to a single system operator.

To illustrate the technical aspects, we briefly describe the example of a telephone card which carries a memory chip. In memory cards, irreversible counters are simply deducted during payment. It is not necessary for the terminal to keep precise track of the number of deducted units, but only to ensure that the card's counter is always deducted when a call is made. Thus the terminal is a kind of "machine for the destruction of electronic funds". Of course, in practice, a balance is kept for each terminal, but the debited units are only accounted for internally by the purse operator. Fraud during debiting of units between the terminal owner and the purse operator is thus impossible in principle, since both belong to the same organization (the telephone company).

In contrast, in open systems the terminal owner and purse operator can be completely different companies. Hence the latter must be able to satisfy himself when totalling the terminals' turnover that these are correct and not manipulated. This must be taken into consideration from the start when finalizing the system concept, otherwise accounting between the terminal owner and purse operator is very difficult or even impossible. In the above example of a memory card, the system concept makes it impossible for the former to guarantee the latter that the amount forwarded is correct. The terminal owner can only submit a certain number of units for accounting, but no forgery-resistant signatures against the amounts paid from the genuine electronic purses.

System structure and terminal connection

The system structure of electronic payment systems with Smart Cards can be either centralized or decentralized. System security is the most important aspect here. Hence one often tends to prefer a centralized system, since then the system operator exercises direct control over all its aspects. Concretely, this means an online system in which every payment is executed directly and online to the background system. If connection cannot be established, payment is impossible. Nevertheless, a centrally operated system offers several benefits. For instance, the incoming transactions can be compared directly and in real-time with the current barring list. Key replacements can be carried out by the background system without any delays. Software modification in the terminals, or that of global parameters in the cards, is possible without complications or great expense since a direct link to the background system must be established for each transaction.

However, these advantages are counterbalanced by several great drawbacks. In many countries telecommunication costs are so high that it is not worth the trader's while to set up a permanent link to the background system or to dial in for each transaction. Nor is the reliability of the telephone network in many regions of such quality that an online link is possible at any time to the superordinate computer. Due to their active nature, Smart Cards are eminently suitable for decentralized systems, since they control part of the system security on site. This is their great advantage compared with the passive magnetic cards, which cannot enforce any processes in the system.

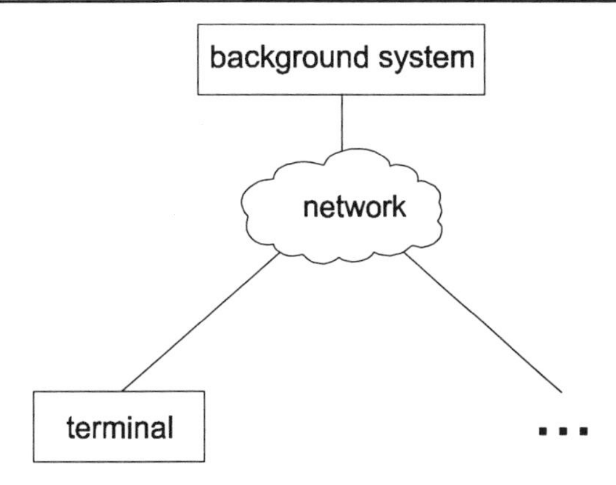

Figure 12.1 Basic structure of centralized electronic payment system (all illustrated connections are permanent)

The implementation of electronic purses in the vending machine industry, in particular, necessarily requires a decentralized system since they may operate quite independently for weeks or months, and offer no possibility of establishing contact with any communication system. All these reasons explain why decentralized systems are often favoured. To this we may add substantially higher reliability. If the centralized background system should crash, the entire electronic payment traffic is blocked. In a decentralized system, in contrast, the consequences of a temporary crash usually do not even spread to the trader's terminal.

However, decentralized systems too suffer from shortcomings, mainly in the area of system administration. The reason is that online links can only be established during certain intervals, and as a rule only by the terminals. But for system security it is vital that the terminals always work with the current barring list. This is one of the reasons why it is usual in many systems to require terminals to talk online to the background system at least once a day. The transaction data is then transmitted to the latter, and the administrative data is transferred to the terminal in return. Examples of such data are new terminal software, new key sets, updated barring lists and data for loading onto customers' cards.

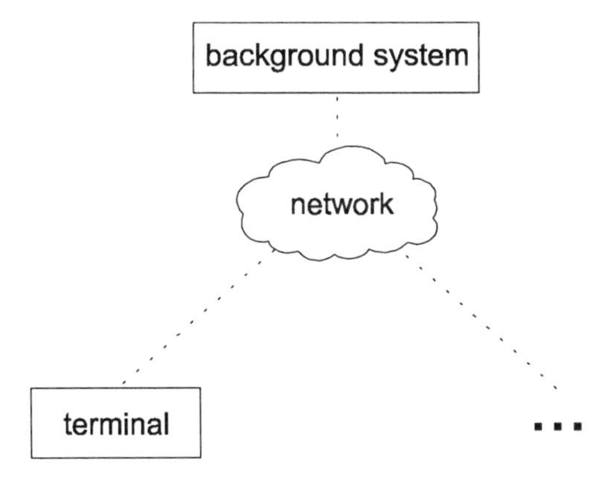

Figure 12.2 Basic structure of decentralized electronic payment system (all illustrated connections can be created as needed)

In practice, a compromise is often chosen between a purely decentralized and centralized system, to combine the advantages offered by both structures without their respective disadvantages. In this case, both the terminal and the Smart Card can force an online connection if certain conditions obtain. If this cannot be established, the payment transaction is not completed.

Typical conditions are listed below. Online authorization must be carried out if a particular amount is exceeded. This amount can usually be set by the system operator individually for each card. The number of off-line transactions and the time expired since the last online transaction can be used as another parameter for deciding to "go online". A random number generator is also sometimes present, whose output causes a certain proportion of all transactions to take place online. Some systems have an additional button on the keyboard, which also forces an online transaction. It can be pressed by the sales staff if they suspect that the customer is using a manipulated card.

These criteria ensure that on average every card is in direct contact with the background system at least once within a given and statistically computable time interval. Thus, the operator recovers the direct control originally lost through the decentralized system. Terminals and vending machines in which only small sums are dealt with can be excluded from the above online requirements, since even in the event of fraud only small losses arise. This saves the need to set up a link to the communications network, since the data exchange is performed manually by the sales staff.

12.1.2 Electronic funds

Electronic funds must have certain properties for them to enjoy the same flexibility as normal money. Where these are wholly or partly absent, the options available to users of electronic funds are necessarily more or less curtailed. Below we describe the essential features for minimizing the difference between the two types of funds.

Processability

An important, though in principle trivial property of electronic funds is complete automatic processability by machine. This ensures that large systems can be run economically.

Transmitability

Electronic funds must not be bound to a particular medium, such as Smart Cards. There must be the possibility of transferring them securely through arbitrary media such as networks or computers.

Divisibility

Electronic money must be divisible, so that arbitrary amounts can be paid without having to fall back on normal money. This property in turn is analogous to normal money, which although not arbitrarily divisible, is available in sufficient denominations so that conventional sums can be paid with a minimum of coins and notes.

Decentralization

Centrally designed payment systems are simple for the operator to supervise, and the possibilities for fraud are very limited. The best example is the online authorization of credit card purchases. However, centralized systems suffer from many drawbacks. They are expensive, susceptible to technical failures, inflexible and difficult to modify or expand. Decentralized systems minimize these disadvantages. This is easy to see in the case of payments made directly between persons by exchange of money, where no central body is involved. Electronic funds should ideally offer the same feature, since they will otherwise be unable to compete with normal money. When applied concretely to payment systems this means that both off-line payments, as well as payments from one electronic purse to another, should be possible. The property of direct payment between purses (purse-to-purse transaction) is also known sometimes as transferability.

System monitoring

Despite the requirement for anonymity, electronic funds must offer the operator the possibility of supervising the system. Only thus can tampering and security defects be detected and corrected. In this aspect too, electronic funds are similar to conventional money, where each citizen is expected to report any discovery of forged money to the appropriate authorities. In the case of electronic money, the operator is able and required to monitor the flow of payments for consistency.

Security

The fundamental property of electronic money must, of course, be its security against forgery. Any system would quickly crash if it were possible somehow to forge money, duplicate it or manipulate the flow of funds. This is the reason for the very extensive use of cryptographic functions in the field of electronic transactions to ensure that the required level of security is achieved.

Anonymity

Anonymity means that it is impossible to link payments and people. This requirement can be assessed very differently, depending on the viewpoint taken. The purse issuer desires, from a technical perspective, a non-anonymous system as far as possible, so that he can perform optimized system monitoring. This approach severely limits the possibilities for fraud, since the criminal can very quickly be identified. The same is true for official institutions such as the police or the Inland Revenue, who have considerably more supervisory options at their disposal in the case of non-anonymous electronic money.

The user's position is diametrically opposed to that. The current payment method, using normal money, is taken to be the ideal arrangement. Complete anonymity and undetectability of payments are regarded as the optimum. However, this feature is often modified by the purse operator for reasons of system security. Payments are mostly anonymous, but electronic purse loading is not. This allows a relatively efficient system supervision to be effected simply and inexpensively.

At first sight, the above features appear somewhat contradictory. For example, complete anonymity and perfect system monitoring are often mutually exclusive. Nevertheless, we are still at the beginning of developments in this field, and many systems are being planned in which these two properties can be simultaneously realized.

Two properties of real money were not mentioned in the preceding discussion, although they are highly significant. Real money is legal tender, which in the relevant country must be accepted by everyone. In almost all countries, the vendors of goods or services are required to accept the legal currency in payment. The second point relates to the currency's value. Except for a few high-inflation countries, the legal currency in circulation retains its value. If this is not the case, the result is a switch to payment in kind or in foreign currency.

12.1.3 Fundamental options for system architecture

An electronic payment system based on Smart Cards can be constructed in a variety of ways. For economic reasons, it is often based on existing systems, mostly those using magnetic cards. However, there is no fundamental model used by all payment systems, since the requirements vary too widely. Hence only the basic principles will be described here, with reference to the components necessary for such a system.

Large Smart Card payment systems fundamentally consist of four different components. These are the background system, the terminals, the network and the cards.

Background system

The background system consists of two parts, Clearing and Administration. The first of these deals with accounting functions relating to all incoming transaction data and the banks, traders and card holders participating in the system. Simultaneously, Clearing takes over the system monitoring tasks. For example, one simple task is a continuous check that the total amounts sent for clearance are never greater than the total in the electronic purse. Should this be discovered to be the case, then an attacker has credited the card without the system's knowledge.

The background system's Administration section deals with the distribution of new warning lists, switching to new key versions, software modifications in the terminals etc. The production of data sets for card personalization also happens here. All the threads involved in electronic transactions join up in the background system. This is completely independent of system structure. Even in the case of systems running completely off-line, the setting of global system parameters as well as the monitoring of security and functionality take place here.

Network

The network links the background system with the terminals. This can be a line-oriented link (e.g. ISDN) or a packet-oriented one (e.g. X.25). As a rule, the network is totally transparent to the messages, which are forwarded from sender to receiver without any modification.

Terminals

The various terminals used in this field can be divided functionally into loading and payment terminals. However, another division is into automatic and attended terminals.

The classic example for the former is cash dispensers. In purse systems they are mostly used only for crediting the cards. It is, of course, conceivable to have an additional function for debiting the purse, and thus dispensing cash. Attended terminals are typically located at supermarket checkouts and in retail shops. They are always used to pay for goods. However, depending on the system, terminals in banks can also be used to credit Smart Cards against payment in cash.

Smart Cards

Smart Cards are used as electronic purses. However, they can also be used as security modules in various types of terminals, and as transfer cards to transport data between various system components. Transfer cards serve for manual transmittance of transaction data between a terminal working completely off-line and one working online (e.g. a cash dispenser).

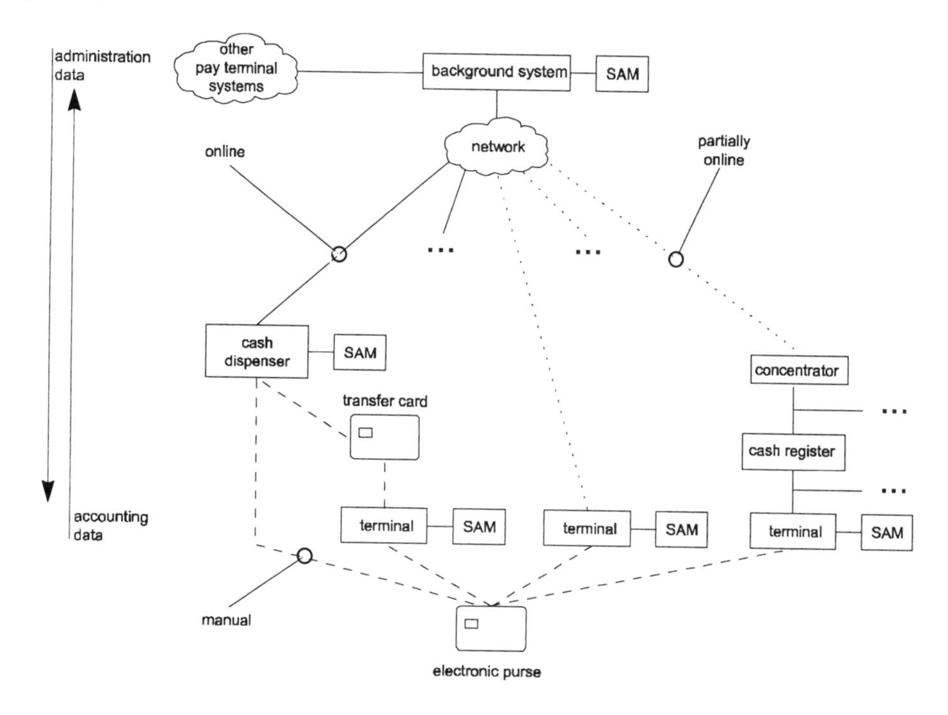

Figure 12.3 Example of system architecture of electronic payment system for electronic purse (SAM – security module)

Figure 12.3 illustrates the system's components and logical links. The background system is connected to the various components via a transparent network. The background system may or may not be part of the same operator's system.

Cash dispensers are mostly used to credit electronic purses and in many cases work online, though in case of a network crash they can also work off-line for a certain time. Therefore they contain their own security module, in which all the keys relevant for operation as well as the derivation procedures are stored.

There are, however, also systems which work completely off-line for payment with the purse. For example, we might mention parking meters and terminals in taxis. A transfer card can be used to transfer the transaction data from the security module to a cash dispenser, from which they reach the background system via the network. In the other direction, the terminal receives current administration data such as warning lists and software modifications.

The second type of payment terminal is connected to the network through an online link set up as necessary. This terminal normally works off-line and periodically connects to the background system, with which it exchanges the accounting and administrative data as necessary. The third type of terminal in the vending field is not connected directly to the network. For example, it might be linked to a supermarket checkout till, which in turn is connected with a concentrator at that store. This concentrator, which is normally a PC functioning as a server, might, for example, be connected to the background system once a day via the network. The necessary data exchange takes place during this connection.

The system just described exists in Austria in a similar form, and is used there with a chip-containing Eurocheque card storing an electronic purse. For larger applications, one may conceive of a distributed system structure with many separate and parallel-running background systems. This would allow the operation of different purse systems with more than one operator, which are compatible with each other.

12.2 PREPAID MEMORY CARDS

One must not neglect memory cards in the context of electronic payment transactions. They are prepared in very large numbers in the form of prepaid electronic purses, and used in many applications. This is unlikely to change over the next few years. Although they will slowly but continuously be replaced by microprocessor Smart Cards, their strength lies in their unbeatably low price. The typical application of debit cards containing a memory chip is the telephone cards widely used in many countries, which are simply discarded once used.

Functionally, memory cards[2] only need a logic unit and an irreversible decrementing counter. In addition, recent versions allow a unidirectional authentication of the card by the terminal. The logic unit is here supplemented by a simple encryption function, whose task is to encrypt a random number received from the terminal via a secret key stored in the card, and subsequently return it to the terminal. Only thus can the terminal ensure that the memory card being used is genuine.

Apart from difficult authentication, memory cards suffer from an additional disadvantage which impedes their implementation as a general electronic payment method. Due to their susceptibility to simple tampering, it is difficult to construct a transaction system which will support various independent application providers (e.g. kiosk, taxi). This derives from the fact that no secure communications are possible between the terminal and the card, and therefore the correct assignment of the paid amounts, and thus system supervision can by definition only be carried out in roundabout ways.

---

[2] Chip structure for memory cards, see also 2.3.1 Memory cards.

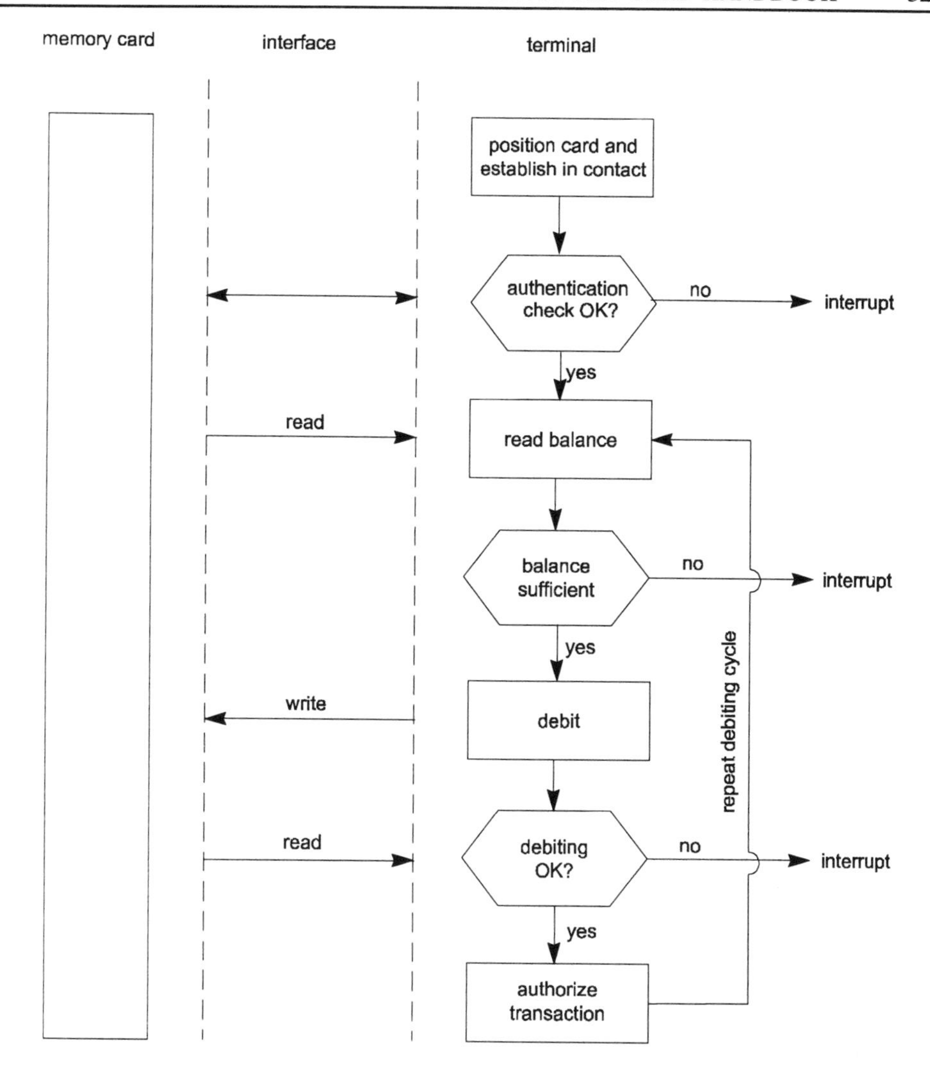

Figure 12.4 Debit procedure of a prepaid memory card seen from the terminal

Memory cards enjoy advantages compared with microprocessor cards in specific and narrowly defined applications due to their low cost, but they are not very suitable for open applications in the field of payment transactions. This is likely to be dominated in the future mainly by Smart Cards, which are significantly more flexible.

12.3 ELECTRONIC PURSES

The idea of implementing an electronic purse on a Smart Card goes back to the very beginnings of Smart Card technology. However, it is only in recent years that this concept has become reality.

If you take a normal purse containing coins and paper money as the basis for an electronic one, it is clear which properties it must possess to satisfy the user. It must be prepaid, and is thus a debit purse. In other words, it cannot be used like a credit card, where payment is only made after receiving the service or goods, but rather like a telephone card which is paid in advance. The process of payment itself must take place quickly and simply, or this method of payment will become unpopular. Furthermore, all purse payments take place anonymously: it is impossible to track down which person bought what and when. The most annoying feature of a purse becomes apparent if it is lost. The money contained in it is irretrievably lost, which in an electronic purse, however, is not necessarily the case.

The greatest advantage of a purse and the cash contained in it, however, is that it is accepted everywhere within a country. It is precisely this factor, though, which is absent in most existing electronic systems. You can only make calls with a telephone card and nothing else. This is typical of closed applications. An ideal electronic purse, on the other hand, is universal, and allows the user to pay for a whole range of services.

12.3.1 CEN standard prEN 1546

The European Commission decided in 1990 to have CEN (Comité Européen de Normalisation) prepare a European standard for a universal electronic purse. Work on the standard started in 1991. Up to now (1996), the separate project teams have spent around 5 work-years on projects connected with this standard. Since very many people have taken part, it is highly unlikely that the standard is free of security defects and thus is semi-certificated. The main parts have been agreed, and after finalization will be published as a European standard.

The CEN standard prEN 1546, entitled "Inter-sector electronic purse", is public, and the processes governing its separate functions are described in great detail. Therefore it is very suitable for illustrating payment and loading routines in an electronic purse. This cannot be demonstrated in such detail in many of the existing systems, since the relevant instructions, processes and interval functions are confidential. Thus this standard is very useful for visualizing the fundamental external and internal routines of an electronic purse.

The typical area of application is apparent in the very first systems based on this standard. The Danish system operator Danmønt has introduced a purse corresponding to this standard into its existing system. In Austria, the widely-issued Eurocheque card carries, besides other applications, an electronic purse which also uses the prEN 1546 standard as a basis.

The standard is divided into four sections. The first, "Definitions, concepts and structures", describes the overall system. The documentation defines and explains in abstract form all the logical components and interrelationships.

The second section uses these concepts to describe the security architecture for both the overall system and its individual components. It covers the mechanisms designed to maintain security, and the possible attacks and necessary countermeasures.

Section three, "Data elements and interchanges", contains the descriptions and definitions of the data needed in the purse. It also covers the instructions and responses used by the card and the security modules.

In the final section, the standard describes the state automata and the states of the devices used. It uses a symbolic representation similar to the well-known flowchart diagram. This form, known as SDL notation, derives from the CCITT Z.100 recommendation[3].

---

[3] See also 4.7 SDL symbolism.

The standard thus contains the complete description of an electronic purse, covering all aspects of the project from Smart Cards through to the terminals with their security modules, and the background and clearing systems. Its objective is the standardization of large electronic purse systems with a very high number of cards and large scale market penetration.

The main advantage of a general standard for electronic purses is that individual and independently operated systems become compatible with each other. As in GSM, the future possibility exists for the user to pay with his card through other purse providers' systems. This is an important precondition for the success of this kind of system. Nevertheless, at this point, a small comment is necessary: the prEN 1546 standard leaves considerable room for manoeuvre as far as the final implementation is concerned, and is more concerned with the framework than with the precise specification of the bits and bytes used. Hence it is perfectly possible for two systems to correspond to the standard in all respects and yet be incompatible with each other, since the chosen crypto-algorithms are different.

The purse providers have overall responsibility, and are also the system administrators. They are comparable to the GSM network operators. The standard specifies the term "purse holders" for the electronic purse users. They are the ones who pay via the application on the card, and in return receive goods or services.

In addition, three other participants are responsible for tasks in the system. The service provider offers goods or services to the purchaser, whilst the acquirer takes over the setting up and administration of the data links between the purse issuer and the service providers. Additionally, acquirers may amalgamate the individual transactions arriving from the payment facility, so that the purse provider only receives collected vouchers. The crediting authority is the service provider's counterpart, in that it can recredit the card against payment.

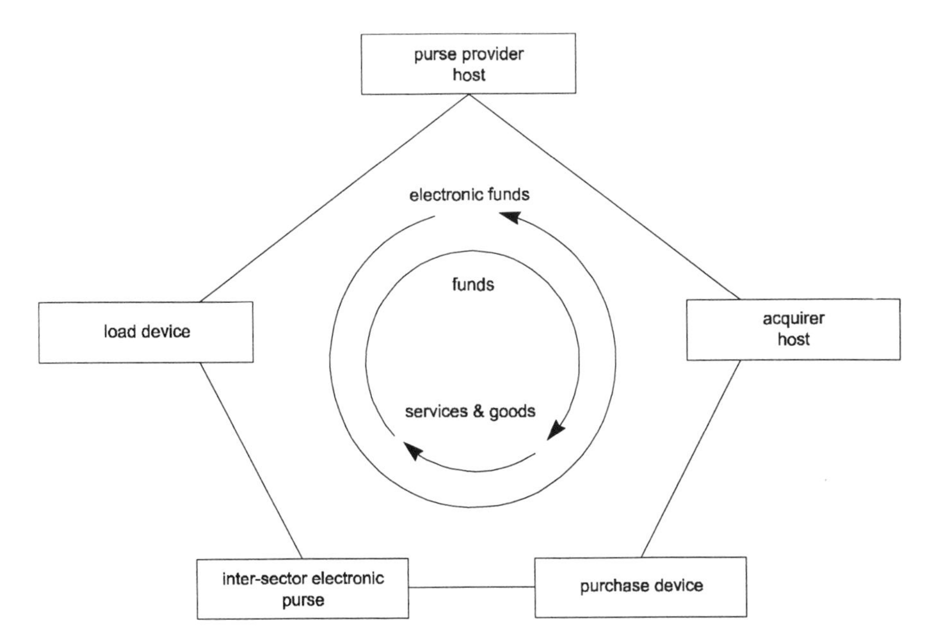

Figure 12.5 Basic structure of inter-sector electronic purse and flow of payments covered by prEN 1546

The five participants listed above need not exist in reality, as their nature is virtual. However, each of them is allocated real and existing technical components, divided by the measure of security they offer. The secure components allow data to be processed and stored in them in such a way that it cannot be manipulated or tampered with from the outside. In the non-secure components such manipulation is possible, at least theoretically. However, the system as a whole is devised so that the manipulation of components identified in Figure 12.6 as non-secure does not compromise overall security.

The initials "IEP", standing for inter-sector electronic purse, denote the standard electronic purse in a Smart Card. The payment facility is used to pay for received goods or services. It is a terminal with keyboard and display, and must have a security module. The standard denotes these uniformly as SAM (secure application module). These secure components contain all secret keys necessary for actions between the IEP and the purse provider's central computer.

Naturally, the keys never leave the security module but are only used internally by the crypto-algorithms. Thus, in many cases, direct links exist only between the system's secure components. The non-secure ones are only used in order to relay the sensitive data transparently .

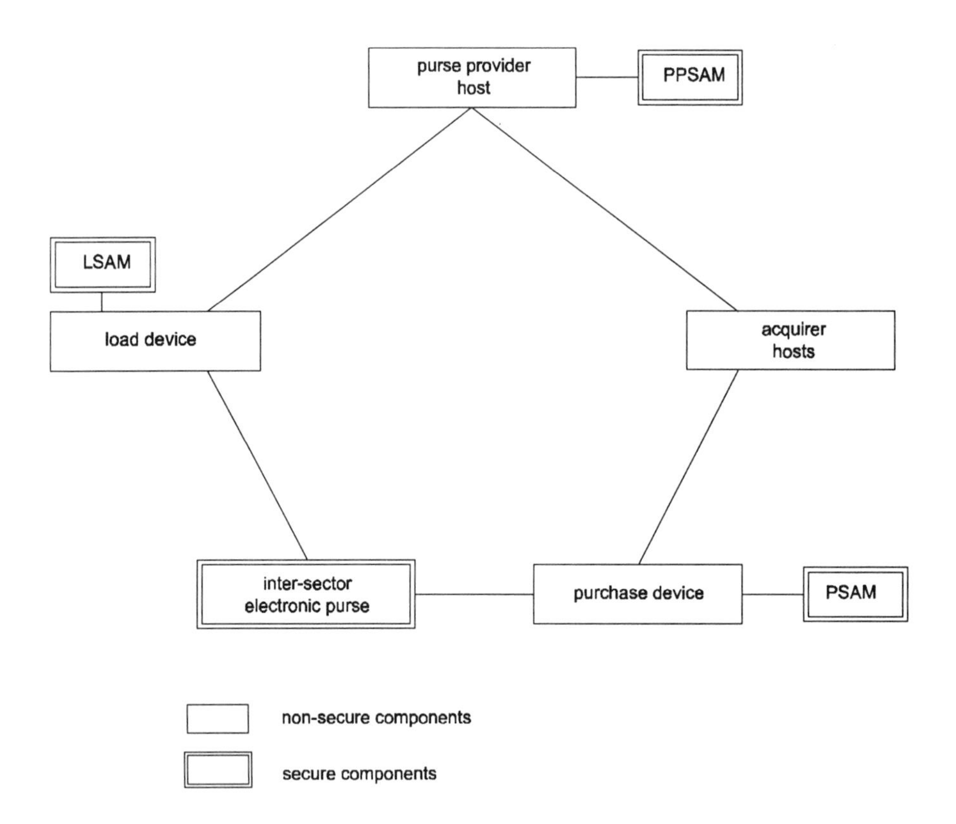

Figure 12.6 Components and connections of electronic purse covered by prEN 1546

Smart Cards made it possible, for the first time, to realize the electronic purse concept. Hence Smart Cards are the central component of the system description in the CEN standard prEN 1546. All the relevant files, instructions, states and routines are defined and described in it. To ensure that not only the Smart Card but the whole of the system is defined, the report contains similar sections encompassing the security modules linked to other components.

However, since the standard is not a specification, the purse provider is naturally allowed a great degree of freedom, and many options. The purse can be constructed to offer the widest variety of functions. For example, it is quite possible to implement a simple system in which only loading and payment with the purse are possible. Further stages may then add payment reversal, alteration of purse parameters and currency conversion. The choice of parameters to be included, among the many options offered by the standard, is largely left to the purse provider. The following describes the main aspects of the IEP application in a Smart Card.

Data elements

An identification of all data elements was introduced to make unambiguous specification possible for the data used system-wide in an electronic purse application. As a result of the very concise designations, data flow and processing can be represented simply and unambiguously in a mathematically correct format. The standard also contains a simple data dictionary, which lists the relevant data contents and corresponding formats for the standardized data elements.

Files

The complete electronic purse application is contained in its own DF in the card. All the files necessary for smooth operation are contained in it. In addition, data concerning the card, the chip, further applications and so on are stored in several files directly under the MF.

The data elements necessary to run the purse are located in six EF files within a purse directory. Figure 12.8 is a summary of the files and the data elements contained in them.

The purse's global parameters are laid down in EF_{IEP}, forming the basis for all transactions taking place. The file EF_{IK} contains specific data for every available key. EF_{BAL} contains the current amount still in the purse which is at the user's disposal. The three protocol files are there exclusively for logging all transactions separately by function. They alone make a transaction balance and error detection possible. There are individual files for loading, paying, alteration of purse parameters and currency conversion. Their structure is always cyclic, thus logging and storing the latest transactions.

Instructions[4]

The files form the basis of the purse, on which the instructions are built. Eight different files are necessary for operation, of which three are covered by the ISO/IEC 7816-4 standard: SELECT FILE, READ BINARY and READ RECORD. They are only used to select the purse application with an AID, and subsequently to read a variety of purse data from the files as required.

---

[4] For a detailed explanation of the instructions see 7.9 Instructions for electronic purses.

Data elements	Description
ALG_{IEP}	cryptographic algorithm used by an IEP
AM_{IEP}	authentication mode required by an IEP
AP_{IEP}	application profile of an IEP
BAL_{IEP}, BAL_{PSAM}, BAL_{PPSAM}	balance of an IEP/PSAM/PPSAM
$BALmax_{IEP}$	maximum balance of an IEP
CC_{IEP}, CC_{PSAM}, CC_{PPSAM}	completion code from the IEP/PSAM/PPSAM
CT	collection status
$CURR_{IEP}$, $CURR_{LDA}$, $CURR_{PDA}$	actual currency for an IEP/LDA/PDA
$DACT_{IEP}$	activation date of an IEP
DD	discretionary data
$DDEA_{IEP}$	deactivation data of an IEP
$DEXP_{IEP}$	expiry date of an IEP
ID_{IEP}, ID_{PSAM}, ID_{PPSAM}	identifier for an IEP/PSAM/PPSAM
IEP	inter-sector electronic purse
IK_{IEP}, IK_{PSAM}, IK_{PPSAM}	key information for an IEP/PSAM/PPSAM
LDA	load device application
LSAM	load SAM
M_{LDA}, M_{PDA}	transaction amount for load/purchase
$MTOT_{IEP}$, $MTOT_{PSAM}$	total transaction amount for a purchase
NC	number of collection
NI	number of individual transactions
NT_{IEP}, NT_{LSAM}, NT_{PSAM}	transaction number for an IEP/LSAM/PSAM
PDA	purchase device application
PP_{IEP}, PP_{PSAM}, PP_{PPSAM}	purse provider identifier for an IEP/PSAM/PPSAM
PPSAM	purse provider SAM
PSAM	purchase SAM
R	random number
S_1	IEP signature
S_2	PSAM/PPSAM signature
S_3	IEP signature
S_4	PSAM signature
SAM	secure application module
TM	total amount
TRT	transaction type and status

Figure 12.7 Overview of important standardized data elements of prEN 1546

Data	Function	Data elements contained
EF_{IEP}	Purse fixed data and parameters	PP_{IEP} identifies IEP purse provider ID_{IEP} IEP identification $DEXP_{IEP}$ IEP expiry date $DACT_{IEP}$ IEP activation date $DDEA_{IEP}$ IEP deactivation date AM_{IEP} IEP authentication mode AP_{IEP} IEP application profile DD user-specific data
EF_{IK}	Information on all keys	ALG_{IEP} IEP cryptographic algorithms IK_{IEP} IEP key information DD user-specific data
EF_{BAL}	Purse holdings	BAL_{IEP} IEP balance $CURR_{IEP}$ IEP currency $BALmax_{IEP}$ IEP maximum balance DD user-specific data
EF_{TFIELD}	Transaction field	NT_{IEP} IEP transaction number
EF_{LLOG}	Protocol file for loading	TRT transaction type and condition NT_{IEP} IEP transaction number BAL_{IEP} IEP balance (new balance) M_{LDA} transaction amount for loading $CURR_{LDA}$ currency for loading ID_{PPSAM} identifies purse provider for PPSAM CC_{IEP} IEP completion code DD user-specific data
EF_{PLOG}	Protocol file for payment	TRT transactin type and condition NT_{IEP} IEP transaction number BAL_{IEP} IEP balance (new balance) M_{PDA} transaction amount for payment $CURR_{PDA}$ currency for payment ID_{PSAM} identifies purse provider for PPSAM NT_{PSAM} PSAM transaction number CC_{IEP} IEP completion code DD user-specific data

Figure 12.8 The necessary files and their data elements for the electronic purse, in accordance with prEN 1546. The protocol files for exchange rate calculations and parameter changes are not represented

The other five instructions were developed specially for use in electronic purses. They are always used in pairs in individual transactions, since their principle of operation is similar to that of mutual authentication. At the same time, the data needed for the transaction are exchanged. The instructions and responses are, of course, so structured that manipulation at the interface between the card and terminal is immediately detected, followed by an immediate interruption of the transaction with appropriate logging.

All purse instructions directly access data elements in the purse files, for both reading and writing. The files are first automatically selected by the operating system. For example, basic purse data is sometimes needed during the execution of instructions. The operating system then selects the EF_{IEP} file, and the required data element is made available to the instruction. All transactions and the most important data relating to them are written to the relevant protocol files as part of the instruction–response cycle.

The instructions listed in Figure 12.9 are defined by prEN 1546, and their internal functions in the card precisely laid down.

Command	Function
INITIALIZE IEP	initialization for following purse command
CREDIT IEP	loading purse
	confirms payment made
	error correction
DEBIT IEP	payment through purse – confirms payment
CONVERT IEP CURRENCY	converts currency
UPDATE IEP PARAMETER	within purse parameters

Figure 12.9 Specific commands for electronic purse covered by prEN 1546

The standard does not provide instructions for PIN checking or modification, since this is unnecessary for the purse's correct functioning. However, if required, additional instructions for PIN checking and administration can be introduced into the purse application, without interference or problems with existing ones.

States

As may already be apparent from the summary of instructions, each transaction consists of an introductory initialization instruction and a further instruction which closes the transaction. In order to fix the sequence of instructions, the state diagrams lay down the necessary states and state transitions in the application. This naturally requires the card to contain a state automaton. Depending on its current state, the card will accept various instructions or block them.

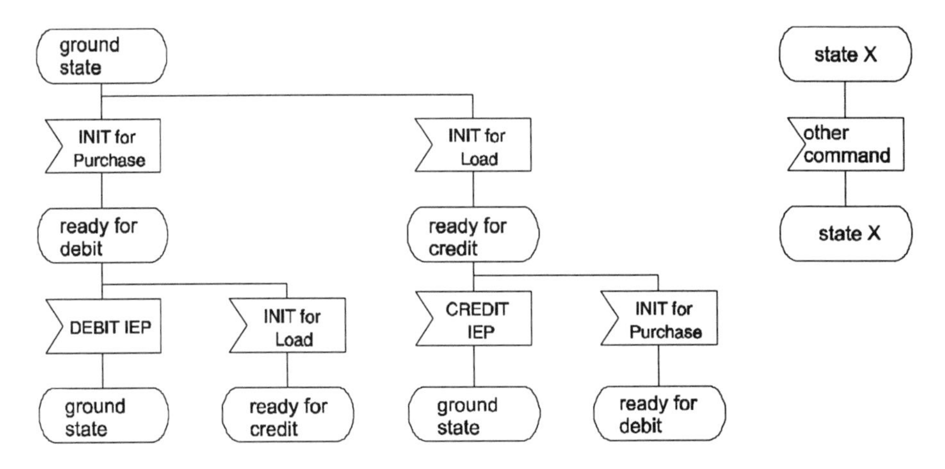

Figure 12.10 Simplified diagram of conditions for loading and payment using electronic purse, in accordance with prEN 1546

Cryptographic algorithms

The system's entire security rests on a cryptographic algorithm. The messages exchanged between the components are all provided with a subsequent signature, so that any tampering can be recognized. This is the only protection for messages which are always exchanged in plaintext[5].

The message exchange is so arranged that every algorithm can be used to generate the signature. The symmetric DES algorithm is currently the most commonly used signature, but the standard also allows asymmetric ones such as RSA or DSS. This independence is a great advantage, since it considerably extends the standard's projected life and flexibility.

Execution

The standard not only specifies files, instructions and states but also describes and explains the relevant execution routines. These are laid down in a BASIC-like pseudo-language, based on the data elements. This was necessary since the security of some processes depends strongly on the order of internal instruction execution. For example, during transactions the appropriate protocol file must always first be changed to the new state before the terminal receives a response. The routines and transactions are precisely defined for all components. Descriptions are available for the following routines in a Smart Card with a purse application:

- Loading
- Payment
- Payment reversal
- Error correction
- Currency conversion
- Change of purse parameters

In addition, the files' read instructions can also, of course, be used for monitoring purposes. The actions involved may, however, vary between purse providers, depending on the purpose of the monitoring.

All runs are basically divided into three phases. In phase 1, complete initialization of the participating components takes place. Phase 2 serves for the actual execution of the function involved, whilst phase 3 is optional, and is used to confirm the previous routines. The successful execution of phases 1 and 2 contains the unidirectional, or optionally mutual, authentication of the two components.

---

[5] In the case of a DES algorithm, the messages are followed by a 4-byte MAC as a signature.

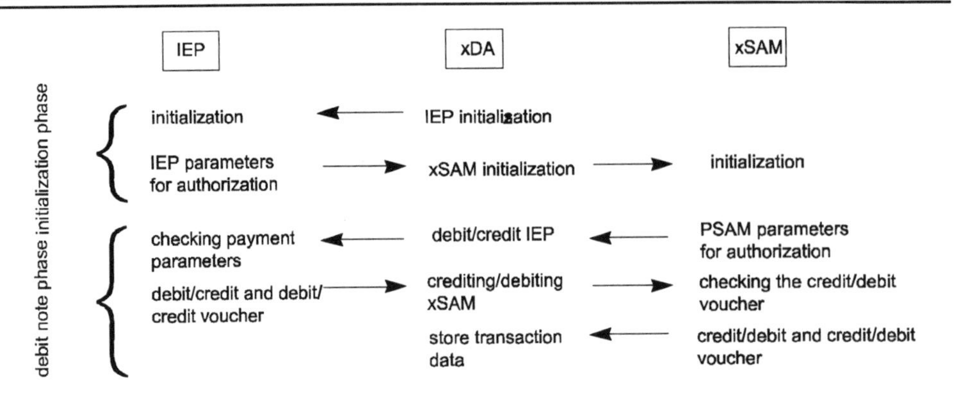

Figure 12.11 Basic process of transaction using electronic purse, in accordance with prEN 1546 (phases 1 and 2)

During all runs, the actual purse function (payment, loading, etc.) and the unidirectional or mutual authentication are nested in each other. This is intended to minimize the time needed for transaction and increase security, since considerably fewer instructions are necessary. This is comparable to the ISO standardized instructions INTERNAL AUTHENTICATE and EXTERNAL AUTHENTICATE, which may be replaced by the non-standardized instruction MUTUAL AUTHENTICATE without losing any functionality or security.

Abbreviation	Meaning
Ax	describes an action
Cx	describes a command
Parameters (...)	request to a participant with the data elements given
Response (...)	response to previous request with the data elements given
Rx	describes a response
Sign (...)	creation of signature for data elements given
Verify (...)	checks data elements given or a function
Write (...)	writes file with data elements given

Figure 12.12 Abbreviations of functions and processes for illustration of payment transactions in accordance with prEN 1546

The process of loading

In order to be able to pay with an electronic purse, it must first be credited, or loaded, as shown below and in Figure 12.13. This is the version in which the IEP purse is loaded directly at the terminal by an online background system with the corresponding PPSAM security module. The standard provides for other options such as loading the purse via an

LSAM security module in a terminal. However, the process described here is common in current systems, since it allows control of loading to rest completely with the system operator.

In this example, an IEP electronic purse is loaded via an LDA terminal by the background system with the PPSAM security module. The user first inserts the card in the terminal, which then resets itself. In the ATR, the IEP notifies the terminal of various global parameters for the subsequent communication process. Then the terminal selects the purse DF on the card. Once this is successfully achieved, the user pays into the terminal the amount to be credited in the acceptable currency. This data is then sent to the PPSAM together with the first purse instruction. The PPSAM checks the specified currency unit and the amount still possible to load. In response, it returns three data elements to the terminal.

The terminal extends the data received from the PPSAM by the loading amount M_{LDA} as well as the corresponding currency $CURR_{LDA}$, and sends them to the IEP card with the instruction INITIALIZE IEP for Load. The card tests, among other things, whether the new purse balance resulting from the amount loaded would exceed the maximum $BALmax_{IEP}$ stored in the card. Should this not be the case, the IEP increments a transaction counter NT_{IEP}, computes the session key $KSES_{IEP}$ and calculates a signature S_1. These are returned to the terminal together with a few other data elements.

The terminal's task after this instruction is only to relay the received data elements to the PPSAM. Here the data is checked against the permitted range of values, and both the card-specific key KD_{PPSAM} and the session key $KSES_{PPSAM}$ are generated. If the subsequent S_1 check is successful, then the card is authenticated as it must know the secret key for calculating S_1. The PPSAM then prepares a signature S_2 and sends it to the terminal together with the key information IK_{PPSAM}. Again, this only relays the data to the card, this time with the instruction CREDIT IEP.

The card now checks the signature S_2. If this test is successful, the PPSAM has also been authenticated by the IEP. The balance BAL_{IEP} is then updated in the purse. The IEP then produces a third signature S_3, which is sent to the terminal to confirm successful updating of the balance. The final instruction to the PPSAM provides this signature, which concludes the whole loading transaction.

The routine just described is one of many possible variations. It is often used in practice, since it is very common to carry out purse loading transactions online. There are prEN 1546 variants which describe loading via a special loading security module, LSAM. It can be built in a decentralized fashion into special loading terminals, such as cash dispensers.

The process of payment

The following example (Figure 12.14) demonstrates payment with the relevant components: electronic purse (IEP), terminal (PDA) and security module in the terminal (PSAM).

After inserting the card in the terminal, the latter performs a reset and requests an ATR from both the PSAM and the IEP. If the ATR does not concord with the expected value, the terminal interrupts the payment procedure. If it does, the terminal selects the purse DF in the IEP. Once again, if this cannot be completed then the procedure is interrupted. However, for reasons of clarity, neither this point nor the general error handling approach are listed.

IEP		**LDA**		**PPSAM**
R1: Response (ATR)	← →	C1: RESET		
R2: Response (CC_{IEP})	← →	C2: SELECT Parameters (DF_{IEP})		
		A1: Input ($M_{LDA} \Vert CURR_{LDA}$)		
		C3: INITIALIZE PPSAM for Load Parameters ($M_{LDA} \Vert CURR_{LDA}$)	→ ←	A2: Verify ($CURR_{LDA}$) Verify ($BAL_{PPSAM} \geq M_{LDA}$) Generate R R3: Response ($PP_{PPSAM} \Vert ID_{PPSAM} \Vert R$)
A3: Verify (PP_{PPSAM}) Verify ($CURR_{LDA}$) Verify ($BAL_{IEP} + M_{LDA} \leq BALmax_{IEP}$) $NT_{IEP} := NT_{IEP} + 1$ $KSES_{IEP} = f(KD_{IEP}, DEXP_{IEP}, NT_{IEP})$ $S_1 := Sign(PP_{IEP} \Vert ID_{IEP} \Vert DEXP_{IEP} \Vert NT_{IEP} \Vert M_{LDA} \Vert CURR_{LDA} \Vert BAL_{IEP} \Vert ID_{PPSAM} \Vert R)$ Write (EF_{LLOG}) R4: Response ($PP_{IEP} \Vert ID_{IEP} \Vert ALG_{IEP} \Vert IK_{IEP} \Vert DEXP_{IEP} \Vert NT_{IEP} \Vert BAL_{IEP} \Vert S_1 \Vert CC_{IEP}$)	← →	C4: INITIALIZE IEP for Load Parameters ($PP_{PPSAM} \Vert ID_{PPSAM} \Vert R \Vert M_{LDA} \Vert CURR_{LDA}$)		

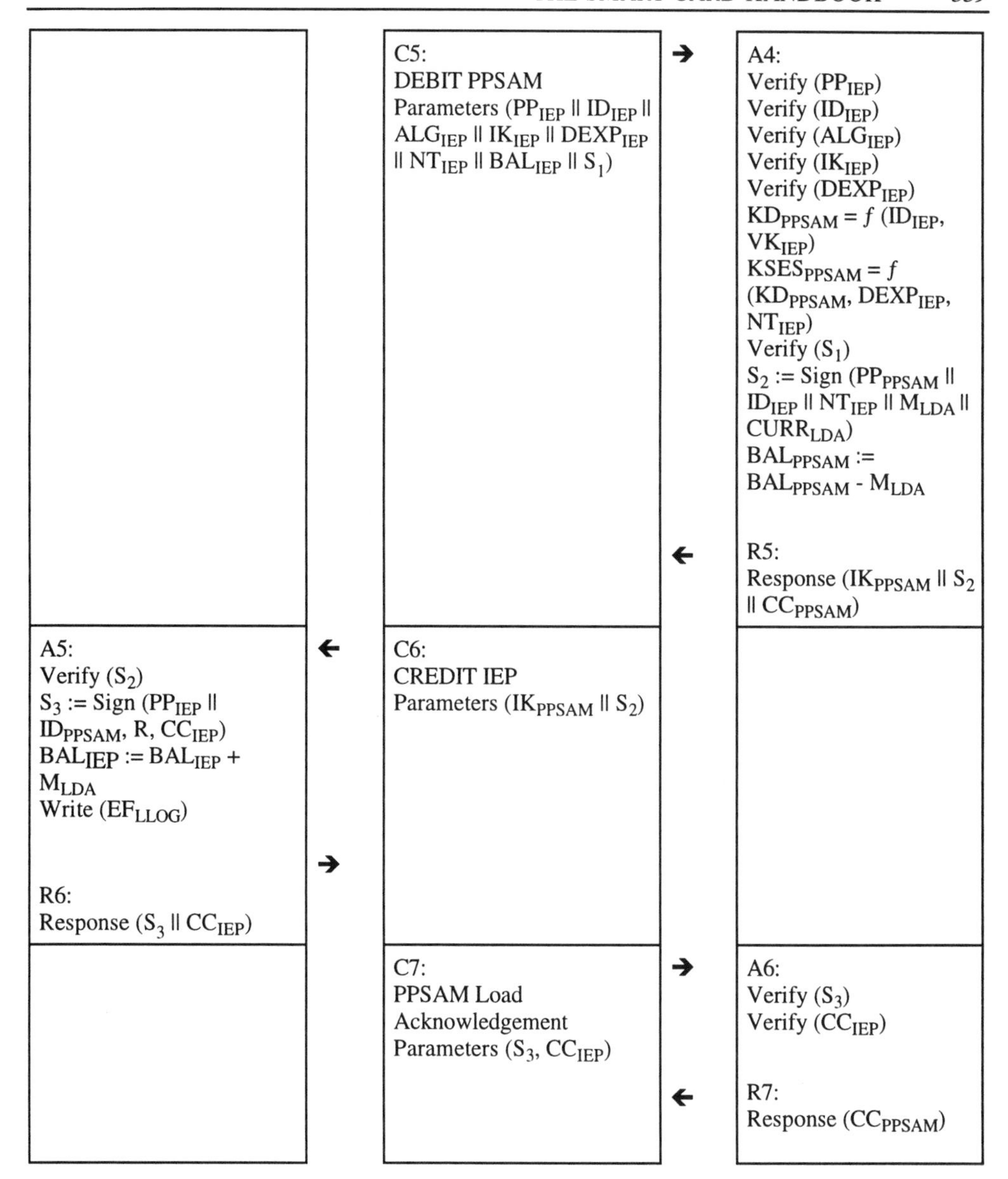

Figure 12.13 Transaction in loading of electronic purse (IEP) to a terminal (LDA) on-line, with a security module (PPSAM)

IEP		**PDA**		**PSAM**
		C1: RESET	→ ←	R1: Response (ATR)
R2: Response (ATR)	← →	C2: RESET		
R3: Response (CC_{IEP})	← →	C3: SELECT (DF_{IEP})		
A1: $NT_{IEP} := NT_{IEP} + 1$ $KSES_{IEP} = f\ (KD_{IEP},$ $DEXP_{IEP}, NT_{IEP})$ $S_1 := \text{Sign}\ (PP_{IEP} \parallel ID_{IEP}$ $\parallel DEXP_{IEP} \parallel NT_{IEP})$ Write (EF_{PLOG}) R4: Response ($PP_{IEP} \parallel ID_{IEP}$ $\parallel ALG_{IEP} \parallel IK_{IEP} \parallel$ $DEXP_{IEP} \parallel CURR_{IEP} \parallel$ $AM_{IEP} \parallel NT_{IEP} \parallel S_1 \parallel$ CC_{IEP})	← →	C4: INITIALIZE IEP for Purchase Parameters ()		
		C5: INITIALIZE PSAM for Purchase Parameters ($PP_{IEP} \parallel ID_{IEP} \parallel$ $ALG_{IEP} \parallel IK_{IEP} \parallel DEXP_{IEP}$ $\parallel CURR_{IEP} \parallel AM_{IEP} \parallel$ $NT_{IEP} \parallel S_1$)	→ ←	A2: Verify (PP_{IEP}) Verify (ID_{IEP}) Verify (ALG_{IEP}) Verify (IK_{IEP}) Verify ($DEXP_{IEP}$) Verify ($CURR_{IEP}$) Verify (AM_{IEP}) $NT_{PSAM} := NT_{PSAM} + 1$ $KD_{PSAM} = f\ (ID_{IEP},$ $VK_{IEP})$ $KSES_{PSAM} = f\ (KD_{PSAM},$ $DEXP_{IEP}, NT_{IEP})$ Verify (S_1) $MTOT_{PSAM} := 0$ $S_2 := \text{Sign}\ (PP_{PSAM} \parallel$ $ID_{PSAM} \parallel NT_{PSAM} \parallel$ $MTOT_{PSAM} \parallel ID_{IEP} \parallel$ $AM_{IEP} \parallel NT_{IEP})$ R5: Response ($ID_{PSAM} \parallel$ $NT_{PSAM} \parallel IK_{PSAM} \parallel S_2 \parallel$ CC_{PSAM})

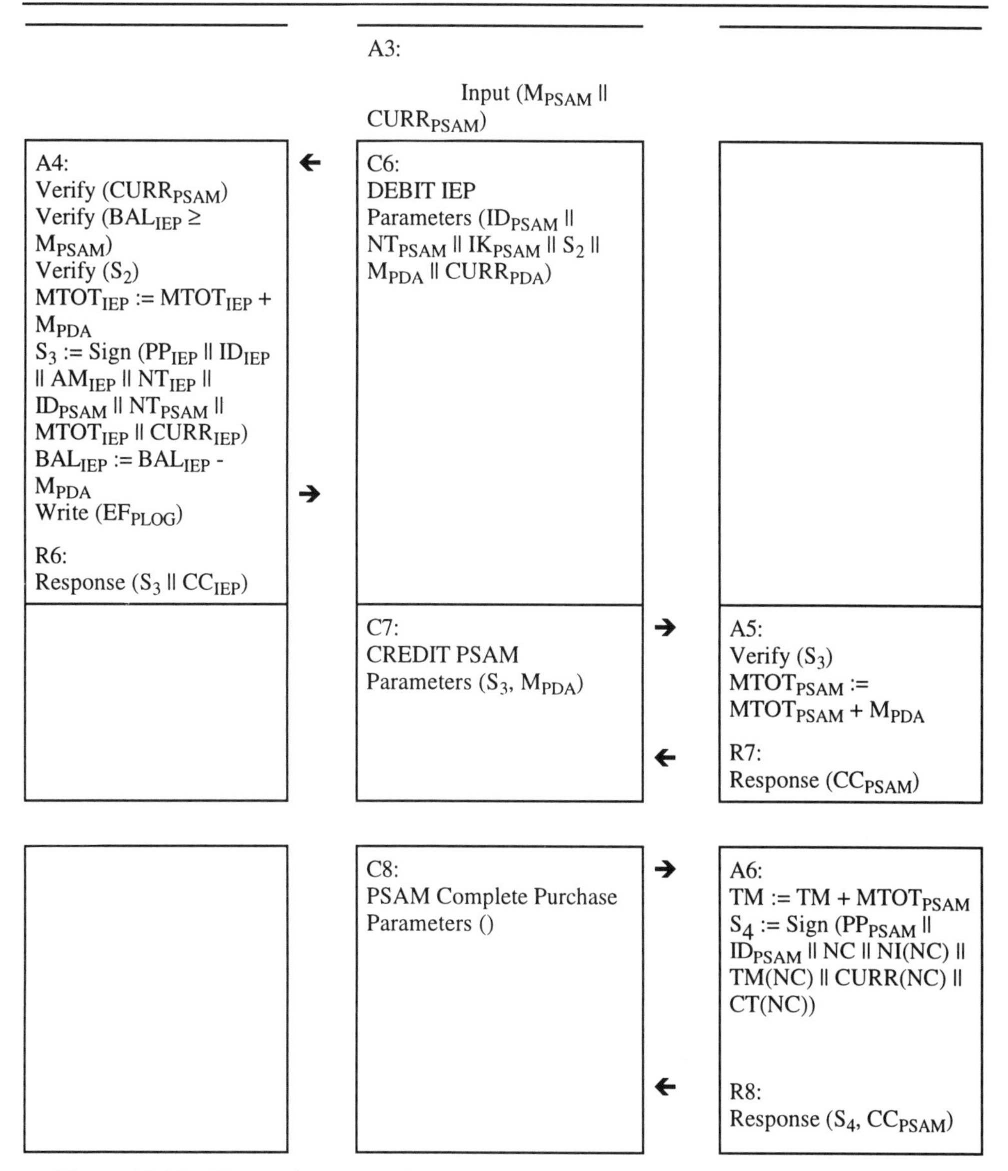

Figure 12.14 Transaction process in payment through an electronic purse (IEP) to a terminal (PDA) and security module (PSAM) in accordance with prEN 1546

After selecting the purse DF in the IEP card, the terminal sends the initialization instruction INITIALIZE IEP for Purchase. The card receives this instruction, increments the transaction counter and finally uses various data elements to produce a signature S_1. It then sends this data and the signature to the terminal.

The terminal then sends the initialization instruction to the PSAM: INITIALIZE PSAM for Purchase. This only involves relaying all the data which arrived from the card in the previous instruction forward to the PSAM. The latter now checks this data (i.e. the execution data $DEXP_{IEP}$, currency $CURR_{IEP}$, cryptographic algorithm used ALG_{IEP}) and the other data received, with the values stored in the PSAM. If all the comparisons are successful, the transaction counter NT_{IEP} is incremented. If even one comparison fails (e.g. the IEP has expired), further processing is immediately interrupted and an appropriate return code is sent to the PDA terminal.

The PSAM now carries out a key derivation with the data sent by the IEP as well as a session key generation, and then checks the signature S_1. If correct, it follows that all the data sent is authentic, and the card is simultaneously authenticated by the PSAM. In other words, the PSAM knows that the card containing the electronic purse is genuine.

In the next step, the PSAM also produces a signature S_2, which is sent to the terminal together with a few additional data elements. The amount to be paid M_{PDA} and the relevant currency $CURR_{PDA}$ are now entered at the terminal. The terminal sends the M_{PDA} and the data previously received from the PSAM to the card, using the instruction DEBIT IEP. The card now checks if there is sufficient money stored in the purse to make the payment. If so, then the signature S_2 is checked. If the signature is correct, then the data has not been tampered with during transmission and the PSAM has also been authenticated by the IEP. This is the case since only a genuine PSAM can have the secret key necessary to generate the signature S_2. The appropriate amount is subtracted from the purse balance, the protocol file is updated and a third signature, S_3, is produced to confirm the debit transaction just performed.

S_3 as well as the debited amount are sent by the terminal to the PSAM, which checks the former. If this signature is correct, the amount debited in the IEP is added to an internal data element $MTOT_{PSAM}$. A further instruction "PSAM Complete Purchase" causes the PSAM balance to be updated, by adding $MTOT_{PSAM}$ to the purse balance TM. Finally the PSAM receives a signature S_4 to confirm correct payment.

The process just described is a very simple variant of the different payment procedures listed in prEN 1546. Other variations exist, including a special faster debiting for phonecards, as well as a variant with an additional acknowledgement option at the end of the transaction.

The files, instructions and routines described above for the card are also specified for all other important system components. This applies above all to the security module, since system security relies solely on it. Statistical methods may be employed to monitor the overall running of the system, which in the case of large applications may consist of tens of thousands of terminals with many hundreds of thousands of Smart Cards. To produce complete accounting every time for each card would conflict with the demand for anonymity, and would also be far too computation-intensive. However, as trials have proved, supervision of the flow of funds based on random samples can continually test overall system security at an acceptable cost.

The European standard prEN 1546 laid the foundation stone for a universal Smart Card electronic purse system. Almost all procedures and functions currently in use and those considered to be plausible were integrated in the standard. Only one function has so far not yet been described, in spite of being very significant for the card user. This is the so-called purse-to-purse transaction, that is, the transfer of electronic funds from one purse to another. At the moment prEN 1546 does not describe such a transfer. Nevertheless, various institutions are attempting to integrate this type of payment transaction into the standard.

12.3.2 The Mondex system

There are currently only a few large payment systems worldwide which use Smart Cards as a central component. Only a fraction of these systems are based on an electronic purse, in which the monetary units are stored directly in the card and not in a background system. Only one of these systems offers electronic payment transactions corresponding to those using normal money. This is the Mondex system.

The idea behind this concept, currently unique in the world, was born in 1990. After five years of R&D, a field trial was launched in July 1995 in Swindon, Wiltshire. Swindon is a town of 190***000 inhabitants about 60 miles west of London, and so far 8000 Smart Cards (of the 40***000 planned) and 700 trader terminals (1000 planned) are in use. The widest variety of retailer categories, ranging from news-stands and snack bars through to supermarkets, travel agents, filling stations and telephones, are taking part in the trial. The maximum sum in the purse was fixed at £500 for the trial, but this value can in principle be increased to any desired level.

Mondex is a consortium of three companies: British Telecom, National Westminster Bank and Midland Bank. Its purpose is to create a means of payment which can be used like cash but without the drawbacks. The results of this technical development will then be sold as a franchise to banks and other companies. Mondex is currently one of the few electronic purse systems offered as a complete service, from the card to the background system. This is also the reason why there are relatively high expectations for this technology.

The system

Since the Mondex purse is designed to be used in the same way as conventional money, purse-to-purse transactions are naturally possible. This allows card holders to pay each other without the intervention or knowledge of a bank or similar organization. The system is completely open and anonymous, and as many participants can be involved as desired. Figure 12.15 shows the possible flow of funds and the participants.

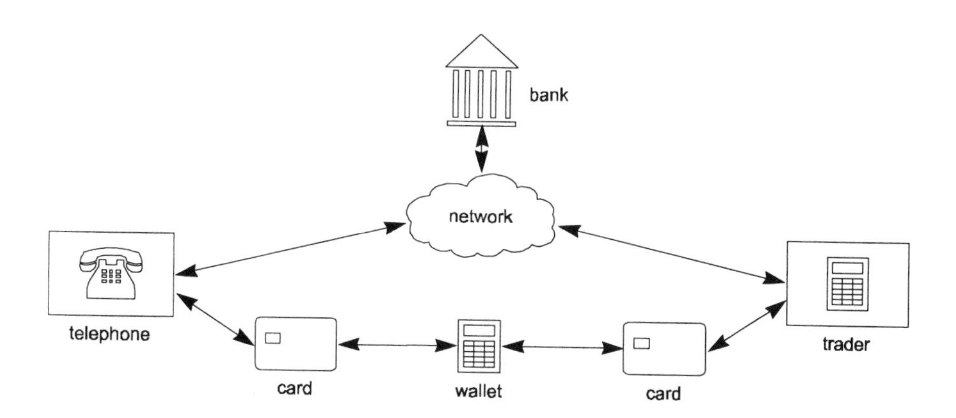

Figure 12.15 Flow of funds and participants in electronic transactions in the Mondex system

The electronic purse is located in the chip of a conventional ID-1 card with contacts. In order to inspect the balance, there is a matchbox-size key ring with a display. After inserting the card into this mini-terminal, the current purse balance and the last ten transactions can be viewed. If one wishes to give another cardholder some money from the purse, then the so-called wallet needs to be used. This is a device similar to a pocket calculator with a small keyboard and display. It also contains a built-in security module and a card reader for the purse. To perform a purse-to-purse transaction, the Smart Card is first inserted in the wallet. The amount to be transferred is input, and it is relayed from the purse to the wallet's security module. The second card is then inserted, and the amount transferred to it from the security module. This completes the whole transaction.

An additional device in this payment system is a telephone with a built-in card reader. It allows money to be transferred over the line during the call. The typical application is the ordering of goods from a mail-order catalogue. Payment is carried out when the order is placed. This technique naturally also allows the loading of the purse over the phone, or a transaction between two cardholders. When loading the card from a bank account, a four-figure PIN must, of course, be entered at the same time for security reasons, to protect the account holder against unauthorized withdrawals.

Figure 12.16 "Wallet" terminal for purse-to-purse transactions, parameter changes and "Balance Reader" last transaction and current balance on Mondex electronic purse (Mondex)

Each electronic purse can hold up to five different currencies. As soon as the balance corresponding to one currency reaches zero, it is possible to replace it with a new one. The purse can be blocked with a simple instruction and unblocked by entering a four-figure PIN, so that no unauthorized use is possible.

The trader's terminal contains, as a security module, the same type of Smart Card as that held by the customer. It is thus also possible to pay for other goods in turn with the trader's security module. Interestingly, this might make the theft of such a card worthwhile, as it can then be used just like a normal purse. However, this problem has been recognized and a solution provided. The traders' cards can be so configured to allow only the receipt of electronic money. Debiting is only possible during an online transaction with the trader's bank. It is evident that electronic money is not necessarily safe from theft. If the trader's terminal is able to set up an online link to the bank, it is possible to arrange for automatic transfer from the card to the trader's account once a particular amount has been reached.

Security mechanisms and the process of payment

All the specifications which describe the process of payment, as well as the Mondex system's security model, are confidential. This makes it very difficult to obtain precise technical details of the system and its individual components. Therefore the following can only provide a general technical summary, and illustrate only a few of the mechanisms and procedures used by the system.

The microprocessor currently used is a Hitachi H8/3102. For mass production, a processor specially developed for Mondex is planned, with a numerical arithmetic unit and appropriate memory size, since the application requires around 5 kbytes of memory in EEPROM. It is probable that the cryptographic algorithm used is a symmetric one, such as DEA. Due to the fact that a special processor with an arithmetic unit will be implemented in future, it is safe to assume that this will be changed to an asymmetric algorithm for increased security, like RSA. In principle, though, the system is independent of the crypto-algorithm used, since it does not rely on special properties of particular algorithms but only uses (digital) signatures to protect data transmissions. In this it differs little from the universal European electronic purse after prEN 1546[6].

Since the Mondex system is operated in a completely decentralized fashion, a special procedure for replacing key versions and algorithms must be present. Each issued card contains at least two totally different crypto-algorithms with several corresponding keys. If it is necessary to switch to another key version or even to use a different algorithm, an appropriate parameter is set in all Smart Cards which link online to the background system. However, these cards in turn have the capability of setting this parameter in all cards with which they execute a payment transaction. This snowballing effect, due to the exponential increase in data redistribution, leads within a very short time to a system-wide switch to the new global parameters. This would even be the case if the background system only modified the parameter in a single card. This is a very effective, fast and simple method of changing global data in a decentralized payment system.

Naturally it must be possible to isolate particular cards in the system. This can be done in three different ways. Suspicious cards may be identified in warning lists, detected at terminals and retained. This, however, is usually only possible at cash dispensers, as they alone possess the technical means needed for retaining cards. The warning lists are loaded into all the terminals which are then able to block cards, so that they can no longer be used for transactions. All electronic purses issued allow, in principle, only a limited number of

---

[6] See also 12.3.1 CEN standard prEN 1546.

transactions, after which they are automatically blocked. They can be reinstated by means of an online query after checking the warning lists, so that the card need not be replaced. This ensures that a card containing an electronic purse cannot be used without any control by the background system.

A typical payment transaction between two Smart Cards in the Mondex system is divided into two phases, which are represented in Figure 12.17. In phase one, a registration of the current transaction takes place, during which all the data required for the subsequent money transfer is exchanged. In the second, "value transfer" phase, Smart Card 2 informs Smart Card 1 of the requested sum. The complete data set is signed digitally, so it cannot be manipulated during transmission. After receipt, Smart Card 1 checks the signature to verify the authenticity of Smart Card 2, and to authenticate the data transmission. If all checks are satisfactorily concluded, the requested amount is debited on Smart Card 1 and sent, together with a digital signature, to Smart Card 2. This signature is checked by Smart Card 2 for the purposes described above. The amount is now credited. Smart Card 2 then produces a confirmation of correct credit, adds to it a digital signature and sends it to Smart Card 1. On receipt of this confirmation of payment and after a final, successful check, the transaction is completed.

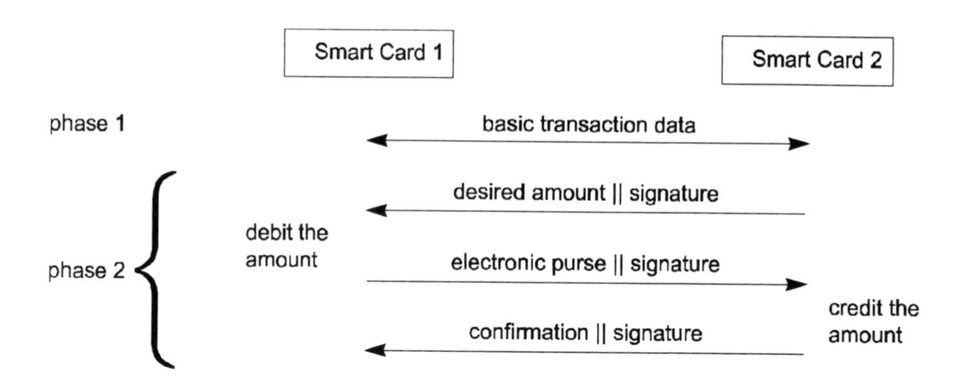

Figure 12.17 Flow of information during transaction between two Smart Cards in the Mondex system

Both cards contain protocol files and have appropriate procedures so that if a transaction is interrupted, it can be correctly continued from the same point. These error recovery procedures are very important, as otherwise it may happen that electronic funds are destroyed by an interrupted transaction. Both participating cards have three separate protocol files each, in which transaction-related data is stored. The transaction log stores data relating the last ten successful payment transactions. The pending log contains all the data which comes up during a transaction and which is needed if an error recovery process is carried out. The exception log stores all transactions which were not completed successfully. Once all the records in this file are written to, the Smart Card is automatically locked, and the card holder needs to unlock it via an online transaction. All the file entries are then loaded into the background system, analysed, and the entries in the protocol file erased.

Summary

The Mondex system is currently the only completely open electronic payment system using an electronic purse. It permits all transactions which are possible with normal cash. Over and above that, it also allows payment transactions across diverse telecommunication media such as the phone. Of course, if the card containing the purse is lost, the money held in it is also lost, just like a real purse containing cash. However, this makes the system completely anonymous, which is sure to bolster user acceptance.

Since the system's principle of operation makes it impossible to demand a fee for each individual transaction, naturally we may ask what income the operator derives from it. The investment needed for setting it up and operating it is not exactly trivial. In the Swindon field trial, each electronic purse user is charged a fairly reasonable fee of £1.50 per month. The same applies, of course, to the traders. Although it is possible to charge the latter transaction clearing fees, the system's completely open nature allows them simply to utilize the electronic money received against purchases from their own suppliers. However, the system operator could also finance it by offering various services to the card holders as well as the traders.

It remains to be seen how the future, real Mondex system will be financed. It is likely to influence the electronic payment market quite markedly in the coming years. In the international arena there are several large banks which are already publicly considering the introduction of such a system. We can expect further interesting developments.

12.4 CHIP-CONTAINING CREDIT CARDS

Smart Card specifications and standards covering the most diverse fields of application have existed for many years. However, traditionally they concentrated on the field of telecommunications, with telephone and GSM cards. This situation has changed markedly in the last two or three years. The European electronic purse standard prEN 1546 is a good example of this development. The most important representative in the area of payment transactions, though, is the EMV specification, so named after its three initiators Europay, Mastercard and Visa. It describes in detail all aspects of credit cards containing microprocessor chips. A corresponding specification for the terminals accepting these cards is also now available.

The three international credit card companies Europay, Mastercard and Visa started work on the specification, known as "IC Card Specifications for Payment Systems", in the autumn of 1993. Version 1 was published relatively quickly, in October 1994. The revised Version 2 was available in mid-1995. The final specification, which will be wholly compatible with Version 2, is scheduled to be published in late June 1996. The final version, in addition to covering credit card functions, will examine the specification of an electronic purse.

The motivation for the three credit card issuers to prepare a specification for chip-containing cards within such a short time is manifold. For one thing, existing credit cards with magnetic stripes can be very easily forged. The only current obstacle is the largely unreproducible hologram. All other card elements can be duplicated at relatively little expense. The second important point is the additional functions which a microprocessor offers. Electronic purses, bonus points and telephone functions represent only a fraction of the possibilities available.

The specification for the chip-containing card is divided into three sections. Section One relies extensively on the ISO/IEC 7816-1/2/3 series of standards, and describes the electromechanical characteristics, logical interface and transmission protocols. According to this section, the Smart Card has the ID-1 format[7] with ISO contact position. Besides other values, it must have a 5 V ± 0.5 V supply voltage, 50 mA maximum current and 1 to 5 MHz clock frequency. The forces must not exceed 0.6 N per contact. Basically, the physical implementation of data transmission is identical to ISO/IEC 7816-3. This applies equally to the time interval for individual bits[8], to the ATR[9] and to the two transmission protocols[10] T=0 and T=1. The specified APDU[11] is identical with the one described in ISO/IEC 7816-4.

Section Two of the specification defines the data elements and instructions. Its contents rely heavily on ISO/IEC 7816-4, with additional instructions specific to EMV. Finally, Section Three, "Transaction Processing", contains the data and routines needed for a payment transaction. An additional global specification deals with the terminals to be used, and lays down the parameters from keyboard layout, to data, to user instructions. The appendix contains a further summary of possible terminal types. The entire specification is kept very general, and only lays down the terminals' fundamental parameters.

Since a typical credit card is a mass-produced article, its manufacturing costs must not be too high. However, since the costs involved in the production of Smart Cards increase with the capacity of the integrated microprocessor, the credit card application was from the start designed to use a minimum amount of memory. Thus, a conventional EMV application without additional functions fits into a processor with 6 kbyte of ROM, 1 kbyte of EEPROM and 128 bytes of RAM. Even in the Smart Card field these values represent the lower end of the range of possible chips, but they do, however, enable economical production.

To an extent, the EMV specifications also form a basic documentation which lays down the minimum standard for other card issuers to follow. The current Version 2 leaves several points open to debate, such as the risk management of terminal and Smart Card during transactions, which remains to be precisely defined. That is why the EMV application is only described here in outline, since many details either are not fully determined, or else are certain to be modified in the future.

Files and data elements

The specification for the chip-containing credit card only states that a tree-structure[12] must be used for the files. The application itself is located in its own DF, which is selected via an AID (application identifier) and which contains all the data necessary for the credit card

---

[7] See also 3.1.1 Formats.

[8] See also 6.1 Physical transmission layer.

[9] See also 6.3 Answer to reset.

[10] See also 6.2 Transmission protocols.

[11] See also 6.5 Message structure.

[12] See also 5.6.2 File hierarchies.

application. These are held in EFs, whose structure may be either linear fixed or linear variable. The latter's FIDs (file identifiers) are restricted to no more than 5 bits, so that all EFs must be selected implicitly by short-FID. The big difference with similar specifications is that no particular EFs and FIDs are laid down; indeed this would not be realized for the payment functions, since all the required data can be processed with ready-made instructions. Directly below the MF there is only a file (DIR file as per ISO/IEC 7816-5) which contains all the data concerning the card's available applications.

All the data in the terminal–card system is identified with clear templates and tags. The specified instructions can address them within the application, without their precise position in the file tree or a particular file having to be known. This makes it possible to leave the division into files to the card issuer, since this does not affect the execution of payment transactions.

Instructions

Technically, only three instructions are necessary for the actual execution of payment functions. However, personalization, administration, special functions and additional applications do need quite a few further instructions. This, though, falls outside the EMV specification's remit. It only states that the following instructions must be available, with all return codes being analogous to ISO/IEC 7816-4[13].

- EXTERNAL AUTHENTICATE (as a subset of ISO/IEC 7816-4)
- GENERATE APPLICATION CRYPTOGRAM (specific to EMV)
- GET DATA (specific to EMV)
- GET PROCESSING OPTIONS (specific to EMV)
- READ RECORD (as a subset of ISO/IEC 7816-4)
- SELECT (as a subset of ISO/IEC 7816-4)
- VERIFY (as a subset of ISO/IEC 7816-4)

Cryptography

By their nature, the cryptographic mechanisms used in an application are highly dependent on the relevant global conditions. This can be particularly clearly seen in the EMV application. It was a fundamental precondition of system design from the start that the terminals did not possess security modules. This makes it impossible to use symmetric crypto-algorithms, since the key confidentiality can no longer be maintained. The absence of security modules can be explained by the considerable increase in cost they would have entailed, since worldwide key management for partly off-line operating terminals is a very complex and expensive process.

---

[13] See also 7.10 Credit card instructions.

In order to make Smart Card authentication by the terminal possible within the framework of these conditions, an asymmetric algorithm must be used. EMV uses unidirectional static authentication with card-specific keys[14]. Although the card is unable to check the terminal's authenticity, the EMV application has no vital need for it since no debiting is carried out on the card; it merely submits a transaction voucher to the terminal. This voucher is not anonymous as far as the card holder is concerned, and can only be sent to the relevant card issuer by a licensed – i.e. known – trader. This largely obviates any possibility of fraud, since a valid transaction voucher can only be "converted" into money by an authorized trader known to the card issuer.

Any cryptographic algorithms can be used in principle, since both the corresponding data and the algorithms themselves are unambiguously identified through TLV-coded data structures. Version 2 of the EMV specification provides for either SHA-1 as per FIPS 180-1, or for ANSI X9.30-2, as the hash algorithm[15]. The cryptographic algorithm used is not DSS, however, as one might expect from the use of SHA-1, but rather the RSA algorithm[16] (ANSI X9.31-1). Key length may vary between 512 and 1024 bits, and is indicated by an appropriate coding (signature indicator). The public keys suggested are numerically small for reasons of computational speed, such as 3 or 2^1+1.

In the event that the Smart Card has established an online link with the background system, it is possible to protect the data transmission by secure messaging[17] as per ISO/IEC 7816-4. A symmetrical cryptographic algorithm is used with this end-to-end communication, namely DEA. In this case it is possible to do this without compromising system security, since both the background system and the card can securely store the secret key.

System and transaction routines

System structure in the credit card industry is traditionally very centralized. Normally there are several background systems, which, either alone or with further systems, are responsible for a particular region (e.g. the United Kingdom). These powerful computer centres are connected to each other over the relevant card issuer's own network. This is used for clearing related data exchanges and also increases operational reliability, since if one computer fails, its tasks can be taken over by another one. Individual terminals are connected to the background system via public telephone and data networks (e.g. ISDN, X.25). There may also be an intermediate acquirer, who routes and bundles the transaction data. This, however, depends largely on the relevant card issuer or country. The following two variants exist at terminal level: direct data exchange with the acquirer, or involvement by the trader's or trading chain's concentrator. Both variants are not only possible but also implemented in practice.

---

[14] See also 8.2.3 Static asymmetric authentication.

[15] See also 4.3 Hash functions.

[16] See also 4.2.2 Asymmetric crypto-algorithms.

[17] See also 6.6 Secure messaging.

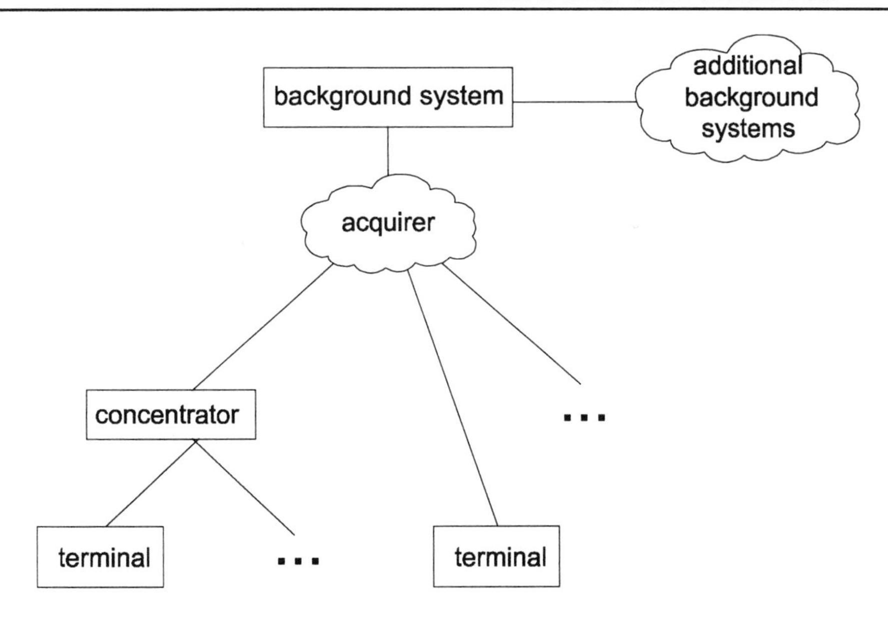

Figure 12.18 Basic structure of transaction system for credit cards with chips

The payment process with a chip-containing credit card does not change substantially in the new system. The customer presents his card at the checkout, and it is inserted into a Smart Card terminal. If a terminal is not available, payment can be made as before by using the normal magnetic stripe or the embossed data. On the other hand, if a terminal is present then it is still possible to check the holder's identity through the signature. The transaction document would then indicate that such identification had taken place. The second option provides for inputting a four-figure PIN. This can be checked online by the background system, but also off-line by the card. The generated transaction voucher then carries an indication of the type of PIN check performed. A summary of this process is contained in Figure 12.19.

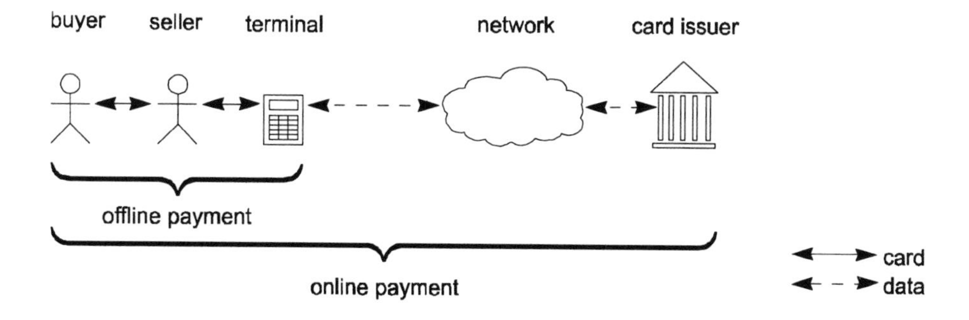

Figure 12.19 Basic infrastructure in Smart Card transactions in accordance with EMV specifications

The details of individual transactions and the process of successful payment with a Smart Card are, of course, somewhat more complicated. An example is shown in Figure 12.20, which illustrates the fundamental mechanisms. Both the card and the terminal compute the exact transaction execution from given data, such as the amount involved. For instance, if the sum to be paid exceeds a particular limit, then the card requests online authorization by the background system. The terminal must then set up a link and have payment confirmed. Only when the background system concurs does the card generate a valid transaction voucher, which the trader can then submit for clearance. The purpose of online authorization is to minimize the card issuer's financial risk. Since a chip-containing credit card has repeatedly to "report" to the background system if these criteria are built into the system, the maximum loss in the event of stolen or manipulated cards can be kept within precisely known limits.

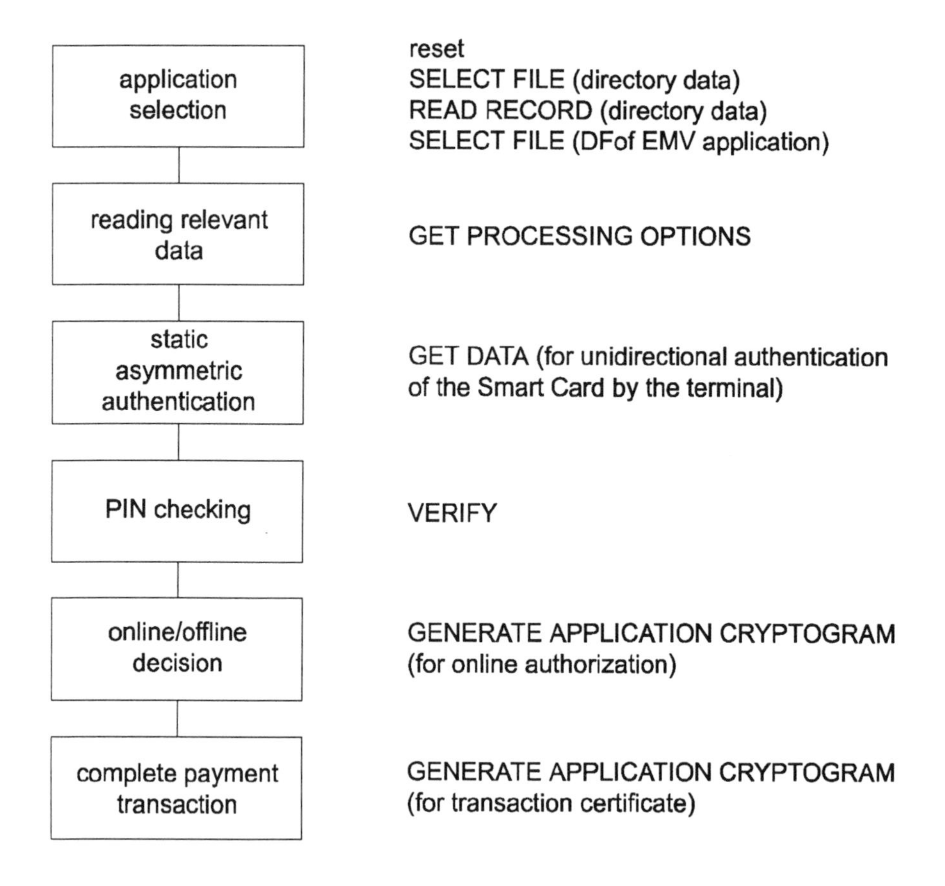

Figure 12.20 Schematic Smart Card transaction procedure in accordance with EMV specifications, seen from the terminal

Future developments

Due to its structure and contents, the EMV specification allows the card issuer great room for manoeuvre in terms of initiative for further developments and variations. This flexibility is certain to be fully exploited in future by the firms involved, and the first amendments are already apparent. Thus, each of the three credit card issuers participating in EMV is already working on an electronic purse which can find its place on the Smart Card as a debit application[18]. A prepaid purse has particular benefits in the area of vending machines or whenever the amounts involved are small, because it avoids the transaction fee that is mandatory with credit cards.

The expanding commercial use of worldwide networks such as the Internet will require in the near future secure, global and acceptable forms of payment. Chip-containing credit cards and a few additional functions would be eminently suitable, since for a long time they have enjoyed international acceptance and widespread use independent of any particular country or currency. R&D work is being carried out in this area. The difference between this and the payment transaction described above is not great, since the Internet can replace the acquirer. It is merely necessary, to begin with, that a Smart Card terminal is available to the customer.

Although the EMV specification still exhibits several inadequacies and also leaves a few points open, it has achieved nearly the same importance in the Smart Card industry as GSM 11.11. There are practically no new Smart Card operating systems on the market which do not claim to be "EMV compatible". Due to the expected number of chip-containing credit cards in circulation in the future, this specification is likely to set the minimum standard for all new applications.

---

[18] See also 12.3.1 CEN standard prEN 1546.

13 Sample applications

Having described all the relevant aspects of Smart Card technology in the previous chapters, we can now introduce a few applications and explain them in general terms. They can serve to illustrate the extent and complexity of large Smart Card applications.

This chapter's objective is also to demonstrate systems in which Smart Cards are implemented, and in which they are only one component among many. The functionality, user friendliness and usually almost the entire security of these systems depend on the Smart Cards used. However, it is always recommended to view these systems as a whole, since they only work satisfactorily when all their components cooperate without conflict.

13.1 CONTACTLESS MEMORY CARDS IN THE AIRLINE INDUSTRY

For many reasons, the system described here is different from the usual applications where Smart Cards are of benefit. This applies to the fundamental system architecture, and also to the card's data transmission and power supply. This is the new ticket-free flight system operated by the German airline Lufthansa. It is based on contactless memory cards which thus do not need to be inserted in a card reader[1]. The basic system concept also differs from all other Smart Card applications described here, in that the card is only used as identification for the user, as the data is stored in the background system.

Lufthansa has been issuing "frequent traveller" and bonus system cards for several years. These cards possess embossing as well as a magnetic stripe for automatic processing. In addition, certain types of card could also be provided with a chip for German public cardphones and for a credit card function. This existing card family was to be supplemented with applications in the areas of boarding and ticketing, and remain compatible with them.

---

[1] See also 2.3.3 Contactless Smart Cards.

In addition to compatibility, another objective was to modify the existing systems as little as possible. Ease of handling by the customer was also important.

In the maximum version, the result is a multi-application card carrying a magnetic stripe, embossing, hologram, and memory chip with and without contacts. A large number of different applications can be served by this card, without problems of compatibility with previously issued cards arising.

Before company-wide introduction, a pilot project was conducted on the Frankfurt/Main–Berlin route. The new cards were issued to 600 totally different customers, and many thousands of flights were carried out on this route between May and December 1995. Appropriate PC-based vending machines were installed at both airports for use by the cards, with a touch-sensitive colour screen and a printer. The machines' PC controller was also connected to a transmitter/receiver for the contactless Smart Cards.

Applications on the card

The most diverse applications are at the customer's disposal on the new card. In this section, attention will be given to the maximum version. There are, of course, simpler variants too, for example a card only containing the contactless memory chip functions. This chip makes a self-service checking-in possible at the machine, which naturally leads to correspondingly higher turnover, a lower waiting time and faster processing. Lufthansa's bonus system is also integrated into the contactless chip, so that no other card or additional data input is necessary.

The checking in and a Boarding Information sheet can be printed out by the machine at the airport. In analogy to current tickets, this carries all the important information concerning the flight booked. If necessary, the seat on the flight can also be selected at the machine. Flights can also be booked by phone, by quoting the number embossed on the card. It is planned to carry this out in the future completely without paper tickets, since all the relevant data can be called up from the background system via the card. Of course, this does not rule out flight confirmation by phone or fax, which will continue to be possible.

In addition to these functions, the new card can also carry the appropriate data on the magnetic stripe and a hologram for a credit card. The possibility also exists of implanting a chip with contacts at the ISO position, which currently is used for a telephone chip.

The overall system

In terms of the Smart Card, all the applications supported by the contactless chip are very simple in structure. The card is only used for identification, using a memory chip which allows dynamic authentication through the background system[2]. After authentication, all the data is read from the card. Currently these are the customer's number, name and profile. With this data the background system can assign the data to a person, and booking data, bonus system points and all other applications are made available. Thus the card is

---

[2] See also 8.2.1 Unidirectional symmetric authentication.

only used as a kind of key, and the data and mechanisms of the various applications are always held in the background system. This system presents considerable benefits, since the background system and all its required databases, programs and interfaces have been available for a long time.

Another advantage of this system is the method used in the event of lost or faulty cards, a problem almost always ignored so far in all other multi-application card systems. Where applications were loaded into the card after personalization, the holder issued with a new card had to contact all the application providers in order for them to reintroduce those applications onto the card. Due to the centralized system architecture, this type of problem is absent in the Lufthansa Smart Card. The holder of a lost or defective card receives a new one, and the old card, identified by its number, is blocked system-wide. Logistically this does not give rise to great expense, since all airports have access to the necessary data through the long-available Lufthansa network.

The contactless card

The new Smart Card is in the internationally common ID-1 format. The memory chip incorporated into the card's body uses an inductive coupling with a coil for power and data transfer. With this technology, data can be both read and written by the card reader at a distance of up to 10 centimetres [3]. If the card is in an ordinary wallet, it is even possible to perform the data exchange with the terminal without removing the card from it: it is sufficient to hold the wallet next to the reader. Since the chip and coil are located inside the card, its graphic layout can be defined totally independently of those components. Furthermore, the contactless technology completely circumvents the problem of worn chip contacts [4].

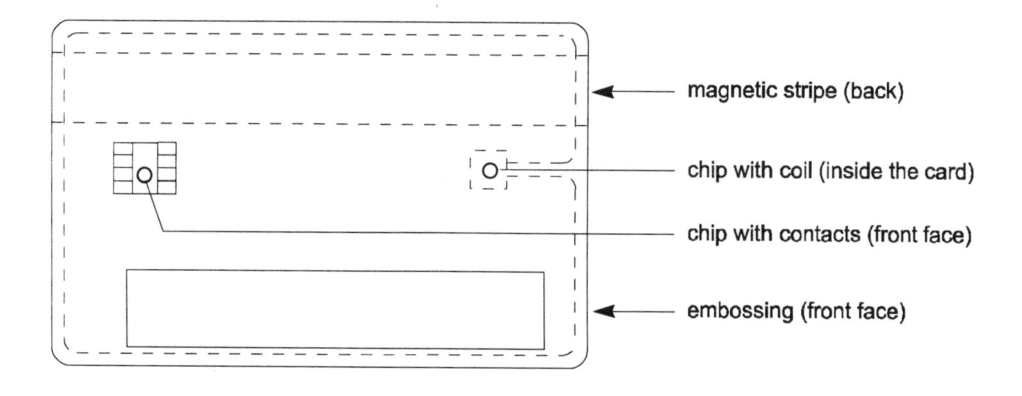

Figure 13.1

[3] See also 3.1.3 Cards without contacts.

[4] See also 6.2 Transmission protocols.

The typical transaction time between terminal and card in the system just described is around 100 ms, and takes place at a frequency of 13.56 MHz. The memory chip used (SLE 44R35) contains a 1 kbyte EEPROM, and is able to perform a unidirectional dynamic authentication of the outside world. Since currently only 48 bytes of data (customer number, name and profile) are held in memory, additional areas of application can be opened up in future, with the added bonus that in this system no further memory is needed in the Smart Card to create a new application.

The future

The field trial on the Frankfurt–Berlin route was very successful. Lufthansa now plan to start issuing series of production cards in March 1996, first to frequent fliers and later to all interested customers. To this end, Lufthansa's check-in areas in all German airports are appropriately being equipped in turn, so that the new contactless card can be used everywhere. Beyond that, it is planned in principle to install this new technology in all airports used by Lufthansa. The new card offers the customer quicker pre-flight handling.

Although this Smart Card application is atypical in its system structure, it does bring a somewhat different and interesting perspective into the Smart Card world. Since all application data is held in the background system, all data protection aspects are located there. This means that data protection law, in whichever country one is, no longer influences the card's data contents. This is also a way of circumventing the well-known multi-application memory limitation problem. The question of who may write which data to the card no longer exists. The separation of applications on the card is also complete, and no interference can exist between the individual applications in terms of system design. Finally we may note that data which is valuable to the system operator is always stored securely in his background system and not on the card, where it may go astray.

Over and above that, the new system represents a model of seamless progress from one card technology to the next generation. This "gentle" migration from one technical level to the next one up is a very important argument, since past investment is not lost and no new system needs to be built up from scratch. True, this benefit must be paid for through a relatively new and expensive card technology. But it is quite conceivable that it will attain very great predominance in the relevant industry, since it makes possible faster and more efficient customer handling than with the current tickets which carry a magnetic stripe.

13.2 ELECTRONIC TOLL SYSTEMS

In some countries it is common to demand a toll for using certain roads. Compared with the fixed charge raised through the sale of a tax disc, the levy of tolls is usage-dependent, i.e., a particular fee is payable depending on the frequency of use and the type of vehicle. So far it has been conventional to pay in cash at toll booths. Electronic road-pricing systems have been appearing here and there for some years, with different types of cards. All existing systems, however, suffer from a disadvantage as the flow of traffic is considerably reduced, due to the need to stop or slow down to walking speed. Furthermore, the toll booths require a lot of space.

For these reasons the Federal German Transport Department decided in 1993 to start a large-scale field trial in automatic toll payment. The road selected was the Federal motorway A555 between Cologne and Bonn. Various systems offered by ten different firms were tested here between May 1994 and June 1995.

The tested systems had to adhere to several noteworthy conditions. Traffic had to be able to flow normally and unhindered through the toll points. However, normal flow in Germany may mean anything up to 250 km/h. The system had to function in such a way that it was not possible to evade payment through excessively fast driving. Furthermore, no toll booths or capture bins were allowed, since only systems were desired whose space requirement was restricted to installation at road bridges or overhead signs. Also, the systems used had to support not only single lane but in addition multi-lane traffic, since for construction reasons separation into single lanes was not desirable; again, it would have considerably reduced traffic flow.

Yet another requirement arose during the course of the project, which came to be seen, publicly, as the ultimate criterion for the whole trial. This was the requirement of complete anonymity: it was to be impossible to produce the movement profiles, or monitor the travelling routes, of specific vehicles. All systems could demonstrate a payment method wherein vehicle anonymity was maintained, but only so long as the fees were paid. As soon as a fee was avoided, the vehicle was captured photographically and the registered keeper traced. This was followed by an appropriate penalty notice. Naturally this was never the case during the trial, as payment was made only with "play" money.

Almost all the automated toll systems used a Smart Card to carry the electronic money units. That is why this trial is described here, since it may well grow in significance for the Smart Card industry in the future.

The systems consisted of a device mounted in the vehicle, known as either OBU (on-board unit) or IVU (in-vehicle unit), as well as additional instrumentation as necessary. These devices, powered from the car battery, contain a Smart Card reader, a display and a simple keyboard. Communication with the outside world is uni- or bidirectional, depending on the system. This is mostly in the microwave range, in the bandwidth of 5.795 GHz to 5.805 GHz as recommended by CEPT for this application. Alternatively, data transmission was implemented by radio waves in the 400 to 500 MHz range or by IR. The latter naturally suffers from the disadvantage that transmission is strongly affected by weather, such as heavy snow or fog. In the participating systems, the OBU's series production price is between 100 and 300 DM, depending on the instrumentation chosen.

The control stations were installed along the motorway, at road bridges or overhead gantries. This makes alterations to the road deck superfluous. The Smart Cards used did not carry complex electronic purses, but only very simple but fast billing instructions. This was due, on one hand, to the fact that no real money was being used, and on the other, to the transaction time available. The optimization process went so far as to even reduce the cards' ATR to four bytes in some cases, to leave sufficient time for the actual debiting. This is understandable when one considers the condition of unrestricted traffic flow. At 250 km/h, a vehicle covers 70 m per second. In the 5.8 GHz frequency spectrum, the control stations had a communications range of ≈ 5 m. This produces a time-interval of ≈ 70 ms during which a vehicle is within the station's range. The following routines must be executed by the card during this time:

- Reset card and send ATR (≈ 10 ms)
- DES encryption to authenticate the card (≈ 12 ms)

- Accessing EEPROM to write the new balance ($\approx$ 2 x 3.5 ms)
- Data transmission to and from card ($\approx$ 30 ms)

Additional time is also needed for data transmission between the OBU and the control station. Hence the time available is measured very strictly. The Smart Cards were supplied with the highest permissible clock frequency (usually 5 MHz). Transmission speeds between OBUs and control stations, at 1 Mbit/s, were substantially higher than between OBU and card, so that this time interval exerted little influence on the calculation.

The following are descriptions of three of the ten systems participating in the field trial. Several systems were almost identical but for a few minor technical details, hence these three are not identified by name and can act, as it were, as representatives of the whole approach.

System 1

This system works with the classic road infrastructure. Two road bridges or gantries mounted closely together across the motorway were used to install the necessary instrumentation. The system is so designed that a change of lanes between the two bridges does not affect the results. There are two different payment types: in the post-paid mode, the accumulating fees are debited to a bank account in the manner of a credit card. Here it is very difficult to guarantee vehicle anonymity. In the prepaid mode, Smart Cards were used in which the appropriate sum was deducted on each occasion from the amount already stored on them.

The system works in the following way: at the first bridge the OBU and the Smart Card are activated; this is followed by the fee being raised, with a general assessment of the vehicle by class, i.e. truck or car. Since the toll is differentiated by vehicle class, it is necessary to check the class as stored in the card. This is done by measuring the vehicle's height when travelling under the bridge, which is sufficient for reliable classification. Within the communications zone at the immediately following bridge, another link to the card is set up via the OBU. There the control station checks whether the initiated toll payment could be successfully executed. If this is not the case, the vehicle is photographed. It is now possible to carry out a fully automated and unambiguous identification based on the number plate. The car's registered keeper then receives an appropriate fine.

The system can be further extended through various procedures, as desired. During payment, for example, the card can be compared with a warning list, in order to exclude stolen cards. It was also considered, in certain special cases (e.g. bank robbery with escape by car) to photograph all passing vehicles in order to obtain data about the escape route. This system's benefits arise from the OBUs' simple construction and resulting low cost, although this must be offset by the necessity for more complex and expensive control stations.

System 2

The second system to be discussed here is not based on a roadside infrastructure. It uses the principle of virtual toll booths, which only exist as databases in the vehicles' OBUs. These contain all the coordinates of existing toll booths, as well as the corresponding charges for motorway sections. As soon as a vehicle enters such a stretch of road, the relevant fee is deducted by the Smart Card. Naturally, this process is far less time-critical than the one described under System 1.

The OBU knows the vehicle's coordinates at any point in time, through an attached GPS model. GPS is short for "global positioning system", and is a worldwide system for pinpointing geographical location. It was developed in the USA for military purposes from 1973, was commissioned in 1993 and consists of 24 satellites which transmit data at 1.6 GHz from an altitude of 20***000 km. The receivers' size has been reduced to no more than that of a cigarette box. In civilian applications, positions can be determined to an accuracy of around 10 to 20 m; in military use the accuracy is substantially better.

Of course, this toll system too has to be controlled, but this is done in the form of random samples, similar to current speed traps. This system offers the great advantage that no installations are needed at any roads, and tolls are raised at virtual points. Control, however, is more difficult, and it is impossible to use post-paid-mode Smart Cards since no data exchange takes place between the OBU and the outside world. On the other hand, this has benefits in terms of anonymity.

System 3

Technically, this is the most ambitious solution for automatic vehicle toll payment. The OBU's equipment consists of a GPS receiver and a GSM mobile phone. The vehicle's position is pinpointed via the GPS unit, and the OBU uses the phone to obtain billing data. Like the two systems described above, the OBU contains a Smart Card and card reader, which are used during the payment procedure.

As in System 2, the toll booths are only virtual points along the roads, determined from coordinates at the earth's surface. In this system the motorway authority is able to change the tariffs at will through bidirectional communication with the OBU. Hence the fees can depend on time, place, vehicle class and external conditions. Calculation of the fee is carried out in the OBU.

In order to achieve a certain level of anonymity, the fees debited to the OBU's Smart Card are not immediately forwarded to the background system, but rather accumulated in the card up to a certain floor limit. Once this value is reached, the total amount due is transmitted by mobile phone for checking and debiting. Due to the bidirectional data exchange, this system can naturally support both debit (prepaid) cards and credit (post-paid) cards. However, anonymity is no longer guaranteed when payment is made in arrears. In prepaid Smart Cards, which are used like electronic purses, the motorway authority is prevented, through the payment of a final total, from building up a profile of the vehicle's movements. Quite independently from that, though, it is of course possible for the GSM network operator to trace the vehicle's coordinates continuously via the permanently activated mobile phone.

In this system too, it is naturally necessary to monitor the traffic flow continuously for offenders. This is achieved by automatic cameras installed at bridges, or by local mobile checkpoints. Their computers can read out the OBU's or Smart Card's ID mark by mobile phone, and compare it to that of the car. If they fail to correspond, or if no connection to the OBU could be established, the car is photographed and a penalty notice produced.

The main advantage of this system is that apart from monitoring, no specific infrastructure needs to be set up. Nevertheless, the OBU's price is necessarily higher than in both previously discussed systems. The two different payment methods (prepaid/post-paid) and the data transmission to the vehicle are also good arguments for this system.

13.3 GSM NETWORK

The GSM network was originally a European standard for mobile phones. Since ever more countries and telecommunication firms joined this standard, it did not remain restricted to Europe but has now been extended throughout the world. The initial meaning of the abbreviation GSM, "Groupe Spécial Mobile", was modified during the entire project's course of internationalization to "Global System for Mobile Telecommunications". The first parts of this network became operational in Europe in July 1991. Currently (1996), a total of 120 mobile phone networks based on this standard operate in 71 countries. Further extension is being pursued by 50 additional network operators in 20 countries.

The GSM network's specification started in 1982 under CEPT (Conférence Européenne des Postes et Télécommunications), and was later continued by ETSI (European Telecommunications Standards Institute). The Subscriber Identification Module Expert Group (SIMEG) started the Smart Card's specification in January 1988.

This specification is divided into two sections. The first, GSM 02.17 (ETS 300509), covers general functional characteristics, whilst the second section, GSM 11.11, deals with the interface description and logical structures. As part of a restructuring process, the GSM specifications were given a new numbering scheme so that the GSM 11.11 standard is now designated prETS 300608.

Both specifications use their own specialized vocabulary. The terms used are defined very precisely in the specification, with their technical meanings restricted to GSM. Further specifications were added in 1994, known together as Phase 2. These contain additional data elements and mechanisms, so that fixed dialling numbers and accumulated call meters are supported. Currently work is being done on the specifications for Phase 2+, known as GSM 11.11 Version 5.0.0. At least another year will pass before final approval of the Phase 2+ standard.

GSM is a cellular mobile phone network, operating in the 900 MHz range. The term "cellular" means that the area covered is subdivided into circular cells with a diameter of between 1 and 30 km. In order to completely cover a country the size of Germany ($\approx$ 360***000 km^2), approximately 3000 such overlapping cells are needed.

The original GSM specifications have been extended, and now also cover an 1800 MHz system known as DCS1800 (digital cellular system) or PCN (personal communication network). Due to the higher frequency and a reduced transmitter power, cell diameter in this system does not exceed 20 km. As a result, its main use is in conurbations rather than in rural areas with low population density. The major difference between GSM and DCS1800 is merely the transmitter and receiver components on either side of the aerial cell interface.

At the centre of each cell is located a so-called base station (BS), whose task it is, on one hand, to establish contact with the mobile phone across the aerial interface, and on the other, to feed into the telephone network. Data transmission across the aerial interface is encrypted and performed at 13 kbit/s, and possesses a loss-bearing compression procedure with error correction mechanisms.

GSM designates the mobile phone as MS (mobile station), consisting of the physically and logically distinct parts ME (mobile equipment) and SIM (subscriber identification module). The ME is the station's radio and encryption section, and SIM is another name for the Smart Card specific to GSM. Together they form the functional mobile phone[5].

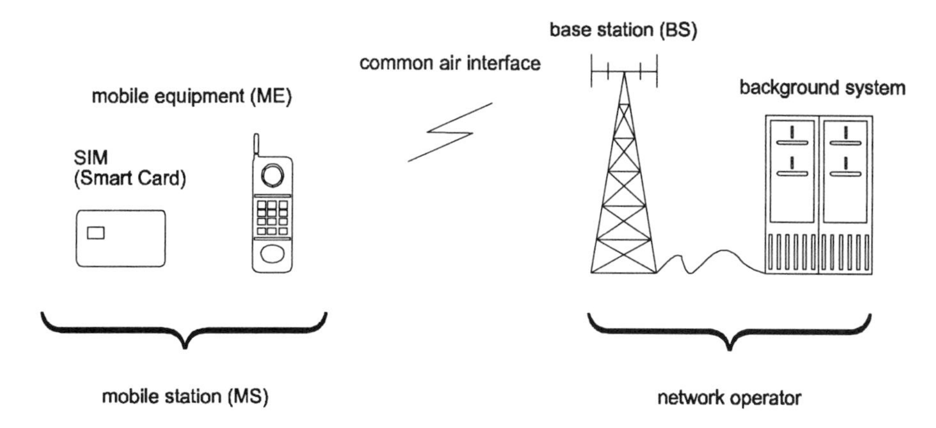

Figure 13.2 Structure of a mobile station in the GSM network

GSM's background system is made up of two parts: the switching station and the base station system. The latter's task is to exchange data with the mobile stations via many base transceiver stations, by aerial broadcasting. Besides some administrative data, this is where the actual telephone conversation is conducted. Several base transceiver stations are managed by a base station controller. If the mobile station leaves the transmitter's range, the controller also ensures handover to another base station system[6].

The management of several base station systems is carried out by a mobile services switching centre. It is responsible, for example, for forwarding calls to other networks such as the country's public telephone system. In order for a subscriber to be reached on his or her phone, the mobile switching services must always know the location of the relevant transceiver station in whose range the subscriber is located. Data concerning the actual location of the subscriber and the mobile station called, is located in the switching system's home location register. Another database, the "visitor location register", holds all the locations of the mobile stations currently within range of the relevant switching system. Here GSM subscribers can also be temporarily recorded by other network operators.

At the highest level in the hierarchy is the authentication centre, which is the sole authority in control of the necessary keys and algorithms for authenticating mobile stations (i.e. the SIMs). If a subscriber is authenticated at the start of a call or during it, this is performed by the authentication centre. This, in rough outline, is the structure of a single operator's GSM system. In reality, of course, several additional authorization points and databases are required, e.g. for billing the calls.

---

[5] See also 6.2 Transmission protocols.

[6] See also 6.2 Transmission protocols.

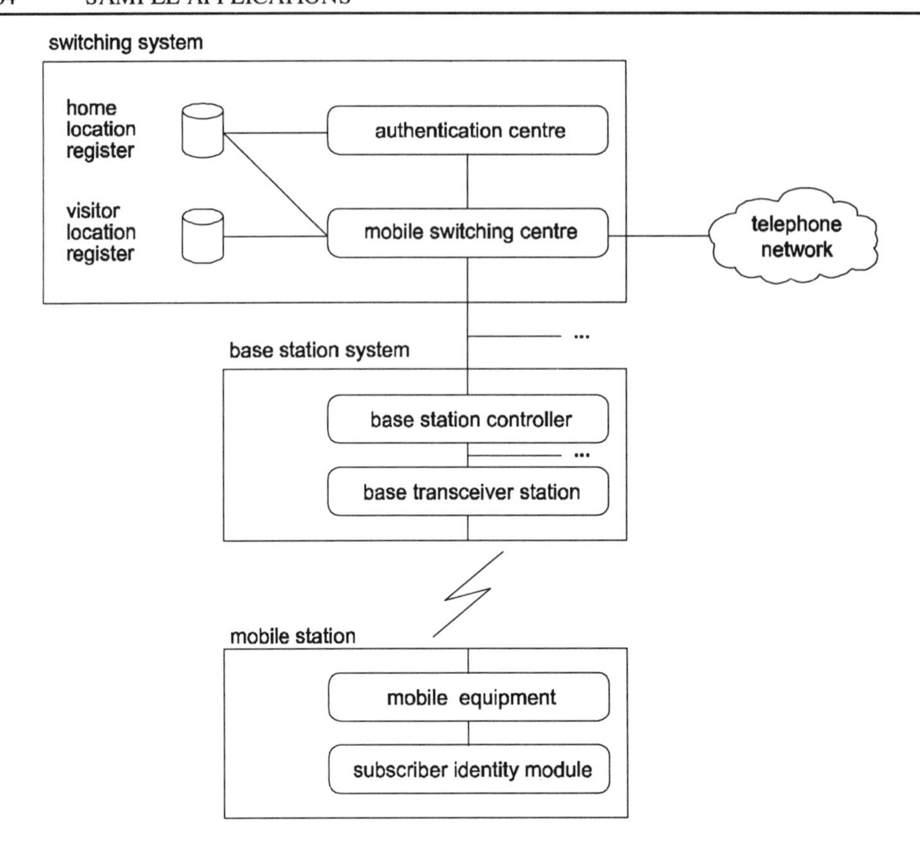

Figure 13.3 Structure of GSM network of single user

GSM allows the SIM to take two different card formats. ID-1 is used in mobile phones which replace the SIM frequently. Mobile phones in which the SIM is only replaced rarely and which are physically very small, employ a plug-in SIM in the ID-000 format. The card's dimensions, though, are the only difference, as they are identical in terms of their physical or logical features. The SIM's task is to permit network access only to authorized persons, thus ensuring reliable billing. The following functions are necessary here: data storage, protection of access to this data and execution of a cryptographic algorithm under secure conditions.

To understand these functions one needs to be aware that the SIM, and thus the mobile equipment, is authenticated by the relevant network operator. This is a unidirectional authentication of the SIM by the background system. Data transmission between mobile station and base station across the aerial interface is encrypted, to make tapping impossible. Short text messages arriving over the network can be stored in the SIM and read out as necessary.

SIM identification is based on a number unique across the whole GSM, which is no more than 8 bytes long and is known as IMSI (International Mobile Subscriber Identity). The subscriber can thus be identified by the system worldwide, on all GSM networks. In order to keep the subscriber's identity as confidential as possible, the network uses a TMSI (Temporary Mobile Subscriber Identity) instead of the IMSI whenever possible, which is only valid within one part of the relevant GSM network.

The card-specific keys for authenticating and encrypting the data at the aerial interface can be derived from the IMSI. However, encryption is not carried out in the Smart Card, since the computational and transmission capacity of a Smart Card is not sufficient for real-time encryption of voice data. Instead, the SIM calculates a temporary key devised for transmission encryption, and relays it to the mobile equipment. The latter contains a powerful encryption unit, which can encrypt and decrypt the voice data in real-time[7].

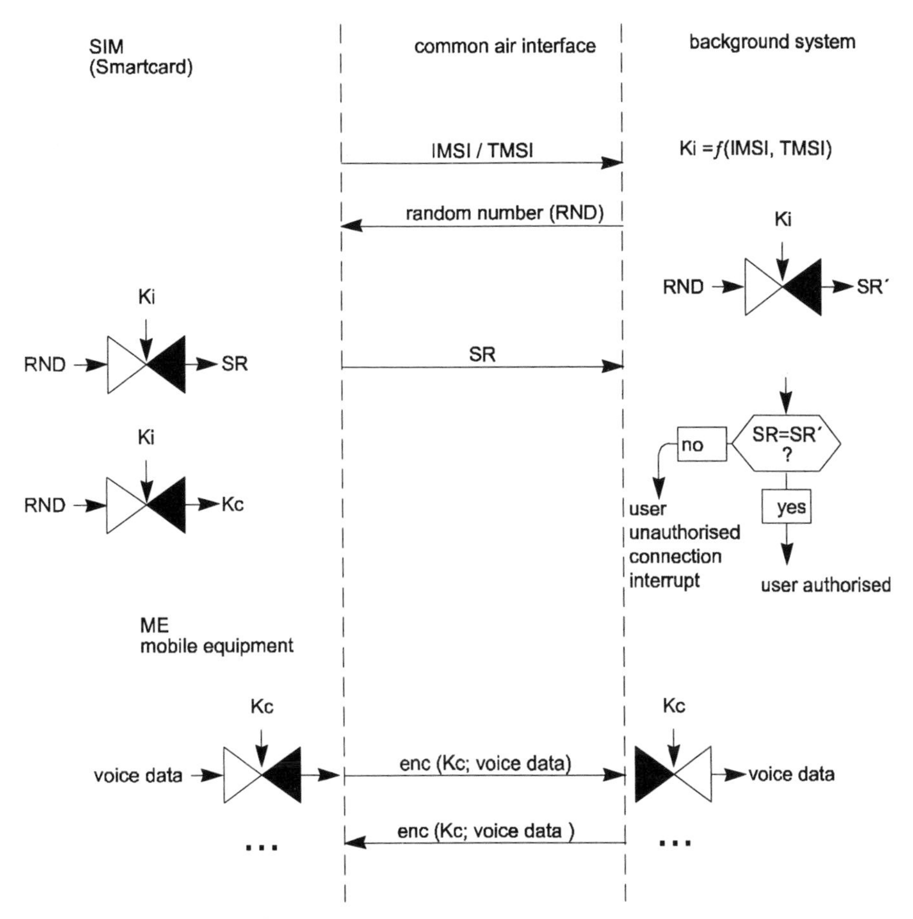

Figure 13.4 SIM cryptographic functions in the GSM network

---

[7] See also 6.2 Transmission protocols.

When a subscriber wants to conduct a conversation, his mobile station sets up a link to the optimal base station, and relays to the latter the IMSI or TMSI from the SIM. If the subscriber's IMSI is registered there, the mobile equipment receives a transmitted random number, which is passed on to the SIM. It uses this number as a plaintext block for encryption, whose key is specific to the card and to the subscriber and is derived from the IMSI. The result is a ciphertext block which is broadcast to the base station by the mobile equipment.

The background system, permanently linked to the base station, derives the card-specific key from the IMSI and then performs the same calculation as the SIM. After the SIM's ciphertext block arrives at the background system, the latter only needs to compare it with the self-computed one to determine if the subscriber is authentic and thus authorized to conduct the call.

Both the SIM and the background system use the random number and the card-specific key to calculate a temporary key, which is used for transmitted data encryption. It is passed on to the mobile equipment where encryption takes place, since the Smart Card is too slow for the required real-time decryption and encryption of voice data.

The GSM network is currently the world's largest Smart Card application, with over 5 million cards in use. It is the first application ever in which Smart Cards had to meet the requirements of different national and international system operators. This global application has set a standard for the Smart Card industry which covers both instructions and the characteristics of the card itself. This normative muscle forces all standardizing groups to remain compatible with GSM at least in all essential respects, if the respective standard is to achieve international acceptance. This is why the GSM field is discussed here at some length.

The SIM

The SIM has a hierarchical file system, with an MF and two DFs which contain the EFs with the application data. The EFs' possible file structures are transparent, linear fixed and cyclic. The GSM 11.11 specification defines 18 Smart Card instructions, identified by Class 'A0'[8].

Input of the 4-figure PIN, which incidentally is designated CHV (chip holder verification), is rather unusual. With a special instruction and the correct PIN, further PIN interrogation by the card can be switched off by the user, so that it is unnecessary to enter them before each call. The user bears the risk that lost cards can be improperly used until they are blocked by the network operator. Another instruction enables the user to switch the PIN interrogation function on again.

Communication between mobile equipment and SIM takes place with the T=0 data transmission protocol, using standard parameters. However, the convention used can be freely selected by the card in ATR. A PTS is allowed for, but currently is not being used due to the restriction to T=0.

The GSM 11.11 specification defines 30 different EFs for the application data, held in two DFs. The file identifiers (FID) are unusual in that the DF's first byte is always '7F'. EFs directly below the MF must contain the value '2F' as the FID's first byte, and EFs below a DF the value '6F'. In addition to the specified files, each network operator can store in the SIM his own files for maintenance and administrative purposes[9].

---

[8] Instructions breakdown: see 15.4.5 Table of class bytes used.

[9] See also 6.2 Transmission protocols.

Data type	FID	Structure, size	Description
MF	'3F00'	—	root description
EF_{ICCID}	'2FE2'	transparent, 10 bytes	contactless card ID number
$DF_{TELECOM}$	'7F10'	—	DF Telekom (German telecom)
EF_{ADN}	'6F3A'	linear fixed, x bytes	abbreviated dialling numbers
EF_{FDN}	'6F3B'	linear fixed, x bytes	fixed dialling numbers
EF_{LND}	'6F44'	cyclic, x bytes	last number dialled
EF_{SMSS}	'6F43'	linear fixed, x bytes	Short Message service status
EF_{SMSP}	'6F42'	linear fixed, x bytes	Short Message service parameters
EF_{SMS}	'6F3C'	linear fixed, 176 bytes	Short Message
DF_{GSM}	'7F20'	—	
EF_{LP}	'6F05'	transparent, x bytes	preferred language
EF_{KC}	'6F20'	transparent, 9 bytes	key Kc
EF_{SPN}	'6F46'	transparent, 17 bytes	Service Provider Name
EF_{PUCT}	'6F41'	transparent, 5 bytes	price per unit and currency table
EF_{SST}	'6F38'	transparent, 4 bytes	SIM service table
EF_{IMSI}	'6F07'	transparent, 9 bytes	IMSI
EF_{LOCI}	'6F7E'	transparent, 11 bytes	TMSI + local information
EF_{PHASE}	'6FAE'	transparent	phase information on GSM

Figure 13.5 GSM card simplified data tree

Immediately below the MF, the Smart Card's ID number (unique throughout the system) is held in a transparent EF. One DF is available as a directory for GSM data, and one for telecommunication data.

The GSM DF contains, for example, an EF (EF_{LP}) in which the preferred language is stored for displaying data to the user. Another EF (EF_{IMSI}) holds the IMSI used to identify the user and to prepare bills. Another file (EF_{TMSI}) stores the current TMSI with further location data. Since this file must be written to with each base station change and for each call, it is particularly protected by the card's operating system. The EEPROM sections, frequently restricted to 10***000 write/erase cycles, would not suffice here, since during a SIM's lifetime this data must be written far more often than that.

EF_{PHASE} stores information concerning the GSM 11.11 phase specified by the SIM. At the moment this value is likely to be 2. The SIM's second DF contains an EF (EF_{ADN}) with abbreviated dialling numbers, so that the user always carries with him in the SIM his personal "important numbers". Another EF (EF_{FDN}) contains the fixed dialling numbers. After activation of a mechanism in the GSM application, only the fixed numbers can be called, all others being blocked. The next three files (EF_{SMS}, EF_{SMSS} and EF_{SMSP}) contain the short messages as well as various related status data which can be received by air and then read out from the SIM at a convenient time. The last file we may mention (EF_{LND}) stores the last dialled number.

Originally it was intended that GSM Smart Cards would be replaced every two years, in order to prevent failures through the EEPROM's limited number of write/erase cycles. However, since few problems have so far come to light in this respect, most application providers only replace the cards in the event of failure. This saves each provider considerable expense, since their logistics department need only bother with replacing defective cards. The number of replaced cards is much lower than expected, as most last rather more than two years. This reduces manufacturing costs quite substantially.

The GSM application has meant an international breakthrough for Smart Cards, and is "the" standard for both cards and operating systems. GSM instructions and mechanisms may already appear outdated compared with the latest developments in the Smart Card sector, but it has been and still is the pioneer for large and international Smart Card applications. In the final analysis, all later applications can only learn from its experiences and problems, and profit thereby. In many respects, the GSM 11.11 specification represents the foundation for all new and eventually more complex Smart Card applications.

Application design

14

The first part of this chapter contains general notes and data for the implementation of Smart Cards. This data has been distilled from many applications and can be directly used for Smart Card application design. It also represents a summary of the latest technology.

The card bodies are not critical at the concept stage, since almost any module can be incorporated into any type of card. The difference often lies in the microprocessor used and its memory size, so these are the aspects which will be dealt with in detail here.

Much of the technical data is hardware-specific, but there are sufficient features in common to give general guidelines about application design, since most of the physical and electrical properties of processors available on the market are identical. The main differences are really only in the memory size.

The second part summarizes the principles of work tools, with which even complex applications can be generated simply and quickly without programming. The last section discusses several possible applications, illustrated by two examples. These are built up systematically, so that the reasons for the various mechanisms and routines can be easily grasped.

14.1 GENERAL NOTES AND DATA

14.1.1 Microprocessor

Production

If a new application requires the programming of a special Smart Card operating system, then this requires a great deal of time. A rule of thumb for the duration of the concept, programming and testing phases for a completely new system is around nine months. If it is possible to use library routines and existing program sections, an individual developer will still require about four to six months. Parallel tasking can only be performed to a limited extent, due to the very complex programming in Assembler. As a result, using current software technology and the maximum of effort and staff, writing a new Smart Card operating system will take at least two to three months.

Once a semiconductor manufacturer starts microprocessor production, it takes between eight to ten weeks before finished dice are available in intermediate numbers (i.e. several tens of thousands). Dice incorporation in the module then requires another week to a fortnight. Hence, turnaround time in microprocessor production is about 12 weeks.

Life expectancy

A Smart Card's life expectancy is effectively governed by the body's bending strength, resistance to corrosion by the contacts in the module and the number of EEPROM write/erase cycles.

The card body's life expectancy depends largely on the area of application. In typical GSM cards, which are permanently inserted in the mobile phone and are never removed, there is practically no limitation on the body. At the other extreme, for example in a company ID card which is also used in the canteen and for access control, the body may fracture after two or three years. Carrying the card in a wallet held in a hip pocket further contributes to the problem.

The second restriction on life expectancy is the maximum number of insertion cycles. The contacts are gold-plated for that reason, and can survive about 10***000 insertions into a card reader. Eventually, the surface becomes so scratched that the adhering dirt and grease prevents reliable electrical contact. In addition, the largely worn gold coating leads to oxidation, which also adversely affects electrical conduction. Naturally, the life expectancy of the contacts and the card is also very dependent on the type of card reader and the forces exerted by it. If optimally designed, life expectancy may be increased by a factor of between two and four.

The greatest limitation on a Smart Card's life expectancy derives from the number of write/erase cycles in the EEPROM[1]. Most semiconductor manufacturers guarantee between 10 000 and 100 000 cycles per EEPROM section. In practice, however, at room temperature and with small supply voltage fluctuations, an average EEPROM will only fail after 200 000 to 400 000 cycles. Such failure is gradual rather than sudden. A first sign of failing performance is when the first write attempt does not set the desired value in the EEPROM, and the written data is no longer stored in memory after a few hours. If such a memory section continues to be used, then after a few thousand further cycles no data can be stored in EEPROM at all, as it loses its contents at once. However, this only affects one section at a time (usually a 4- or 32-byte section). Other sections remain unaffected by this failure. This fact can be used as an error recovery strategy as part of memory organization.

Data storage in EEPROM is based on electric charges held in tiny capacitors. Like all such devices, these also suffer from current leakage which results in a charge loss with time and the stored data is thus no longer available. This effect is exacerbated by high temperatures. Hence the EEPROM's data content is not unlimited, and is only guaranteed by the manufacturer for 10 years.

[1] See also 3.4.2 Memory types.

The problem when calculating data lifetime is that measurement cannot be performed directly, otherwise one would need to wait 10 years. Discharge time can only be calculated from a measurement of the current leakage. Since this is technically almost impossible to achieve, the maximum data retention is obtained through variation of environmental parameters such as temperature and charging potential, from which deductions can then be made. Furthermore, data retention depends on the number of write/erase cycles. The 10 years specified apply to maximum loading so that, as a rule, much higher figures can be expected. For application design this means that data retention can be assumed to last for 10 years but, if possible, this maximum value should not be reached.

A general statement concerning a Smart Card's life expectancy can only be made if all the restrictions are taken into consideration. A practical guideline is three to five years for a conventional application which does not make undue demands on the card. In some fields, such as GSM, many firms are changing over to card replacement only as necessary, say in the event of failure, which is economically beneficial.

Data transmission

In addition to the internal speed for executing instructions, the Smart Card's operating speed depends mainly on data transmission rates at the terminal–card interface[2].

Function	Transmission speed at 3.5 MHz clock	Transmission speed at 4.9 MHz clock
Data transmission at 372	9600 bit/s	13212 bit/s
Data transmission at 512	6975 bit/s	9600 bit/s
Data transmission at 372, T=0 / T=1	≈ 7680 bit/s	≈ 10570 bit/s
Data transmission at 372, T=1 and Secure Messaging (authentic mode)	≈ 3800 bit/s	≈ 5200 bit/s
Data transmission at 372, T=1 and Secure Messaging (combined mode)	≈ 1900 bit/s	≈ 2600 bit/s

Figure 14.1 Typical data transmission speeds in contactless cards

Due to the asynchronous serial interface, a transmitted byte must be extended to 12 bits, since start bit, parity bit and stop bits need to be transmitted besides the 8 bits of actual data. A value of 1.25 ms/byte and a transmission speed of 9600 bit/s can be used as a rule of thumb. If 20% is added for the necessary transmission protocol, a value of 1.5 ms/byte is obtained. This is sufficiently accurate to be able to estimate data transmission speeds.

---

[2] See also 6.2 Transmission protocols.

Algorithm execution speeds

Since Smart Cards often serve as protected computers for the processing of algorithms, a few typical values are given in Figure 14.2. These are average figures which vary from one implementation to the next. The table assumes a 3.5 MHz or 4.9 MHz clock.

All values, other than the writing of data in EEPROM, are directly proportional to the applied clock frequency. Turnaround time can be multiplied by a factor of 1.4 when increasing the clock frequency from 3.5 to 4.9 MHz, which is possible in every microprocessor currently on the market. Writing or erasing of EEPROM data is independent of the clock, and cannot be accelerated.

Function	Absolute value		Transmission	
	3.5 MHz clock	4.9 MHz clock	3.5 MHz clock	4.9 MHz clock
XOR-calculations	1 μs/byte	0.7 μs/byte	1 Mbyte/s	1.4 Mbyte/s
CRC-calculations	0.2 ms/byte	0.14 ms/byte	5000 byte/s	7100 byte/s
Writing or deleting data in the EEPROM (page size 4 bytes)	3.5 ms/4 byte	3.5 ms/4 byte	1142 byte/s	1142 byte/s
Creation of 8 byte random number	26 ms	18.5 ms	—	—
DEA-calculation (8 byte block, ECB-/CBC-mode)	17 ms/8 byte	12 ms/8 byte	470 byte/s	660 byte/s
DEA-calculation of MAC (8 byte block)	17 ms/8 byte	12 ms/8 byte	470 byte/s	660 byte/s
IDEA-calculation (8 byteblock, ECB-/CBC-mode)	12.3 ms/8 byte	8.8 ms/8 byte	650 byte/s	910 byte/s
RSA-calculation (512 bit key length with NPU)	308 ms/64 byte	220 ms/64 byte	208 byte/s	290 byte/s
RSA-calculation (768 bit key length with NPU)	910 ms/64 byte	650 ms/64 byte	105 byte/s	148 byte/s

Figure 14.2 Typical transmission speeds of algorithms in contactless cards

14.1.2 Application

Besides the specific factors relating to the microprocessor, further aspects need to be considered during the concept and development stages of an application. As far as the former is concerned, this applies mainly to key distribution, administration of application data and the fundamental principles involved in data exchange.

Key administration

In all applications which use cryptographic algorithms, security relies on the confidentiality of the appropriate key. If a secret key is revealed for any reason, then all the security mechanisms based on it will crash. Since this eventuality cannot, in principle, be ruled out with complete certainty, precautions must be taken against it[3].

The most trivial and expensive option is to replace all the cards, but this is economically unacceptable in large applications. In practice, therefore, other methods are used to minimize the consequences of revealed keys.

In principle, only card-specific keys should be held in Smart Cards. Each card then holds individual keys, so that should this key be discovered, it can only be used to clone this one card. The corresponding main key, however, must remain secret under all circumstances.

In order to keep the main key secure, several generations of card-specific keys can be stored in the card, so that if necessary it is possible to switch over to a new generation.

The third principle governing keys is their separation by function. The consequence of this policy is that, for example, an authentication key may not be used for encryption. This allows the repercussions of the discovery of a key to be minimized.

The three principles discussed above must be observed in all large applications, as they guarantee the application provider a significantly more secure system, and in certain circumstances will save him a great deal of money that would otherwise be spent on replacing all the cards. However, in this context we must address one important point: each of these requirements leads to an increase in the number of keys. This means that possible memory problems resulting from this application will have to be taken into consideration, and that the system will also require a more or less complex key management system. In practice, therefore, it is necessary to choose very carefully whether all three criteria really need to be met, or whether it is possible to compromise over the full implementation of certain of these.

Data

The principles that are standard in the IT industry should be adhered to when storing data in Smart Cards. For an application this means that classificatory number systems should be avoided as far as possible, since they often "crash" even in the case of small modifications or extensions. Identificatory numbering systems, however, are frequently too abstract, so that in practice mixed numbering is the rule.

A good example for this effect is provided by telephone numbers. The first part, the area code, is classificatory. Knowledge of this number allows one to say with certainty in which geographical region the number is located. The second part, the subscriber number, is purely identificatory, since at least in small towns it provides no information about the subscriber's precise location. The two parts together result in a mixed number, which enables fairly straightforward extensions.

The ASN.1-coded data object is very suitable for dealing flexibly with data elements in many different versions[4]. As a rule of thumb based on current practice, the additional

[3] See also 8.4 Key management.

[4] See also 4.5 Data structuring.

coding overheads needed represent approximately 25% of the data. This only applies to small data sets, which are common in the Smart Card domain.

We must make another comment in respect of Smart Card memory size. Applications with over 2 kbytes of user data are extremely rare, since as a rule the available memory is very restricted in size.

When planning a new application it is necessary to calculate, at least in approximate terms, the amount of memory which needs to be reserved. Not only the size of the user data is needed, but also the required volume of administrative data. In newer operating systems which allow for several applications to be present simultaneously on one card and possess object-oriented file management, the proportion of administrative data is relatively high. The size of file headers is usually between 16 and 32 bytes, which is a fixed figure, independent of the amount of user data held in the file: it makes no difference whether one byte or 200 bytes of user data are present. As a result, it is preferable not to set up an individual file for each data element, since otherwise too much memory is wasted on administration.

Data exchange

Compared with the performance of modern computer systems, the card's serial interface is very slow. The conventional 9600 bit/s renders data transmission very robust against interference, but means that data exchange takes a great deal of time. It is therefore recommended to ensure that communication between terminal and card is limited, in every case, to essentials. Data which the terminal has already obtained once during a session, should not be requested again. Naturally, this does not need to be a consideration in applications in which time is a low priority. However, where people are involved in the process, minimizing the data exchange at the interface must always be a priority.

The following points are also relevant here. With secure messaging[5], all data can in principle be very effectively protected at the terminal/card interface. An attack at this interface is all but impossible. Nevertheless, this sharply reduces the effective transmission speed. Hence data should only be transmitted with secure messaging when this is unavoidable for system security.

With the exception of secret keys, it is sufficient for almost all user data to be protected by a MAC (authentic mode) only. This does not reduce transmission speed as severely as the additional encryption in the combined mode. Also, transmission is transparent at the interface and can easily be checked from outside. This may be necessary in the context of data protection legislation, since in this way it is possible, at any time and without knowledge of the secret keys, to check which data passes across the interface.

14.1.3 System

Security

DEA is used as a cryptographic algorithm in many systems. The reasons for this are manifold, the primary one being that DEA is very well known. However, besides DEA

---

[5] See also 6.6 Secure messaging.

there are many other good data encryption and decryption algorithms. Due to its familiarity, DEA is also subjected to the most attacks, which may sometimes be successful. Hence in the case of large Smart Card projects, the option of using an alternative to DEA to minimize the risks should be considered. The current GSM algorithms provide an excellent illustration of this philosophy, as they were developed specially for this application and are totally independent of DEA.

User interface

User acceptance is critical for a successful Smart Card project. Although this has much to do with the terminal's man–machine interface, the interface between the terminal and the card is also of importance. Experience shows that user acceptance decreases when a transaction lasts for more than one second, as the user will often suspect a technical fault and attempt to withdraw their card, leading to an uncontrolled interruption of the session. This can be prevented by optimizing all procedure executions to take less than one second.

Above all, it is paramount to remember, when designing the man–machine interface, that if user–acceptance is insufficient there will be considerably more technical problems through unexpected interventions (e.g. withdrawal of the card from the terminal).

Concept

When designing a Smart Card IT system, Kerckhoff's principle, that system security relies solely on the secret keys, should always be considered. This is because testing a system based on the confidentiality of the data is very difficult, since it can only be carried out by a very limited number of people. Since this group have usually also designed most of the system, such testing cannot be very effective. Total revelation of all the internal data is, however, also problematic, since it would allow a potential attacker to gain free access to them. In practice, therefore, a compromise has to be chosen, in which the system's fundamental design is open but specific data is kept confidential.

If, in contradiction to Kerckhoff's principle, certain items relating to a Smart Card project must remain secret, then logically security should not rely on one single person. It is better to distribute knowledge between several people who would each be familiar with only part of the system.

During the design of Smart Card-based systems, the entire system should always be looked at and not merely the card and its immediate environment. This is the most common mistake made during a project, and one which should be carefully avoided. Due to the Smart Card's active character, these systems are always distributed, their components must react autonomously and there is no pure client–server relationship between them. After integration, the whole system must cooperate smoothly, and not just the card and terminal. Therefore it is necessary to maintain an overall perspective from the early stages of the design process onwards.

14.2 AIDS TO THE GENERATION OF APPLICATIONS

Several PC-based programs exist nowadays for designing Smart Card applications. They allow quick and simple production of complete applications, without internal knowledge of the relevant operating system.

These tools are usually first employed to construct a file tree, in which the various applications (i.e. DFs) and the corresponding EFs are then loaded. It is of course necessary with the EFs to establish both the file structure and the relevant access conditions. If the card's operating system also possesses a state automaton for the instructions, this can also be defined on the PC's graphical user interface. Some of these development programs also permit a consistency check for the defined automata. Since the application needs various data and keys in its EFs, this can be followed by setting up a link to a database. Thus the contents of the EFs in the individual cards are connected to the database's data sets.

Once the complete application is defined, various global parameters can be established for the card's operating system. This might include the transmission protocol and its divider. The application can then be experimentally loaded into one or several Smart Cards with the appropriate memory size. When a few trial cards have been produced, they are tested in a terminal. If it appears that modifications are necessary, it is possible to erase the application and subsequently load a revised version.

If all tests are concluded satisfactorily and a larger number of cards becomes necessary, a regular production setup can be used to produce the required number. The application data generated via the PC program (files, instructions, states, etc.) forms the basis for the card's completion, initialization and personalization, so that the turnaround time remains very low despite the high degree of flexibility.

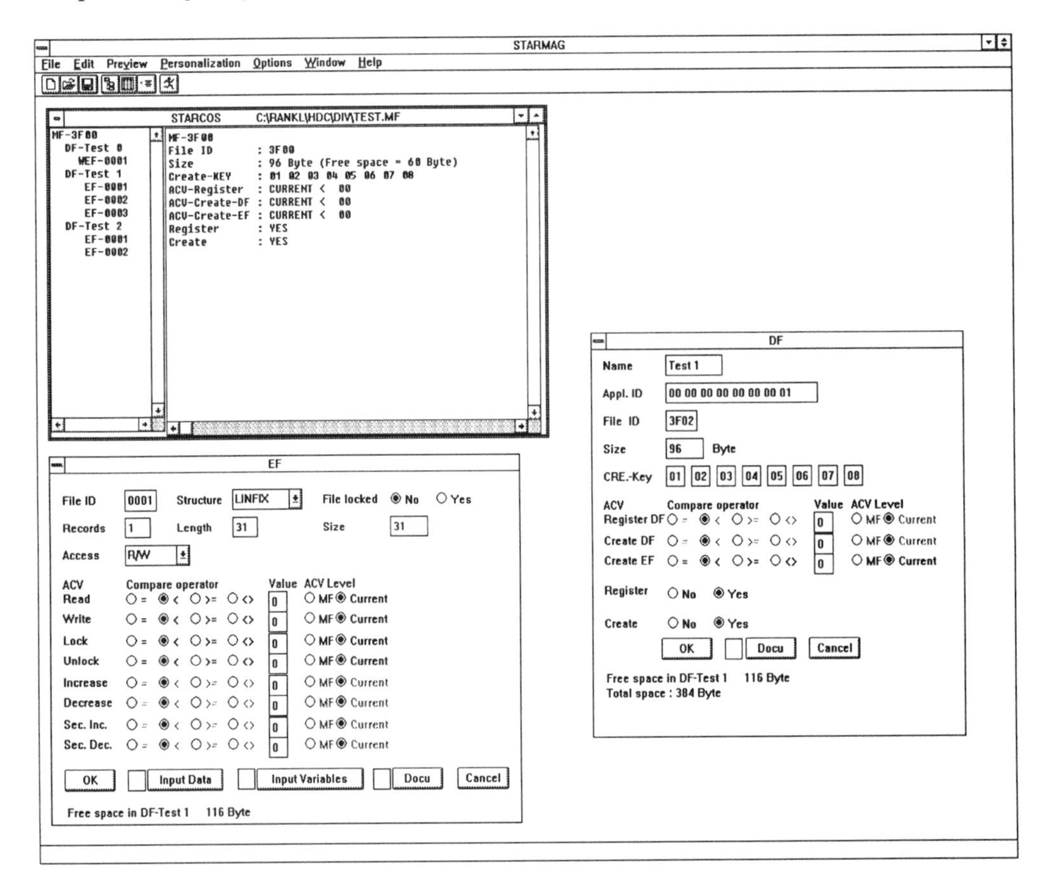

Figure 14.3 Screen output for PC-based software tool for the creation of complete applications for contactless cards (Giesecke und Devrient)

Alongside these tools for application design and production, there are also Smart Card simulators. They behave in an identical manner to normal cards, but consist only of a dummy Smart Card linked to a PC interface. Suitable software on the PC simulates the card in real-time. Naturally, all the applications can also be generated and tested on the PC in the way described above. However, a PC is always needed for simulation which is often problematic due to the size.

14.3 EXAMPLES OF APPLICATION DESIGN

The following examples illustrate two typical Smart Card applications. These are mid-range DP applications, and do not require large-scale system design. They may be implemented in medium-sized firms. The background system employed may be a PC in a secure environment, in which case installation and running costs are fairly low.

These simple examples demonstrate very effectively how a typical Smart Card application is constructed. This will be explained step by step, and the gradual construction up to the finished application in the card will be demonstrated. The terminal and background system aspects are only treated in outline, but can be derived from the available data.

The basic principle behind these examples is the distribution of information among many individual systems. This contrasts with the common centralized mainframe solutions, in which all the information is concentrated in one point. When translated to Smart Card applications, this means that the card is only a kind of ID document, and all the data is stored in an omniscient background system. The extension of such an application, then, regularly leads to the necessity to redesign the all-controlling background system, a very expensive and time-consuming process.

Here an attempt was made to find an alternative solution. The background system is only responsible for the overall system's administration and consistency. All the other data is held locally in the cards. A global database is necessary to administer the system, carrying out such tasks as reproducing lost or faulty cards by using the stored data. However, this database must not be central to the system's operation.

A branched Smart Card system may be thought of as a large tree, deriving its energy from photosynthesis in the numerous leaves. Energy extraction is thus located simultaneously in the tree's many leaves. Figuratively, an effective and efficient Smart Card application behaves in the same way. The data is stored in decentralized cards and is thus safe from attack. In this way the cards do not put heavy demands on the background system, which only deals with the centralized administrative tasks. The actual routines are decentralized, just like the process of photosynthesis in the tree's leaves, and take place in parallel in the terminals and cards. This makes it very simple to extend the overall system through additional terminals and cards, without needing to worry about large-scale consequences for the background computer.

The opposite approach to the system just sketched out is one where the entire work is dealt with centrally in the background system. In our analogy, photosynthesis would be transferred from the tree's leaves to its trunk, which would lead to a huge trunk. The overall system would then become not merely very large and expensive, but also extremely susceptible to computer failures. This solution, then, should be avoided if possible.

In their ignorance of the characteristics of Smart Cards, many new system operators make the mistake of designing a whole system top-down. The operators then often specify that the security-sensitive areas of the system, the cards and terminals, should be made secure in a more or less unknown way.

The outlines described here represent, in this sense, a bottom-up strategy, in which by starting with the lowest link in the chain – the Smart Card – and moving upwards, the system and its required properties are all properly defined. The risk of security gaps can thus be efficiently minimized, since the system is constructed securely from the smallest unit towards the top.

14.3.1 Purse for a gaming machine

Starting point

The aim of this Smart Card application is to make it possible to pay small amounts into the games machine. The coins used conventionally are replaced by Smart Cards in order to reduce operational costs.

Two types of terminal exist. The loading terminal has a slot for coins and a testing unit, and can load electronic money units onto the card. These units are then deducted from the card by the largely autonomous billing terminal.

Requirements

The entire system should be anonymous, whilst still allowing the monitoring of all fund movements. If fraud is suspected, it is necessary to be able to identify cards and selectively block them.

Due to the field of application, where monetary transactions are made and the machines are fully automated rather than supervised by people, an intermediate security level is indicated.

Proposed solution

The solution is based on a simple closed purse, designed specially for this one application. Naturally, it could easily be modified for similar applications by changing the files and routines. Implementation of an electronic purse after CEN prEN 1546 was avoided here, as it would be more expensive for the application provider than the proposed solution. This explanation is also intended to demonstrate the principle of a simple closed purse.

A payment machine is placed in a central position. It accepts both coins and notes, and transfers them to the card as electronic money units. Neither a PIN nor other user input is necessary, as the purse is anonymous. The only additional book-keeping of the credited amounts relies on the card's number, which is unique in the system.

All data and fund movement administration is carried out by a PC. The PC also runs a database, which stores all the issued cards with their parameters. Daily or weekly balancing ensures that the system's monetary transactions are still closed.

The loaded units are debited to the electronic purse at the billing terminal. A display shows the user the amount debited. To keep the cost of the terminals as low as possible, data transmission is secured not by a shutter but by secure messaging. Each terminal possesses a security module for the secret key and the book-keeping of the amounts, sorted by card numbers. At regular intervals, the data obtained is transferred by cable or via a special transfer card to the administration PC, which then undertakes an evaluation. The solution proposes the file tree shown in Figure 14.4 in the card. Depending on the card's operating system, the application requires around 100 bytes in EEPROM.

File type		FID	Structure	Description
MF		'3F00'	—	root directory of contactless card
DF		—	—	directory for "arcade games" application
	EF 1	'0001'	transparent	date of issue card number
	EF 2	'0002'	cyclic	amount
	EF 3	'0003'	linear fixed	key 1 key 2 key 3 key 4

Figure 14.4 File tree for the "arcade games" application

Since a data exchange takes place at regular intervals between the billing terminal and the administration computer, this can also be used to update a warning list in the terminal. If the terminal discovers that an inserted card is listed, it blocks file EF2 which contains the electronic money. This makes payment with this card impossible, and the user needs to unblock the card at the administration terminal, where a check can be conducted into why the card appears on the list.

Key	Application	Function	Change of condition
Key 1	MUTUAL AUTHENTICATION	mutual authentication of terminal and contactless card – payment through purse – blocking purse	x ➔ 1
Key 2	MUTUAL AUTHENTICATION	mutual authentication of terminal and contactless card – unblocking purse – administration of card	x ➔ 2
Key 3	MUTUAL AUTHENTICATION	mutual authentication of terminal and contactless card – loading purse	x ➔ 3
Key 4	Secure Messaging	securing data transmission – authentic mode	—

Figure 14.5 Necessary key for the "arcade games" application

The implementation of derived and thus card-specific keys was waived in favour of a simpler overall system (Figure 14.5).

The proposed solution is very suited to payment for received services at an automatic machine. Human supervision is unnecessary. However, the funds need not always be loaded in the card automatically at a special machine: this can take place manually at a cash till after payment of the appropriate sum.

The system outlined here can be used, with minor modifications, at a launderette or canteen.

File	Read	Write	Block	Unblock	Increase amount	Lower amount
EF 1	≥ 0	$= 2$	< 0	< 0	< 0	< 0
EF 2	$\geq 0 \perp$ SM	< 0	$= 1$	$= 2$	$= 3$	$= 1$
EF 3	< 0	< 0	$= 1$	$= 2$	< 0	< 0

Figure 14.6 Access conditions for "arcade games" application (≥ 0: always, < 0: never, SM: Secure Messaging)

During the loading of electronic monetary units, the generalized steps shown in Figure 14.7 occur. Figure 14.8 shows how payment takes place.

Smart Card		Terminal	User
	→	SELECT FILE (DF)	*Message* "insert coins"
	→	SELECT FILE (EF 1)	
	→	READ BINARY	
	→	SELECT FILE (EF 2)	
	→	READ BINARY	
	→	ASK RANDOM	
	→	MUTUAL AUTHENTICATION	
		Switch on Secure Messaging	
	←	INCREASE	
Payment [... ǁ Returncode]	→	IF (Returncode=OK) THEN command successfully completed ELSE cancelled	*Message* "£ x.xx loaded onto the Smart Card"

Figure 14.7 Theoretical execution of commands at loading of electronic monetary units in the "arcade games" application

Smart Card		Terminal	User
	→	SELECT FILE (DF)	
	→	SELECT FILE (EF 1)	
	→	READ BINARY	
	→	SELECT FILE (EF 2)	
	→	READ BINARY	
	→	ASK RANDOM	
	→	MUTUAL AUTHENTICATION	
		Switch on Secure Messaging	
	←	DECREASE	
Response [... ‖ Returncode]	→	IF (Returncode = OK) THEN command successfully completed ELSE cancelled	*Message* "£x.xx debited on Smart Card"

Figure 14.8 Theoretical execution of commands when debiting electronic monetary units in the "arcade games" application

14.3.2 Access control

Starting point

The purpose of this application is to create a Smart Card-based, graduated access control system to various rooms and computers. Terminals are therefore fitted on certain doors and computers, which can unlock them after communicating with the card. It is important to define the security levels and establish the groups and users who can access each level. Access is granted on the basis of authentication and user identification. Identity is proved, on the one hand, by possession of a genuine card, and on the other by knowledge of the corresponding PIN. If both criteria are met, access is permitted. The terminals must be able to contain simple warning lists to prevent access via "lost" cards, and if necessary, to block such cards irreversibly.

Requirements

In order to maximize user acceptance of the solution, the process of communication between the terminal and the Smart Card, and the subsequent lock release, must not significantly exceed one second. Any longer, and the solution's acceptability would sooner or later diminish, and incite users to circumvent security by the use of various methods such as jamming the door open. In order to obviate the need to write the PIN on the card as an aid to memory, it must be possible for the user to change it.

The system's security level should be designed to be basic, as it is unlikely to be subjected to complex attacks or analysis. The cost of introduction and operation of the system should not significantly exceed that of good-quality locks, as otherwise the latter would be a more financially viable option. The system and card should be laid-out in such a way that time-logs and canteen bills could be integrated at a later stage.

Proposed solution

Simple terminals with numeric keypads are permanently installed on the required doors and computers. The terminals can work autonomously; are fitted with economical and replaceable security modules (e.g. a Smart Card in the plug-in format); and can release the lock on the relevant door or computer when appropriate. All terminals located at central and sensitive areas can, if necessary, establish an autonomous online link to the PC acting as a main computer via a dual cable. The simple structure ensures that the cost of this solution remains low.

The main computer is equipped with a simple multi-tasking operating system, so that several functions can be performed in parallel. It is connected to an additional terminal responsible for overall system management.

The Smart Card must possess an operating system capable of managing several applications, and of subsequently generating files (DFs and EFs) in the card's file tree, thus being able to load further applications whilst connected, if necessary. The EEPROM, required both for the original application and for those projected for the future, does not exceed 1 kbyte. The cards are obtained from the manufacturer complete with operating system, and appropriately adapted for use with the management computer.

All cards are initialized with a uniform and easily remembered PIN, '0000' being the simplest. This obviates the complication of PIN generation and preparation of PIN correspondence to users. Each user must alter the uniform PIN at the management terminal on receipt of the card, since the terminals reject a PIN input of '0000'.

If the user forgets the PIN or the error counter reaches its limit, a new PIN can be input at the system terminal after authentication, and the error counter reset.

Since the required security level is intermediate and the administration must not be too costly, a sharply reduced key management system is recommended. Hence, there is no need in this solution for derived keys or for several key generations. The keys should only be separated by function, which leads to the breakdown shown in Figures 14.9 and 14.10. Figure 14.11 shows the access conditions defined for the card's file tree.

Key	Application	Function	change of condition
Key 1	MUTUAL AUTHENTICATION	application administration – creating new files – writing in files –unblocking application	x → 1
Key 2	MUTUAL AUTHENTICATION	mutual authentication of terminal and Smart Card –right of access – blocking application	x → 2
PIN	VERIFY CHV	user identification	2 → 3

Figure 14.9 Necessary keys for the "access control" application

File type	FID	Structure	Description
MF	'3F00'	—	Smart Card root directory
DF	—	—	directory for the "access control" application
EF 1	'0001'	transparent	name, first name, department
EF 2	'0002'	linear fixed	reasons
EF 3	'0003'	linear fixed	key 1 key 2 PIN

Figure 14.10 File tree for the "access control" application

File	Read	Write	Block	Unblock	Creation of DFs and EFs
DF	—	—	—	—	= 1
EF 1	≥ 0	= 1	< 0	= 1	—
EF 2	= 3	= 1	= 2	= 1	—
EF 3	< 0	< 0	= 2	= 1	—

Figure 14.11 Access conditions for the "access control" application

The file EF2 with the authorization data for access to rooms and computers is record-oriented in structure. The records are all of uniform length (linear fixed). Each record has one entry which indicates to which rooms the card holder has access. Security levels may be defined, so that each room does not need to be explicitly specified. Access can then be restricted globally to certain areas.

Since experience shows that in access control, particularly, changes and restructuring are often needed, the records' contents should have a TLV structure. This allows extensions and modifications to be carried out in a technically elegant manner.

Only standard instructions are used, incorporated into an ISO- or ETSI-based operating system available on the market. This means that nothing need be programmed in the card, which considerably reduces acquisition costs. The instructions needed are the following:

- ASK RANDOM
- CHANGE CHV
- CREATE
- INVALIDATE
- MUTUAL AUTHENTICATION
- READ BINARY
- REHABILITATE

- SELECT FILE
- UNBLOCK CHV
- VERIFY CHV
- WRITE BINARY

Figure 14.12 shows the typical execution which occurs for access functions during Smart Card use.

Smart Card		Terminal	User
	←	Reset	
ATR	→	IF ATR = OK THEN continue ELSE cancel	
	→	SELECT FILE (DF)	*Message* "Please put in PIN"
	→	VERIFY CHV	
	→	SELECT FILE (EF 2)	
	→	READ RECORD	
		evaluate file contents	
		IF (permission = yes) THEN activate door opener	*Message* "Please enter"

Figure 14.12 Progress of commands during access control in the "access control" application

If necessary, the terminal can (directly after the ATR) read the user's name from the file with READ BINARY, and compare it with a warning list. If the name is listed, it is possible, through the blocking of all EFs with INVALIDATE, to prohibit further use of the card for access. At the management terminal one may then, if desired and after mutual authentication, reactivate the application with REHABILITATE.

Access to computer systems is completely analogous to the process described above. The only difference is that instead of a door release mechanism being activated, a signal is sent to the computer authorizing access.

If the system operator also decides to carry out canteen payments through the card, a new application with DF and EFs must be generated. This can be done in two ways. Either all personnel must go to the management terminal with their cards, or the files are automatically loaded during an access procedure. The second approach is certainly the cheaper and more user friendly, since it does not give rise to further administrative costs.

Appendix

15

15.1 GLOSSARY

Acquirer

An entity which operates and administers the transfer of data from the operator of a payment system to the individual service provider. The acquirer can consolidate all the individual transactions, so that the payment system operator receives more organized data.

Administrative data

Data which is only administrative and has no other function in an application.

AFNOR

The Association Française de Normalisation is a French Standards Organization based in Paris.

ANSI

The American National Standards Institute is an American Standards Organization based in New York.

Application

Data, commands, procedures, conditions, mechanisms and algorithms for use with an integrated circuit(s) card, allowing it to be used within the framework of a certain system.

Application operator

An entity using an application on an integrated circuit(s) card. Generally identical with the application provider.

ASN.1

Abstract syntax notation is a data description language. It allows data to be clearly defined and viewed independently by computer systems. ASN.1 is defined by ISO/IEC 8824 and ISO/IEC 8825.

Authentication

Procedure to prove the authenticity of an entity (e.g. machine) via a cryptographic process.

Authenticity

Truth and consistency of a message.

Auto-eject reader

Card reader which can automatically eject an inserted card.

Background system

A computer system which takes over the processing and administration of data from the terminal hierarchy.

Brute force attack

Attack on a cryptographic system through calculating all the possibilities of a key.

Buffer

Attack on magnetic stripe cards, in which they are read, data is changed via a terminal Reading of a magnetic card and writing back, after a terminal had changed data on it (e.g. fault presentation counter).

Card body

Plastic card into which further functional elements can be incorporated following certain production methods (e.g. implanting chip).

Card issuer

An entity which issues users with cards. Usually the card issuer is also the application provider, but this need not be the case.

Card manufacturer

An entity which produces the actual cards and incorporates modules in them.

Card owner

Owner of the card, usually also the user.

Card reader

Largely mechanical device, which receives data from an integrated circuit(s) card via galvanized contacts.

Card user

Person using the card. This does not have to be the owner.

CCITT

The Comité Consultatif International Télégraphique et Téléphonique is an international organization for telephone and telegram services based in Geneva. The work of the organization includes making suggestions for the international standardization of data and telephone transmissions.

CCS

This is a cryptographic checksum of data, which recognizes manipulation of data during storage. If data is protected during transmission by a CCS, the term MAC (message authentication code) applies.

CEN

The European Standards Organization Comité Européen de Normalisation, in Brussels, Belgium, works with all national European standards organizations and is the "official institution of the EU for European Standardization".

CEPT

The Conférence Européenne des Postes et Télécommunications is a European Standards Organization composed of the national telecommunications organizations.

Clearing system

A computer background system which, in an application, assumes the task of central calculations for electronic financial transactions.

Cloning

Attack on circuit card system through complete copying of ROM and EEPROM of microcontroller.

Closed application

IC card application which is only available to a proprietor and cannot be used generally.

Closed purse

Closed application for an electronic purse. Can only be used within the framework set by the user and not for general payment transactions.

Contactless credit card

Card in which energy and data transmission is achieved without contacts via electro-magnetic fields.

Credit card

Card which is not prepaid. Payment occurs after purchase of goods or services.

Debit card

Card which is pre-paid and debited on receipt of goods or services.

Die, dice

Silicon crystal, on which a single half-duplex microcontroller is built.

Digital signature

Often referred to as an electronic signature, it is used to establish the authenticity of electronically transmitted messages or electronic documents.

Download

Transmission of data from a master system (e.g. Host) to a slave system (e.g. Terminal).

DRAM

Dynamic RAM (dynamic random access memory) requires a constant power supply and periodic refreshing of the contents to preserve them. DRAM takes up less space on the chip than SRAM, but SRAM is faster to access.

Duplicate

Transfer of data onto a second card.

ECC

The error correction code is a checksum which can to a degree detect and correct errors in the data.

EDC

The error detection code is a checksum which can to a degree detect errors in the data. A typical example of an EDC is the XOR- or CRC-checksum used in various data transmission protocols.

EEPROM (electrical erasable programmable read only memory)

A non-volatile memory form, used in IC cards. An EEPROM is divided into memory pages and its content can be altered or deleted, but there is a physically determined limit to the number of write operations.

Embossing

Part of the physical personalization process, where raised characters are imprinted onto the plastic body of the card.

EMV

General specification for Europay, Mastercard and Visa IC credit cards and terminals.

EPROM (erasable programmable read only memory)

A non-volatile memory form, which is seldom used now in IC cards. EPROM can only be deleted through UV light and can therefore only be used for WORM storage (write once read many) with IC cards.

ETS

The European Telecommunication Standard is the description of a standard provided by ETSI, which is mainly concerned with European telecommunications.

ETSI

The European Telecommunications Standards Institute based in Sophia Antipolis, France, is the standards institute for European telecommunication companies.

Finalization

Completion of operating system through loading EEPROM parts. This enables subsequent changes and modifications to be made without having to create a new ROM mask.

FIPS

The term Federal Information Processing Standard describes the American standard, supplied by NIST.

GSM

The Global System for Mobile Communications is a specification for an international, terrestrial mobile telephone system. Originally intended to cover a few countries in central Europe, it is increasingly developing into an international standard for mobile telephones.

Hash algorithms

Process for the compression of data through an irreversible function, so that the original data cannot be calculated backwards. These processes are created in such a way that any changes to the original data can affect the hash value calculated.

Hybrid card

Describes a card with two contrasting card technologies. A typical example is a card with a magnetic stripe and an additional chip.

ID-l card

Standard format for integrated circuit(s) cards (length = 85.6 mm, width = 54 mm, thickness = 0.76 mm).

Identification

Process for establishing the authenticity of a machine or a person using a password.

Initialization

Loading independent application data into the EEPROM.

Irreversible function

Mathematical function for which there is no inverse function or whose inverse function can only be calculated with great difficulty.

ITSEC

The Information Technique System Evaluation Criteria were released in 1991 and represent the catalogue of criteria for the evaluation and certification of security information technical systems in Europe.

Key fault presentation counter

Fault counter which decides whether or not a certain key may be used again. If the fault counter reaches the maximum value, the key is locked and can no longer be used. The fault counter is normally reset to zero upon completion of a successful action.

Laser engraving

Process in which special plastic layers are burnt with a laser, creating a blackened effect.

MAC

The message authentication code (data security code) is a cryptographic checksum for data, which recognizes if data has been manipulated during transmission. If data is protected with a MAC while being stored, the term CCS (cryptographic checksum) is applied.

Magnetic stripe card

Card with a magnetic stripe serving as a data carrier.

Memory card

Card with a chip, which has a simple logic and additional read and write memory.

Module

Base for a die and its contact elements.

Module manufacturer

A company which builds the dice into the module and creates an electronic connection by bonding it with the contact elements.

Multi-application IC card

Integrated circuit(s) card which is used in numerous applications, e.g. bank card with telephone card function.

NBS

Before 1988 the NIST was known as the National Bureau of Standards (NBS).

NCSC

The American National Computer Security Center is a subsidiary of the NSA and responsible for testing security products and publishing criteria for safety in computer systems, including the TCSEC.

Negative file

A file which lists all IC cards which can no longer be used with particular applications.

Negative result

The case in which a logical decision leads to a less advantageous or unwanted result.

Nibble

The four most significant or least significant bits of a byte.

NIST

The American National Institute of Standards and Technology is a part of the US Department of Commerce and responsible for the national standardization of information technology. Known as NBS until 1988. The NIST publishes the FIPS codes of standards.

Noiseless

Property of a cryptographic algorithm, which means that encrypting and decrypting will always require the same amount of time, regardless of coding, or whether the text is simple or encoded. If a cryptographic algorithm is not noiseless, the key possibilities can be limited by an analysis of the calculation time.

Non-volatile memory

A kind of memory (e.g. ROM, EEPROM), which retains its contents even without power supply.

NSA

The American National Security Agency is the official institute for communication security of the US government. It reports directly to the Department of Defense and one of its functions is to tap in to foreign communications and decode them. They also develop new crypto-algorithms and limit the use of existing ones.

Open application

Application on an IC card, which is open to many application holders and can be widely used.

Open purse

Open application for an electronic purse. Can be used for general payment transactions.

Operating system manufacturer

A company which carries out programming and testing of an operating system.

Optical memory card

Card in which information is burnt into a reflective surface (similar to a CD).

Paging

Compressing numerous memory bytes into so-called pages, which can only be written or deleted as a whole. Paging is only available as EEPROM with circuit card microcontrollers. The standard page size today is 4 bytes, i.e. 32 bytes.

Personalization

Process whereby a card is assigned to a person. This can occur through physical personalization (e.g. embossing, laser engraving) or through electronic personalization (i.e. loading personal data in the circuit card's memory).

Personalizer

Entity which carries out the personalization.

Plug-in

Circuit card in very small format, used especially in the GSM field (length = 25 mm, width = 15 mm, thickness = 0.76 mm).

Positive result

A case in which a logical decision leads to a more favourable or intended result.

Processor

The most important functional unit on a microcontroller. It carries out the commands and memory accesses incorporated in the program.

Protection

Protective surface on a half-duplex, protecting it from oxidation and other chemical processes.

Purse-to-purse transaction

Transfer of electronic monetary units from one purse directly to another, without having to go over a third, slave system. As a rule, this function means that the purse system has to operate anonymously and the electronic purse has to use a main key for this function.

RAM

Volatile memory, used as working storage in circuit cards (random access memory). The RAM loses its contents in a power failure.

Record

A piece of data, usually one of a number of pieces of data, each with a similar internal structure.

Reset

Resetting a computer (in this context, an IC card) to a clearly defined original state. One speaks of a cold reset or a power-on reset if, to reset, the power supply voltage is switched off and on again. A warm reset is done via a signal on the reset cable to the circuit card, the power supply voltage remaining untouched.

ROM (read only memory)

A non-volatile memory system, used in IC cards. Mainly used to store programs, as ROM contents cannot be changed.

ROM mask

Mask for the creation of ROM with half-duplex. The expression is also used to describe the data contents of ROM with circuit card microcontrollers.

Scrambling

Mixed arrangement of buses (address, data and control bus) on the chip of a microcontroller, so that assignment according to function is no longer possible without background information.

Secure messaging

Data transmission on an Interface, secured against manipulation (secured with MAC, i.e. authentic mode) or against tapping in (encoding, i.e. combined mode).

Security module

A building block secured mechanically by IT, for the storage of confidential data (SAM – secure application module).

Session

Time between switching on and off of an IC card, in which the whole data exchange and necessary IT mechanisms run their course.

Shutter

Mechanical element in terminals, which cuts off all wires leading from the IC card to the card reader, in order to prevent manipulation of communication. If a separation is impossible, the inserted circuit card will not be electronically activated.

SIMEG

The Subscriber Identification Module Expert Group is a group of experts who, within the framework of ETSI, lay down the specifications for the Interface between circuit card and mobile telephone (GSM 11.11). Composed of representatives of card and mobile phone manufacturers and networks.

Smart Card

Storage card with expanded logic for additional security functions which control memory access.

SRAM

Static RAM (static random access memory) requires a constant power supply but does not need to be periodically refreshed to maintain the store contents. Access time to the SRAM is less than DRAM, but SRAMs take up more space on the chip.

TCSEC

The Information Technique System Evaluation Criteria were published in 1985 by the NCSC and are a catalogue of criteria for the evaluation and certification for the security of information technology systems in the United States.

Terminal

Complement to circuit cards. A device, with keyboard and display, which enables power supply and data exchange with the IC card.

TLV format

A data format conforming to ASN.1, which describes a Value through a chosen Tag and Length.

Trapdoor

Mainly a mechanism in software or algorithms used to circumvent security functions or protective mechanisms.

Trojan horse

Historically, the wooden horse which enabled Odysseus to gain entry into the heavily fortified city of Troy. It is now a program which primarily fulfils a defined job, but also has additional, unknown effects. It is introduced purposely into a program, and unlike a virus it cannot multiply.

Upload

Transfer of data from a slave system (e.g. terminal) to a master system (e.g. host).

Useful data

Such data as is necessary to an application.

User

Person using an integrated circuit(s) card. This need not be the card owner.

Volatile memory

A form of memory (e.g. RAM), which only retains its contents with constant power supply.

15.2 LITERATURE

Beutelsbacher, A., Kersten, A. and Pfau, A.: *Chipkarten als Sicherheitswerkzeug*, Berlin: Springer Verlag, 1991

Beutelsbacher, A.: *Kryptologie*, 3rd Ed., Braunschweig: Vieweg Verlag, 1993

Europäische Gemeinschaften-Kommission: *Kriterien für die Bewertung der Sicherheit von Systemen der Informationstechnik (ITSEC)*, Version 1.2, June 1991

Fumy, W. and Ries, H. P.: *Kryptographie*, 2nd Ed., Munich, Vienna: R. Oldenbourg Verlag, 1994

Knuth, D. E.: *The Art of Computer Programming*, 2nd Ed., Reading, Massachusetts: Addison-Wesley, 1973

Massey, J. L.: An Introduction to Contemporary Cryptology, *IEEE*, Vol. 76, No. 5, 1988

Massey, J. L.: *Cryptography, Fundamentals and Applications*

Meyer, C. H. and Matyas, S. M.: *Cryptography*, New York: John Wiley & Sons, 1982

Myers, G. J.: *The Art of Software Testing*, 5th Ed., New York: John Wiley & Sons, 1995

Schief, R.: *Einführung in die Mikroprozessoren und Mikrocomputer*, 10th Ed., Tübingen: Attempto Verlag, 1987

Schneier, B.: *Applied Cryptography*, 2nd Ed., New York: John Wiley & Sons, 1996

Simmons, G. J. (Ed.): *Contemporary Cryptology*, New York, IEEE Press, 1992

Sommerville, I.: *Software Engineering*, Wokingham: Addison-Wesley, 1990

Tannenbaum, A. S.: *Betriebssysteme – Entwurf und Realisierung*, Munich, Vienna: Carl Hanser Verlag, 1990

Tietze, U. and Schenk, C.: *Halbleiter-Schaltungstechnik*, 10th Ed., Berlin: Springer Verlag, 1993

15.3 STANDARDS LIST WITH COMMENTARY

The standards as published by institutions and their codes, in alphabetical order ignoring prefixes (e.g."pr") or description of condition (e.g."DIS", etc.).

ANSI X 9.9 : 1986
Financial Institution Message Authentication

ANSI X 9.17 : 1985
Financial Institution Key Management

ANSI X 9.19 : 1986
Financial Institution Retail Message Authentication

ANSI X 9.30 Part 1 : 1993
Public Key cryptography using irreversible algorithms for the financial services industry – Part 1: The Digital Signature Algorithm (DSA)

ANSI X 9.30 Part 2 : 1993
Public Key cryptography using irreversible algorithms for the financial services industry – Part 2: The Secure Hash Algorithm (SHA-1)

ANSI X 9.31 Part 1 : 1993
Public Key cryptography using irreversible algorithms for the financial services industry – Part 1: The RSA Signature Algorithm

ANSI X 3.92 : 1981
Data Encryption Algorithm
Describes DEA.

ANSI / IEEE 1008 : 1987
IEEE Standard for Software Unit Testing
Comprehensive description of basic processes in software testing.

CCITT Z.100 : 1994
Specification and Description Language (SDL)

DIN 44300 : 1988
Information processing terminology

EMV-1 Version 2.0 : 1995
Integrated Circuit Card Specifications for Payment Systems – Part 1: Electromechanical Characteristics, Logical Interface and Transmission Protocols
Specification of the mechanical and electrical properties of integrated circuit cards and terminals. Defines techniques for switching on and off, physical data transfer, ATRs and their relevant data elements. The two transfer protocols T=0 and T=1 are also specified.

EMV-2 Version 2.0 : 1995
Integrated Circuit Card Specifications for Payment Systems – Part 2: Data Elements and Commands
Specification of the APDUs, logical channels, secure messaging, commands and return codes as well as data elements and their relevant TLV codes for integrated credit cards with chips.

EMV-3 Version 2.0 : 1995
Integrated Circuit Card Specifications for Payment Systems – Part 3: Transaction Processing
Specification of transactions and functions for credit cards with chips. The appendix contains examples of the authentication methods for integrated circuit cards.

EMV-ICCT Version 1.0 : 1995
Integrated Circuit Card Terminal Specification for Payment Systems
Determines possible configurations, functions and security requirements of terminals for credit cards with chips. Also lists possible user messages and acquirer interfaces. The appendix contains, amongst other things, the codes of data elements used and characters for messages to the user.

EN 726-1 : 1994
Identification card systems – Telecommunications integrated circuit(s) card and terminals – Part 1: System overview

EN 726-2 : 1995
Identification card systems – Telecommunications integrated circuit(s) card and terminals – Part 2: Security framework

EN 726-3 : 1994
Identification card systems – Telecommunications integrated circuit(s) card and terminals – Part 3: Application independent card requirements
Defines data structures, commands, return codes, data for particular functions and the basic mechanisms of telecommunications integrated circuit cards. The standard used is the ETSI – ISO/IEC 7816-4.

EN 726-4 : 1994
Identification card systems – Telecommunications integrated circuit(s) card and terminals – Part 4: Application independent card related terminal requirements

EN 726-5 : 1995
Identification card systems – Telecommunications integrated circuit(s) card and terminals – Part 5: Payment methods
Defines various methods of payment via related data structures, data elements and processes of integrated circuit cards. Designed for use in telecommunications.

EN 726-6 : 1995
Identification card systems – Telecommunications integrated circuit(s) card and terminals – Part 6: Telecommunication features

EN 726-7 : 1996
Identification card systems – Telecommunications integrated circuit(s) card and terminals – Part 7: Security Module

EN 1038 : 1995
Identification card systems – Telecommunication applications – Integrated circuit(s) card payphone
Defines basic requirements for using integrated circuit(s) cards in public card telephones. Lists existing standards.

prEN 1105 : January 1995
Identification card systems – General concepts applying to systems using IC cards in inter-sector environments – Rules for inter-application consistency
Defines basic requirements governing the use of integrated circuit(s) cards in inter-sector environments. Lists existing standards and regulations regarding chip cards and terminals.

prEN 1292 : 1995
Additional Test Methods for IC Cards and Interface Devices
Defines tests for electronic frame values and basic data transfer. This standard is a supplement to ISO/IEC 10373.

prEN 1332-1 : 1995
Identification card systems – Man-Machine Interface – Part 1: Design principles and symbols for the user interface

prEN 1332-2 : 1995
Identification card systems – Man-Machine Interface – Part 2: Definition of a Tactile Identifier for ID-1 cards
Specification of tactile cut-out for ID-1 cards to identify the cards' orientation.

prEN 1332-3 : 1995
Identification card systems – Man-Machine Interface – Part 3: Keypads

prEN 1332-4 : 1995
Identification card systems – Man-Machine Interface – Part 4: Coding of user requirements for people with special needs

prEN 1545-1 : 1995
Identification card systems – Surface transport applications – Part 1: General

prEN 1545-2 : 1995
Identification card systems – Surface transport applications – Part 2: Transport payment

prEN 1545-3 : 1995
Identification card systems – Surface transport applications – Part 3: Tachograph

prEN 1545-4 : 1995
Identification card systems – Surface transport applications – Part 4: Vehicle and driver licensing

prEN 1545-5 : 1995
Identification card systems – Surface transport applications – Part 5: Freight

prEN 1546-1 : 1995
Identification card systems – Inter-sector electronic purse – Part 1: Definition, concepts and structures
Defines standard terms and description of basic concepts and structures for inter-sector electronic purse.

prEN 1546-2 : 1995
Identification card systems – Inter-sector electronic purse – Part 2: Security architecture
Describes notation used for security mechanisms, security architecture and relevant processes and mechanisms of the inter-sector electronic purse.

prEN 1546-3 : 1995
Identification card systems – Inter-sector electronic purse – Part 3: Data elements and interchanges
Describes data elements, files, commands and return codes between all components of inter-sector electronic purse.

prEN 1546-4 : 1995
Identification card systems – Inter-sector electronic purse – Part4: Devices
Describes TLV mechanism for reading data from files, and gives detailed description of components and conditions for status devices for inter-sector electronic purse. There is also a list of tags for all existing data elements.

prETS 300 331 : 1993
Radio Equipment and Systems; Digital European Cordless Telecommunications; Common interface; DECT Authentication Module *Description of circuit card for DECT System. It encompasses all relevant commands, files, access conditions, and authentication procedures. The measurements for Mini ID cards and Plug-in cards are also defined. This is a GSM 11.11 standard.*

prETS 300 608 : 1995 (identical to GSM 11.11 version 4.16.0)
European digital cellular telecommunications system (Phase 2) – Specification of the Subscriber Identity Module – Mobile Equipment (SIM – ME) interface (GSM 11.11)
Complete specification of interface between circuit card and mobile telephone for GSM. This contains definition of ID-1 and Plug-in card size, physical dimensions for cards and contacts. Defines all electronic frame values such as those of the ATR and PTS. Definition of file structures, security mechanisms, commands and return codes. Definition of all data elements and files necessary for GSM, as well as relevant procedures for the individual functions.

FIPS Pub 46 : 1977
Data Encryption Standard (DES)

FIPS Pub 81 : 1980
DES Modes of Operation

FIPS Pub 180-1 : 1994
Secure Hash Standard (SHA-1)

FIPS Pub 186 : 1994
Digital Signature Standard (DSS)

GSM 11.11 Version 4.15.0 : 1995, see prETS 300 608

GSM 11.12 Version 2.0.0 : 1995
European digital cellular telecommunications system (Phase 2+) – Specification of the 3V Subscriber Identity Module – Mobile Equipment (SIM – ME) interface
Specification for 3 volt SIMs including a compatibility list for SIMs, which were programmed according to the previous specifications. This standard is supplementary to GSM 11.11 and notes the differences.

ISO 639
Codes for the representation of names of languages

ISO 3166 : 1993
Codes for the representation of names of countries

ISO 3309 : 1984
Information processing systems – Data communication – High-level data link control procedures – Frame structure
Defines news structures for data transfer and their protection through an error recognition code (CRC).

ISO/IEC 4217 : 1990
Codes for the representation of currencies and funds

ISO 4909 : 1987
Bank cards – Magnetic stripe data contents for track 3

ISO 7810 : 1995
Identification cards – Physical characteristics
Describes important physical characteristics of a card (without chip) and card dimensions ID-1, ID-2 and ID-3.

ISO 7811-1 : 1995
Identification cards – Recording technique – Part 1: Embossing
Exact definition of the ten numbers as well as the basic writing style for the embossing of cards.

ISO 7811-2 : 1995
Identification cards – Recording technique – Part 2: Magnetic stripe
Defines dimensions and location of magnetic stripe on a card. A standard exists for the physical characteristics of the magnetic material and the coding of the symbols on the magnetic stripe.

ISO 7811-3 : 1995
Identification cards – Recording technique – Part 3: Location of embossed characters on ID-1 cards
Defines possible embossing locations on ID-1 cards.

ISO 7811-4 : 1995
Identification cards – Recording technique – Part 4: Location of read-only magnetic tracks – Tracks 1 and 2
Defines position of read-only tracks 1 and 2 on an ID-1 card.

ISO 7811-5 : 1995
Identification cards – Recording technique – Part 5: Location of read–write magnetic track – Track 3
Defines position of read–write track 3 on ID-1 cards.

ISO DIS 7811-6 : 1995
Identification cards – Recording technique – Part 6: High coercivity magnetic stripe

ISO 7812-1 : 1993
Identification cards – Part l: Numbering system
Specification of numbering scheme for distributors of ID cards.

ISO 7812-2 : 1993
Identification cards – Part 2: Application and registration procedures
Sets down registration procedures and formula for the registration of applications. Also contains algorithm for the checksum according to Luhn (Modulo 10 checksum).

ISO 7813 : 1995
Identification cards – Financial transaction cards
Defines basic physical characteristics, dimensions and embossing of ID-1 cards according to ISO 7810 for financial transaction cards. The data contents of tracks 1 and 2 of the magnetic stripe are also defined.

ISO 7816-1 : 1987
Identification cards – Integrated circuit(s) cards with contacts – Part 1: Physical characteristics
Defines physical characteristics of integrated circuit(s) cards as well as test methods to be used in conjunction with these.

ISO 7816-2 : 1988
Identification cards – Integrated circuit(s) cards with contacts – Part 2: Dimensions and location of the contacts
Defines dimensions and location of card contacts, and possible arrangement of chip, magnetic stripe and embossing. The measuring method for the position of integrated circuit(s) card contacts is also described.

ISO/IEC 7816-3 : 1989
Identification cards – Integrated circuit(s) cards with contacts – Part 3: Electronic signals and transmission protocols
The most important ISO standard for electronic frame parameter definition. Defines basic electronic characteristics of an integrated circuit(s) card as well as power needed for data transfer. Specifies build-up and data elements of ATR and PTS. Basics of data transfer on physical level (e.g. components) and definition of transmission protocol T=0. This standard is presently being rewritten to include a new function and will also include the previous appendices 1 and 2.

ISO/IEC 7816-3 Amd. 1 : 1992
Identification cards – Interrelated circuit(s) cards with contacts – Part 3: Electronic signals and transmission protocols – Amendment 1: Protocol type T=l, asynchronous half duplex block transmission protocol
Complete definition of transmission protocol T=1 with numerous examples of protocol procedures.

ISO/IEC 7816-3 Amd. 2 : 1994
Identification cards – Integrated circuit(s) cards with contacts – Part 3: Electronic signals and transmission protocols – Amendment 2: Protocol type selection
Description of both possible PTS processes (Negotiable/Specific Mode).

ISO/IEC 7816-4 : 1995
Information technology – Identification cards – Integrated circuit(s) cards with contacts – Part 4: Inter-industry commands for interchange
The most important ISO standards for users of IC cards. Definition of data organization, file structure, security architecture, TPDUs, APDUs, secure messaging, return codes, logical channels. A thorough description of the commands for circuit cards. The basic mechanisms of circuit cards for inter-industry use are also described.

ISO/IEC 7816-5 : 1994
Identification cards – Integrated circuit(s) cards with contacts – Part 5: Numbering system and registration procedure for application identifiers

Defines the numbering system for the clear identification of national and international applications in IC cards. The exact data structure of the AIDs is defined and the application procedure explained.

ISO/IEC CD 7816-5 Amd. 1 : 1995
Identification cards – Integrated circuit(s) cards with contacts – Part 5 – Amendment 1: Registration of identifiers

ISO/IEC DIS 7816-6 : 1995
Identification cards – Integrated circuit(s) cards with contacts – Part 6: Inter-industry data elements
Definition of data elements and relevant TLV markings for inter-industry use. Also the relevant TLV structures are explained as well as procedures for the retrieval of data elements in integrated circuit(s) cards.

ISO/IEC WD 7816-7 : 1995
Identification cards – Integrated circuit(s) cards with contacts – Part 7: Enhanced inter-industry commands
Defines additional commands for IC cards to comply with ISO/IEC 7816-4. This standard describes the commands for SQL-access in IC cards, mutual authentication and encrypting. In addition, to supplement Part 4 of the standard series, further mechanisms for secure messaging are defined.

ISO/IEC WD 7816-8 : 1995
Identification cards – Integrated circuit(s) cards with contacts – Part 8: Inter-industry security architecture
Definition of detailed security architecture of integrated circuit(s) cards.

ISO 8372 : 1987
Modes of Operation for a 64-bits Block Cipher Algorithm

ISO 8730 : 1990
Banking – Requirements for message authentication
Basic requirements for securing data during transmission, i.e. creation and checking of MACs. The appendix contains detailed numeric examples, as well as a description of a pseudo-random number generator with DEA.

ISO 8731-1 : 1987
Banking – Approved algorithms for message authentication – Part 1: DEA
Very short standard, in which the DEA appropriate to MAC calculations is described. There is also a short description of parity calculations for DEA keys.

ISO 8731-2 : 1992
Banking – Approved algorithms for message authentication – Part 2:
Message authenticator algorithm
Definition of a fast algorithm for DEA calculations in banking. The appendix contains numeric examples, as well as an exact description of the algorithm.

ISO 8732 : 1988
Banking – Key management
Comprehensive standard for the basic requirements and procedures of key management between two or more entities.

ISO/IEC 8824 : 1990
Information technology - Open Systems Interconnection - Specification of Abstract Syntax Notation One (ASN.1)
Defines basic encoding rules for ASN.1.

ISO/IEC 8825 : 1990
Information technology – Open Systems Interconnection – Specification of Basic Encoding Rules for Abstract Syntax Notation One (ASN.1)
Defines data description language ASN.1.
ISO 9564-1 : 1991
Banking – Personal Identification Number management and security – Part 1: PIN protection principles and techniques
Principles for PIN selection, PIN management and PIN protection in banking. The appendix contains general definitions of PIN requirements for card acceptance devices. There are also suggestions for layout of appropriate keyboards. Furthermore the appendix contains tips for deleting sensitive data on various kinds of media, such as magnetic tape, paper or half-duplex.

ISO 9564-2 : 1991
Banking – Personal Identification Number management and security – Part 2: Approved algorithm(s) for PIN encipherment
Very short standard, defining the DEA as an algorithm for PIN encipherment.

ISO/IEC 9646-3 : 1992
Information technology – Open Systems Interconnection – Conformance testing methodology and framework – Part 3: The Tree and Tabular Combined Notation (TTCN)
Comprehensive standard, describing general terminology for tests.

ISO/IEC 9796 : 1991
Information technology – Security techniques – Digital signature scheme giving message recovery
Defines procedure for creation and testing of a digital signature. The appendix contains numerous numeric examples for key creation, signature creation and checking.

ISO/IEC 9797 : 1994
Information technology – Security techniques – Data integrity mechanism using a cryptographic check function employing a block cipher algorithm
Defines length of message authentication codes using a block cipher algorithm.

ISO/IEC 9798-1 : 1991
Information technology – Security techniques – Entity authentication - Part 1: General model
Description and notation of three further parts of the standard series.

ISO/IEC 9798-2 : 1994
Information technology – Security techniques – Entity authentication – Part 2: Mechanisms using symmetric encipherment algorithms
Definition of authentication procedures using symmetric encipherment algorithms. Describes the basic requirements for single and mutual authentication procedures employing two or three entities.

ISO/IEC 9798-3 : 1993
Information technology – Security techniques – Entity authentication mechanisms – Part 3: Entity authentication using a public key algorithm
Defines asymmetrical authentication procedures. Describes the basic mechanisms of single and mutual authentication procedures using two entities.

ISO/IEC DIS 9798-4 : 1994
Information technology – Security techniques – Entity authentication mechanisms – Part 4: Mechanisms using a cryptographic check function
Basic requirements for authentication and introduction of various mechanisms.

ISO 9807 : 1991
Banking and related financial services – Requirements for message authentication (retail)

ISO 9992-1 : 1990
Financial Transaction Cards – Messages between the Integrated Circuit Card and the Card Accepting Device – Part 1: Concepts and structures

ISO CD 9992-2 : 1994
Financial Transaction Cards – Messages between the Integrated Circuit Card and the Card Accepting Device – Part 2: Functions, messages (commands and responses), data elements and structures
Definition of commands, processes and data elements for IC cards in financial transactions. Contains definitions of tags for data elements and many existing standards in the ISO/IEC 7816 series.

ISO/IEC 10116 : 1991
Information technology – Security techniques – Modes of operation for an n-bit block cipher algorithm
Description of the four most common modes (e.g. ECB, CBC) with which block cipher algorithm can be used. The appendix contains detailed notes for use of the four encryption modes, and additional relevant numeric examples are given.

ISO/IEC 10118-1:1994
Information technology – Security techniques – Hash functions – Part 1: General
General information about hash functions, as well as relevant padding methods.

ISO/IEC 10118-2:1994
Information technology – Security techniques – Hash functions – Part 2: Hash functions using an n-bit block cipher algorithm

Definition of two hash functions, based on a block cipher algorithm. An algorithm with simple and double cryptographic key(s) is described. The appendix contains relevant individual numeric examples for use with DEA.

ISO/IEC DIS 10118-3 : 1995
Information technology – Security techniques – Hash functions – Part 3: Dedicated hash functions

ISO/IEC DIS 10118-4 : 1995
Information technology – Security techniques – Hash functions – Part 4: Hash-functions using modular arithmetic

ISO/IEC CD 10170-1 : 1995
Information technology – Security techniques – Key Management, Part 1: Key Management Mechanisms Using Asymmetric Techniques

ISOIIEC CD 10170-2 : 1995
Information technology – Security techniques – Key Management, Part 2: Key Management Mechanisms Using Symmetric Techniques

ISO 10202-1 : 1991
Financial Transaction Cards – Security Architecture of Financial Transaction Systems using Integrated Circuit Cards – Part 1: Card life cycle

ISO DIS 10202-2 : 1995
Financial Transaction Cards – Security Architecture of Financial Transaction Systems using Integrated Circuit Cards – Part 2: Transaction process

ISO DIS 10202-3 : 1995
Financial Transaction Cards – Security Architecture of Financial Transaction Systems using Integrated Circuit Cards – Part 3: Cryptographic key relationship

ISO DIS 10202-4 : 1995
Financial Transaction Cards – Security Architecture of Financial Transaction Systems using Integrated Circuit Cards – Part 4: Secure application modules

ISO DIS 10202-5 : 1995
Financial Transaction Cards – Security Architecture of Financial Transaction Systems using Integrated Circuit Cards – Part 5: Use of algorithms

ISO 10202-6 : 1994
Financial Transaction Cards – Security Architecture of FinancialTransaction Systems using Integrated Circuit Cards – Part 6: Card holder verification

ISO DIS 10202-7 : 1994
Financial Transaction Cards – Security Architecture of Financial Transaction Systems using Integrated Circuit Cards – Part 7: Key Management

Definition of general mechanisms of key management and key management techniques. Both symmetrical and asymmetrical procedures are described.

ISO DIS 10202-8 : 1994
Financial Transaction Cards – Security Architecture of Financial Transaction Systems using Integrated Circuit Cards – Part 8: General principles and overview

ISO/IEC 10373 : 1993
Identification cards – Test methods
General principles of card testing. Exact description of methodology for testing cards alone and in connection with the implanted chip. Descriptions are supplemented by detailed drawings.

ISO/IEC 10536-1 : 1992
Identification cards – Contactless integrated circuit(s) cards – Part 1: Physical characteristics
Definition of physical characteristics of contactless IC cards, and relevant testing methods.

ISO/IEC 10536-2 : 1994
Identification cards – Contactless integrated circuit(s) cards – Part 2: Dimension and location of coupling areas
Specifies dimension and location of coupling areas for IC cards and use in card terminals and card swipes or on the surface.

ISO/IEC 10536-3 : 1995
Identification cards – Contactless integrated circuit(s) cards – Part 3: Electronic signals and reset procedures
Definition of electronic signals for inductive and capacitive coupling elements between terminal and circuit card.

ISO/IEC CD 10536-4 : 1995
Identification cards – Contactless integrated circuit(s) cards – Part 4: Answer to reset and transmission protocols
Defines data transmission and storage on physical level and data elements of ATR and PTS for contactless IC cards. Definition of transmission protocol T=2 with numerous examples of protocol structures.

ISO 11568-1 : 1994
Banking – Key management – Part 1 : Introduction to Key Management

ISO 11568-2 : 1994
Banking – Key management – Part 2: Key management techniques for symmetric ciphers

ISO 11568-3 : 1994
Banking – Key management – Part 3: Key life cycle for symmetric ciphers

ISO DIS 11568-4 : 1994
Banking – Key management – Part 4: Key management techniques using asymmetric ciphers

ISO DIS 11568-5 : 1995
Banking – Key management – Part 5: Key life cycle for asymmetric ciphers

ISO WD 11568-6 : 1995
Banking – Key management – Part 6: Key management schemes

ISO/IEC 11693 : 1995
Optical memory cards

ISO/IEC DIS 11694-1 : 1994
Optical memory cards – Linear recording method – Part 1: Physical characteristics

ISO/IEC 11694-2 : 1995
Optical memory cards – Linear recording method – Part 2: Dimensions and location of the accessible optical area

ISO/IEC 11694-3 : 1995
Optical memory cards – Linear recording method – Part 3: Optical properties and characteristics

ISO/IEC DIS 11770-1 : 1995
Information technology – Security techniques – Key management – Part 1: Key management framework

ISO/IEC DIS 11770-2 : 1995
Information technology – Security techniques – Key management – Part 2: Mechanisms using symmetric techniques

ISO/IEC DIS 11770-3 : 1995
Information technology – Security techniques – Key management – Part 3: Key management mechanisms using asymmetric cryptographic techniques

ISO CD 13491 : 1995
Banking – Secure cryptographic devices

ISO/IEC WD 14443-1 : 1996
Remote coupling communication cards – Part 1: Physical characteristics

ISO/IEC WD 14443-2 : 1996
Remote coupling communication cards – Part 2: Radio frequency interface

ISO/IEC WD 14443-3 : 1996
Remote coupling communication cards – Part 3: Transmission protocols

ISO/IEC WD 14443-4 : 1996
Remote coupling communication cards – Part 4: Transmission security features

15.4 CHARACTERISTIC VALUES AND TABLES

15.4.1 Time interval for ATR

This table defines the interval during which an ATR must be sent after a reset.

Clock frequency	Minimum 400 clock pulses	Maximum 40 000 clock pulses
1.0000 MHz	0.400 ms	40 000 ms
2.0000 MHz	0.200 ms	20 000 ms
3.0000 MHz	0.133 ms	13 333 ms
3.5712 MHz	0.112 ms	11 201 ms
4.0000 MHz	0.100 ms	10 000 ms
4.9152 MHz	0.081 ms	8 138 ms
5.0000 MHz	0.080 ms	8 000 ms
6.0000 MHz	0.067 ms	6 667 ms
7.0000 MHz	0.057 ms	5 714 ms
8.0000 MHz	0.050 ms	5 000 ms
9.0000 MHz	0.044 ms	4 444 ms
10.0000 MHz	0.040 ms	4 000 ms

Figure 15.1 The time interval during which the ATR must be sent by the Smart Card

15.4.2 Conversion table for ATR data elements

These tables are based on the ISO/IEC 7816-3 definition of the ATR data elements CWT and BWT. The values assume a clock frequency of 3.5712 MHz.

CWI	CWT	CWT (for 93 divider)	CWT (for 186 divider)	CWT (for 372 divider)
0	12 etu	0.313 ms	0.625 ms	1.250 ms
1	13 etu	0.339 ms	0.677 ms	1.354 ms
2	15 etu	0.391 ms	0.781 ms	1.563 ms
3	19 etu	0.495 ms	0.990 ms	1.979 ms
4	27 etu	0.703 ms	1.406 ms	2.813 ms
5	43 etu	1.120 ms	2.240 ms	4.479 ms
6	75 etu	1.953 ms	3.906 ms	7.813 ms
7	139 etu	3.620 ms	7.240 ms	14.479 ms
8	267 etu	6.953 ms	13.906 ms	27.813 ms
9	523 etu	13.620 ms	27.240 ms	54.479 ms
10	1 035 etu	26.953 ms	53.906 ms	107.813 ms
11	2 059 etu	53.620 ms	107.240 ms	214.479 ms
12	4 107 etu	106.953 ms	213.906 ms	427.813 ms
13	8 203 etu	213.620 ms	427.240 ms	854.479 ms
14	16 395 etu	426.953 ms	853.906 ms	1 707.813 ms
15	32 779 etu	853.620 ms	1 707.240 ms	3 414.479 ms

Figure 15.2 CWT conversion table (all entries relate to a 3.5712 MHz clock frequency)

BWI	BWT	BWT (for 93 divider)	BWT (for 186 divider)	BWT (for 372 divider)
0	1 011 etu	26.328 ms	52.656 ms	105.313 ms
1	2 011 etu	52.370 ms	104.740 ms	209.479 ms
2	4 011 etu	104.453 ms	208.906 ms	417.813 ms
3	8 011 etu	208.620 ms	417.240 ms	834.479 ms
4	16 011 etu	416.953 ms	833.906 ms	1 667.813 ms
5	32 011 etu	833.620 ms	1 667.240 ms	3 334.479 ms
6	64 011 etu	1 666.953 ms	3 333.906 ms	6 667.813 ms
7	128**011 etu	3 333.620 ms	6 667.240 ms	13 334.479 ms
8	256**011 etu	6 666.953 ms	13 333.906 ms	26 667.813 ms
9	512**011 etu	13 333.620 ms	26 667.240 ms	53 334.479 ms

Figure 15.3 BWT conversion table (all entries relate to a 3.5712 MHz clock frequency)

15.4.3 Calculation table for transmission speed

Clock frequency	F=372, D=1/64 ⇒8	F=372, D=1/32 ⇒16	F=372, D=1/16 ⇒32	F=372, D=1/8 ⇒64
1.0000 MHz	125 000 bit/s	62 500 bit/s	31 250 bit/s	15 625 bit/s
2.0000 MHz	250 000 bit/s	125 000 bit/s	62 500 bit/s	31 250 bit/s
3.0000 MHz	375 000 bit/s	187 500 bit/s	93 750 bit/s	46 875 bit/s
3.5712 MHz	446 400 bit/s	223 200 bit/s	111 600 bit/s	55 800 bit/s
4.0000 MHz	500 000 bit/s	250 000 bit/s	125 000 bit/s	62 500 bit/s
4.9152 MHz	614 400 bit/s	307 200 bit/s	153 600 bit/s	76 800 bit/s
5.0000 MHz	625 000 bit/s	312 500 bit/s	156 250 bit/s	78 125 bit/s
6.0000 MHz	750 000 bit/s	375 000 bit/s	187 500 bit/s	93 750 bit/s
7.0000 MHz	875 000 bit/s	437 500 bit/s	218 750 bit/s	109 375 bit/s
8.0000 MHz	1 000 000 bit/s	500 000 bit/s	250 000 bit/s	125 000 bit/s
9.0000 MHz	1 125 000 bit/s	562 500 bit/s	281 250 bit/s	140 625 bit/s
100000 MHz	1 250 000 bit/s	625 000 bit/s	312 500 bit/s	156 250 bit/s

Figure 15.4 Common transmission speeds at 3.5712 MHz, resulting from employing standardized values for the divider F and the transmission adjustment factor D, Part 1

Clock frequency	F=372, D=1/4 ⇒93	F=372, D=1/2 ⇒186	F=372, D=1 ⇒372	F=512, D=1 ⇒512
1.0000 MHz	10 753 bit/s	5 376 bit/s	2 688 bit/s	1 953 bit/s
2.0000 MHz	21 505 bit/s	10 753 bit/s	5 376 bit/s	3 906 bit/s
3.0000 MHz	32 258 bit/s	16 129 bit/s	8 065 bit/s	5 859 bit/s
3.5712 MHz	38 400 bit/s	19 200 bit/s	9 600 bit/s	6 975 bit/s
4.0000 MHz	43 011 bit/s	21 505 bit/s	10 753 bit/s	7 813 bit/s
4.9152 MHz	52 852 bit/s	26 426 bit/s	13 213 bit/s	9 600 bit/s
5.0000 MHz	53 763 bit/s	26 882 bit/s	13 441 bit/s	9 766 bit/s
6.0000 MHz	64 516 bit/s	32 258 bit/s	16 129 bit/s	11 719 bit/s
7.0000 MHz	75 269 bit/s	37 634 bit/s	18 817 bit/s	13 672 bit/s
8.0000 MHz	86 022 bit/s	43 011 bit/s	21 505 bit/s	15 625 bit/s
9.0000 MHz	96 74 bit/s	48 387 bit/s	24 194 bit/s	17 578 bit/s
10.0000 MHz	107 527 bit/s	53 763 bit/s	26 882 bit/s	19 531 bit/s

Figure 15.5 Common transmission speeds at 3.5712 MHz, resulting from employing standardized values for the divider F and the transmission adjustment factor D, Part 2

15.4.4 Table for sensing points

This table is based on data transmission after ISO/IEC 7816-3. The values assume a clock frequency of 3.5712 MHz.

	Start	Lower tolerance	Centre	Upper tolerance	End
Start bit	0 pulses 0.000 μs	112 pulses 31.250 μs	186 pulses 52.083 μs	260 pulses 72.917 μs	372 pulses 104.167 μs
Data bit 1/8	372 pulses 104.167 μs	484 pulses 135.417 μs	558 pulses 156.250 μs	632 pulses 177.083 μs	744 pulses 208.333 μs
Data bit 2/7	744 pulses 208.333 μs	856 pulses 239.583 μs	930 pulses 260.417 μs	1 004 pulses 281.250 μs	1 116 pulses 312.500 μs
Data bit 3/6	1 116 pulses 312.500 μs	1 228 pulses 343.750 μs	1 302 pulses 364.583 μs	1 376 pulses 385.417 μs	1 488 pulses 416.667 μs
Data bit 4/5	1 488 pulses 416.667 μs	1 600 pulses 447.917 μs	1 674 pulses 468.750 μs	1 748 pulses 489.583 μs	1 860 pulses 520.833 μs
Data bit 5/4	1 860 pulses 520.833 μs	1 972pulses 552,083 μs	2 046 pulses 572.917 μs	2 120 pulses 593.750 μs	2 232 pulses 625.000 μs
Data bit 6/3	2 232 pulses 625.000 μs	2 344 pulses 656.250 μs	2 418 pulses 677.083 μs	2 492 pulses 697.917 μs	2 604 pulses 729.167 μs
Data bit 7/2	2 604 pulses 729.167 μs	2 716 pulses 760.417 μs	2 790 pulses 781.250 μs	2 864 pulses 802.083 μs	2 976 pulses 833.333 μs
Data bit 8/1	2 976 pulses 833.333 μs	3 088 pulses 864.583 μs	3 162 pulses 885.417 μs	3 236 pulses 906.250 μs	3 348 pulses 937.500 μs
Parity bit	3 348 pulses 937.500 μs	3 460 pulses 968.750 μs	3 534 pulses 989.583 μs	3 608 pulses 1 010.417 μs	3 720 pulses 1041.667 μs
Guard time/ stop bit 1	3 720 pulses 1 041.667 μs	3 832 pulses 1 072.917 μ	3 906 pulses 1 093.750 μs	3 980 pulses 1 114.583 μs	4 092 pulses 1 145.833 μs
Guard time/ stop bit 2	4 092 pulses 1 145.833 μs	4 204 pulses 1 177.083 μs	4 278 pulses 1 197.917 μs	4 352 pulses 1 218.750 μs	4464 pulses 1 250.000 μs

Figure 15.6 Data transmission with 372 divider

15.4.5 Table of class bytes used

Class	Use
'0X'	For standardized instructions after ISO/IEC 7816-4
'80'	For electronic purse after prEN 1546-3
'8X'	For application- and firm-specific instructions
'8X'	For chip-containing credit cards after EMV-2
'A0'	For GSM mobile phones after prETS 300 608 / GSM 11.11 and for standardized instructions after EN 726-3

Figure 15.7 Summary of the allocation of class bytes to applications

15.4.6 Table of the most important Smart Card instructions

Instruction	Function	INS	Standard
APPEND RECORD	Add a new record to a file structured as linear fixed. In a cyclic file, the highest numbered record is replaced	'E2'	ISO/IEC7816-4
ASK RANDOM	Request a random number from the card.	'84'	EN 726-3
CHANGE CHV	Change the PIN.	'24'	GSM 11.11, EN 726-3
CLOSE APPLICATION	Reset all the satisfied access conditions.	'AC'	EN 726-3
CONVERT IEP	Convert currency.	'56'	prEN 1546-3
CURRENCY			
CREATE FILE	Generate a new file.	'E0'	EN 726-3
CREATE RECORD	Generate a new record in a record-oriented file.	'E2'	EN 726-3
CREDIT IEP	Load the purse (IEP).	'52'	prEN 1546-3
CREDIT PSAM	IEP payment against the PSAM.	'72'	prEN 1546-3
DEBIT IEP	Pay with the purse.	'54'	prEN 1546-3
DECREASE	Decrement a file counter.	'30'	EN 726-3
DECREASE STAMPED	Decrement a counter in a file secured by a crytographic checksum.	'34'	EN 726-3
DELETE FILE	Delete a file.	'E4'	EN 726-3
DISABLE CHV	Disable PIN interrogation.	'26'	GSM 11.11 EN 726-3
ENABLE CHV	Enable PIN interrogation.	'28'	GSM 11.11

			EN 726-3
ENVELOPE PUT/ENVELOPE	Embed a further instruction in an instruction for cryptographic security.	'C2'	EN 726-3, ISO/IEC 7816-4
ERASE BINARY	Set the contents of a transparently structured file to the erased state.	'0E'	ISO/IEC 7816-4
EXECUTE	Execute a file.	'AE'	EN 726-3
EXTEND	Extend a file.	'D4'	EN 726-3
EXTERNAL AUTHENTICATE	Authenticate the outside world by the card.	'82'	EN 726-3, ISO/IEC 7816-4
GENERATE	Generate a signature for a payment transaction.	'AE'	EMV-2
AUTHORIZATION			
CRYPTOGRAM			
GET CHALLENGE	Request a random number from the card.	'84'	ISO/IEC 7816-4
GET DATA	Read TLV-coded data objects.	'CA'	ISO/IEC 7816-4
GET PREVIOUS IEP SIGNATURE	Repeat calculation and output of most recent signature received from IEP.	'5A'	prEN 1546-3
GET PREVIOUS PSAM SIGNATURE	Repeat calculation and output of most recent signature received from PSAM.	'86'	prEN 1546-3
GET RESPONSE	T=0 instruction for requesting data from the Smart Card.	'C0'	GSM 11.11, EN 726-3, ISO/IEC 7816-4
GIVE RANDOM	Send a random number to the card.	'86'	EN 726-3
INCREASE	Increment a file counter.	'32'	GSM 11.11,
INCREASE STAMPED	Increment a counter in a file secured by a cryptographic checksum.	'36'	EN 726-3
INITIALIZE IEP	Initialize an IEP for a subsequent purse instruction.	'50'	prEN 1546-3
INITIALIZE PSAM	Initialize a PSAM for a subsequent purse instruction.	'70'	prEN 1546-3
INITIALIZE PSAM for off line collection	Initialize a PSAM for off line billing.	'7C'	prEN 1546-3
INITIALIZE PSAM for online collection	Initialize a PSAM for online billing.	'76'	prEN 1546-3
INITIALIZE PSAM for	Initialize a PSAM for parameter amendment.	'80'	prEN 1546-3
update			
INTERNAL AUTHENTICATE	Authenticate Smart Card by the outside world.	'88'	EN 726-3, ISO/IEC 7816-4
INVALIDATE	Block a file reversibly.	'04'	GSM 11.11,

			EN 726-3
ISSUER AUTHENTICATE	Check card-issuer's signature.	'82'	EMV-2
LOAD KEY FILE	Cryptographically secured loading of keys into files.	'D8'	EN 726-3
LOCK	Lock a file irreversibly.	'76'	EN 726-3
MANAGE CHANNEL	Control a Smart Card's logic channels.	'70'	ISO/IEC 7816-4
PSAM COLLECT Acknowledgement	Complete PSAM online billing.	'7A'	prEN 1546-3
PSAM COLLECT	Execute PSAM online billing.	'78'	prEN 1546-3
PSAM VERIFY COLLECTION	Complete PSAM off line billing.	'7E'	prEN 1546-3
PSAM COMPLETE	Complete IEP payment against the PSAM.	'74'	prEN 1546-3
PUT DATA	Complete IEP payment against the PSAM.	'DA'	ISO/IEC 7816-4
READ BINARY	Read from a transparently structured file.	'BO'	GSM 11.11
READ BINARY STAMPED	Read from a transparently structured file secured by a cryptographic checksum.	'B4'	EN 726-3
READ RECORD/ READ RECORD(S)	Read from a linear fixed, linear variable or cyclic file.	'B2'	GSM 11.11, EN 726-3, ISO/IEC 7816-4
READ RECORD STAMPED	Read from a linear fixed, linear variable or cyclic file secured by a cryptographic checksum.	'B6'	EN 726-3
REHABILITATE	Unblock a file.	'44'	GSM 11.11, EN 726-3
RUN GSM ALGORITHM	Execute the GSM-specific cryptographic algorithm.	'88'	GSM 11.11
SEEK	Search text in a linear fixed, linear variable or cyclic file.	'A2'	GSM 11.11, EN 726-3
SELECT/SELECT (FILE)	Select a file.	'A4'	GSM 11.11, EN 726-3, ISO/IEC 7816-4
SLEEP	Now-redundant instruction to set the card to an energy-saving state.	'FA'	GSM 11.11
STATUS	Read various data to current file.	'F2'	GSM 11.11, EN 726-3
UNBLOCK CHV	Reset expired PIN error counter.	'2C'	GSM 11.11, EN 726-3

UPDATE BINARY	Write to transparently structured file.	'D6'	GSM 11.11, EN 726-3, ISO/IEC 7816-4
UPDATE IEP PARAMETER	Amend purse global parameters.	'58'	prEN 1546-3
UPDATE RECORD	Write to linear fixed, linear variable or cyclic file.	'DC'	GSM 11.11, EN 726-3, ISO/IEC 7816-4
UPDATE PSAM parameter (off-line)	Execute off line amendment of PSAM parameters.	'84'	prEN 1546-3
UPDATE PSAM parameter (online)	Execute online amendment of PSAM parameters.	'82'	prEN 1546-3
VERIFY	Check received data (e.g. PIN).	'20'	ISO/IEC 7816-4, EMV-2
VERIFY CHV	Check PIN.	'20'	GSM 11.11, EN 726-3
WRITE BINARY	Write via logical AND/OR concatenation in a transparently structured file.	'D0'	EN 726-3, ISO/IEC 7816-4
WRITE RECORD	Write via logical AND/OR concatenation in a linear fixed, linear variable or cyclic file.	'D2'	EN 726-3, ISO/IEC 7816-4

Figure 15.8 Summary of standardized Smart Card instructions

15.4.7 Summary of instruction bytes in use

The grey columns (odd numbers) must not be used for the coding of instructions, since the T=0 transmission protocol uses this range for controlling programming voltage[1].

- EMV-2 *
- EN 726-3 +
- prEN 1546-3 $
- GSM 11.11 I
- ISO/IEC 7816-4 #

---

[1] See also 6.2.2 Transmission protocol T=0.

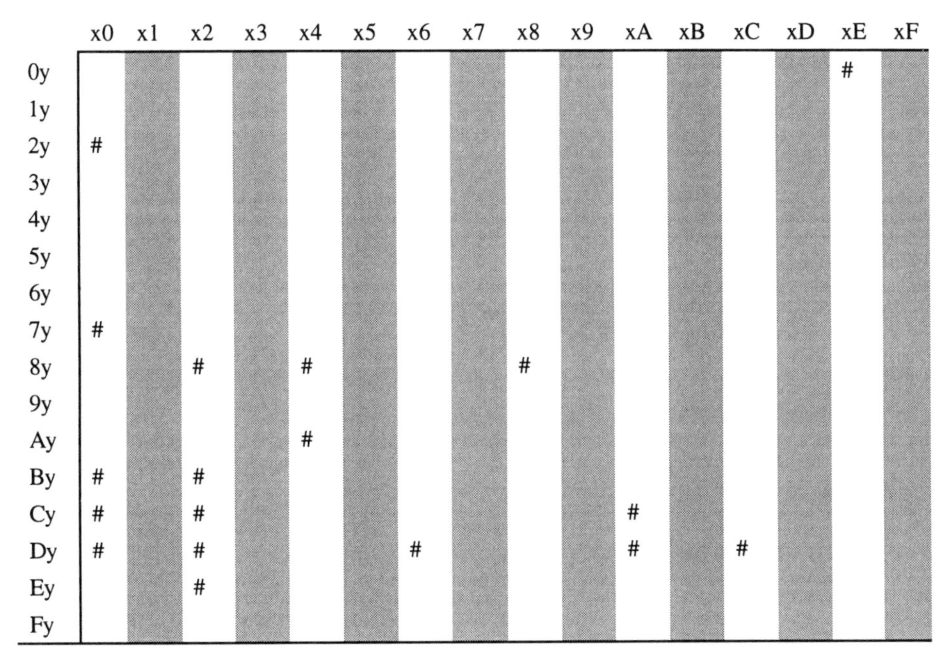

	x0	x1	x2	x3	x4	x5	x6	x7	x8	x9	xA	xB	xC	xD	xE	xF
0y															#	
1y																
2y	#															
3y																
4y																
5y																
6y																
7y	#															
8y			#		#				#							
9y																
Ay					#											
By	#		#													
Cy	#		#								#					
Dy	#		#				#				#		#			
Ey			#													
Fy																

Figure 15.9 Summary of instruction byte (INS) coding with a ‘00’ class byte (CLA)

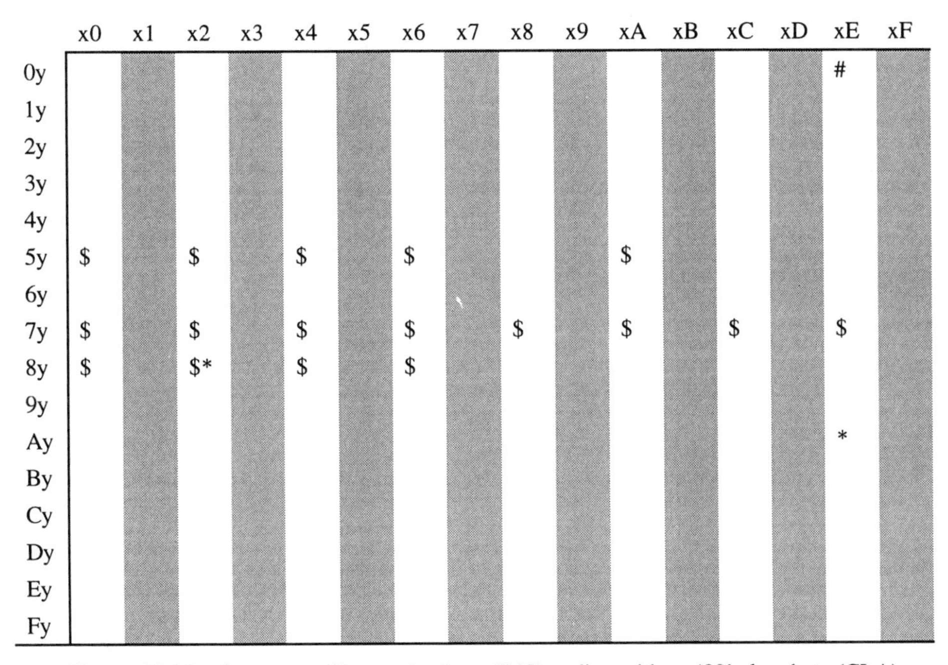

	x0	x1	x2	x3	x4	x5	x6	x7	x8	x9	xA	xB	xC	xD	xE	xF
0y															#	
1y																
2y																
3y																
4y																
5y	$		$		$		$				$					
6y																
7y	$		$		$		$		$		$		$		$	
8y	$		$*		$		$									
9y																
Ay															*	
By																
Cy																
Dy																
Ey																
Fy																

Figure 15.10 Summary of instruction byte (INS) coding with an ‘80’ class byte (CLA)

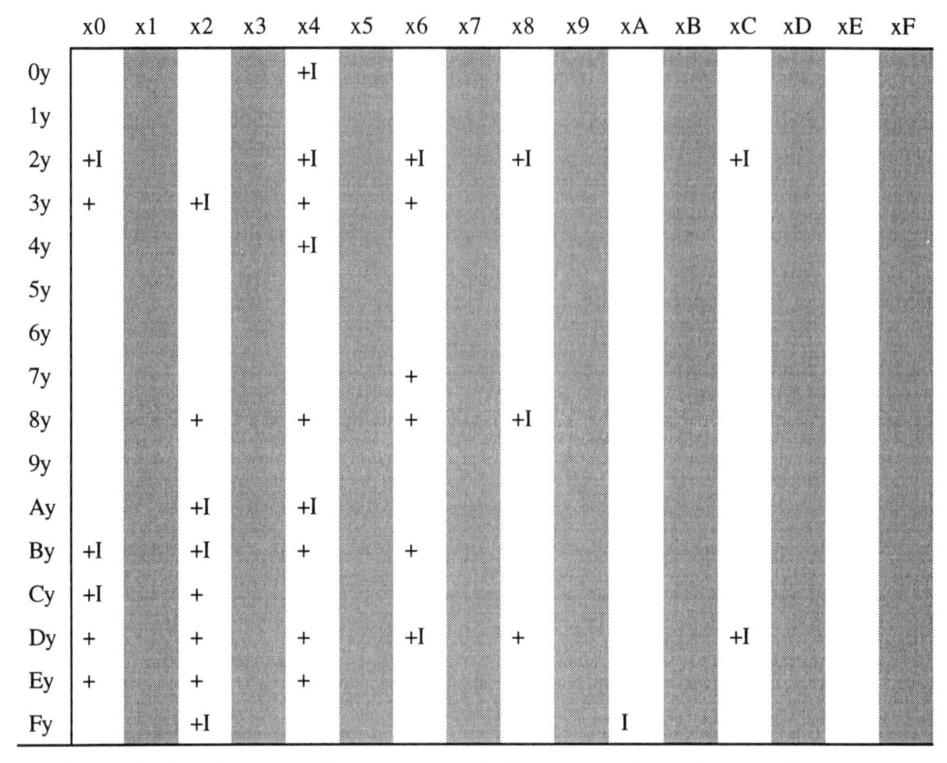

	x0	x1	x2	x3	x4	x5	x6	x7	x8	x9	xA	xB	xC	xD	xE	xF
0y					+I											
1y																
2y	+I				+I		+I		+I				+I			
3y	+		+I		+		+									
4y					+I											
5y																
6y																
7y							+									
8y			+		+		+		+I							
9y																
Ay			+I		+I											
By	+I		+I		+		+									
Cy	+I		+													
Dy	+		+		+		+I		+				+I			
Ey	+		+		+											
Fy			+I								I					

Figure 15.11 Summary of instruction byte (INS) coding with an 'A0' class byte (CLA)

15.4.8 Important Smart Card return codes

The instructions described below are subdivided according to the scheme for return codes[2] in ISO/IEC 7816-4. The following abbreviations are used:

- NP Process completed, normal processing
- WP Process completed, warning processing
- EE Process aborted, execution error
- CE Process aborted, checking error

[2] See also 6.5.2 Response APDU structure.

Return code	Status	Meaning	Standard
'61xx'	NP	Only in T=0: xx additional data can be requested with GET RESPONSE.	ISO/IEC 7816-4
'6281'	WP	The returned data may be faulty in some cases.	ISO/IEC 7816-4
'6282'	WP	Fewer than **Le** bytes may be read, as the file end has previously been reached.	ISO/IEC 7816-4
'6283'	WP	The selected file is reversibly blocked.	ISO/IEC 7816-4
'63Cx'	WP	The counter has reached the value x (the precise meaning depends on the particular instruction).	ISO/IEC 7816-4
'6581'	EE	Memory error (e.g. during a write operation).	ISO/IEC 7816-4
'6700'	CE	Length wrong, parameter P3 wrong.	GSM 11.11, EN 726-3
'6800'	CE	Class byte functions are not supported (general).	ISO/IEC 7816-4
'6881'	CE	Logical channels are not supported.	ISO/IEC 7816-4
'6882'	CE	Secure messaging is not supported.	ISO/IEC 7816-4
'6900'	CE	Instruction not permitted (general).	ISO/IEC 7816-4
'6981'	CE	Instruction incompatible with file structure.	ISO/IEC 7816-4
'6982'	CE	Security status not satisfied.	ISO/IEC 7816-4
'6983'	CE	Authentication method blocked.	ISO/IEC 7816-4
'6985'	CE	Conditions of use not satisfied.	ISO/IEC 7816-4
'6986'	CE	Instruction not permitted (no EF selected).	ISO/IEC 7816-4
'6A00'	CE	Wrong parameter P1/P2 (general).	ISO/IEC 7816-4
'6A80'	CE	Wrong parameters in data section.	ISO/IEC 7816-4
'6A81'	CE	Function not supported.	ISO/IEC 7816-4
'6A82'	CE	File not found.	ISO/IEC 7816-4
'6A83'	CE	Record not found.	ISO/IEC 7816-4
'6A84'	CE	Insufficient memory.	ISO/IEC 7816-4
'6B00'	CE	Parameter 1 or 2 wrong.	GSM 11.11, EN 726-3, ISO/IEC 7816-4
'6D00'	CE	Instruction not supported.	GSM 11.11, EN 726-3, ISO/IEC 7816-4
'6E00'	CE	Class not supported.	GSM 11.11, EN 726-3, ISO/IEC 7816-4
'6F00'	CE	Instruction interrupted – precise diagnosis impossible (e.g. operating system error).	GSM 11.11, EN 726-3,

			ISO/IEC 7816-4
'9000'	NP	Instruction successfully executed.	GSM 11.11, EN 726-3, ISO/IEC 7816-4
'920x'	NP	Writing to EEPROM successful after xth attempt.	GSM II.II, EN 726-3, ISO/IEC 7816-4
'9210'	CE	Insufficient memory.	GSM 11.11, EN 726-3
'9240'	EE	Writing to EEPROM failed.	GSM 11.11, EN 726-3
'9400'	CE	No EF selected.	GSM 11.11, EN 726-3
'9402'	CE	Address range exceeded.	GSM 11.11, EN 726-3
'9404'	CE	FID not found, record not found, comparison pattern not found.	GSM 11.11, EN 726-3
'9408'	CE	Selected file type incompatible with instruction.	GSM 11.11, EN 726-3
'9802'	CE	No PIN defined.	GSM 11.11, EN 726-3
'9804'	CE	Access conditions not satisfied, authentication failed.	GSM 11.11, EN 726-3
'9835'	CE	ASK RANDOM / GIVE RANDOM not executed.	GSM 11.11, EN 726-3
'9840'	CE	PIN verification failed.	GSM 11.11, EN 726-3
'9850'	CE	INCREASE / DECREASE cannot be executed as limit reached.	GSM 11.11, EN 726-3
'9Fxx'	NP	Instruction successfully executed and xx bytes of data available as response.	GSM 11.11, EN 726-3

Figure 15.12 Selected standardized Smart Card return codes after ISO/IEC 7816-4

15.4.9 Typical instruction execution times

This table is based on average successful Smart Card instruction execution times. A variety of cards were used in the measurements, with various operating systems. The figures below are mean values, and may vary sharply in a given case, depending on the operating system. The following conditions applied: transmission protocol T=1, clock frequency 3.5712 MHz, 372 divider and a DEA with a speed of 17 ms/8 bytes.

Execution times of instructions denoted by "*" depend critically on the particular implementation and the range of supported functions. Hence the times given can only be regarded as a rough guide.

Instruction	Execution time with data transmission	Execution time without data transmission
ASK RANDOM (8 bytes random)	55 ms	26 ms
CREDIT *	222 ms	175 ms
DEBIT *	270 ms	235 ms
EXTERNAL AUTHENTICATE	51 ms	22 ms
GET CARD DATA (8 bytes)	33 ms	4 ms
INITIALIZE IEP for Load *	173 ms	89 ms
INITIALIZE IEP for Purchase *	201 ms	135 ms
INTERNAL AUTHENTICATE	65 ms	26 ms
INVALIDATE	34 ms	15 ms
MUTUAL AUTHENTICATE	163 ms	95 ms
READ BINARY (5 bytes)	27 ms	2 ms
READ BINARY (10 bytes)	34 ms	2 ms
READ BINARY (20 bytes)	46 ms	3 ms
READ BINARY (50 bytes)	84 ms	3 ms
READ BINARY (100 bytes)	149 ms	5 ms
READ RECORD (5 bytes)	27 ms	2 ms
READ RECORD (10 bytes)	34 ms	3 ms
READ RECORD (20 bytes)	46 ms	3 ms
READ RECORD (50 bytes)	84 ms	3 ms
READ RECORD (100 bytes)	149 ms	5 ms
REHABILITATE	33 ms	15 ms
SEEK	22 ms	3 ms
SELECT FILE (with 8 bytes AID)	32 ms	3 ms
SELECT FILE (with 2 bytes FID)	24 ms	3 ms
UPDATE BINARY (5 bytes)	54 ms	29 ms
UPDATE BINARY (10 bytes)	84 ms	53 ms
UPDATE BINARY (20 bytes)	157 ms	113 ms
UPDATE BINARY (50 bytes)	376 ms	295 ms
UPDATE BINARY (100 bytes)	741 ms	598 ms
UPDATE RECORD (5 bytes)	54 ms	29 ms
UPDATE RECORD (10 bytes)	84 ms	53 ms
UPDATE RECORD (20 bytes)	157 ms	113 ms
UPDATE RECORD (50 bytes)	376 ms	295 ms
UPDATE RECORD (100 bytes)	741 ms	598 ms
VERIFY PIN (8 bytes PIN) *	56 ms	27 ms

Figure 15.13 Execution times of typical Smart Card instructions

15.4.10 Sample codings of Smart Card instructions

The following tables illustrate the most important codes for several sample Smart Card instructions. For simplicity, it was assumed that no secure messaging or addressing via logical channels was employed. For complete coding of these and further instructions, the reader is referred to the ISO/IEC 7816-4 standard[3].

CLA	INS	P1	P2	Lc	DATA	Meaning
'00'	'A4'	'00'	'00'	'00'	---	Direct MF selection.
...	...	'00'	'00'	'02'	FID	Direct MF/DF/EF selection by specifying the 2-byte FID.
...	...	'04'	'00'	'10'	AID	Direct DF selection by specifying the complete 16-byte FID.

Figure 15.14 Coding for the instruction SELECT FILE after ISO/IEC 7816-4 (only the main options are given)

CLA	INS	P1	P2	Le	Meaning
'00'	'B0'	0XXXXXXX	Y	n	Read n bytes with the offset X||Y from a preselected transparently structured EF (X=MSB, Y=LSB).
...	...	100XXXXX	Y	n	Read n bytes with the offset Y from a transparently structured EF implicitly selected by this instruction. The EF's short FID is XXXXX.
...	...	...	...	'00'	Read all bytes with one of the above access options.

Figure 15.15 Coding for the instruction READ BINARY after ISO/IEC 7816-4 (only the main options are given)

CLA	INS	P1	P2	Lc	DATA	Meaning
'00'	'D0'	0XXXXXXX	Y	n	Z	Write n bytes whose contents are Z with the offset X||Y to a preselected transparently structured EF (X=MSB, Y=LSB).
...	...	100XXXXX	Y	n	Z	Write n bytes whose contents are Z with the offset Y to a transparently-structured EF implicitly selected by this instruction. The EF's short FID is XXXXX.

Figure 15.16 Coding for the instruction UPDATE BINARY after ISO/IEC 7816-4 (only the main options are given)

---

[3] See also 6.5.1 Instruction APDU structure.

CLA	INS	P1	P2	Le	Meaning
'00'	'B2'	Y	'04'	n	Read the first n bytes from the record numbered Y (Y≠0) in a preselected linear fixed/variable or cyclic EF.
...	...	Y	XXXXX100	n	Read the first n bytes from the record numbered Y (Y≠0) in a linear fixed/variable or cyclic EF implicitly selected by this instruction. The EF's short FID is XXXXX.
...	...	...	...	'00'	Read all bytes from the relevant record with one of the above access options.

Figure 15.17 Coding for the instruction READ RECORD after ISO/IEC 7816-4 (only the main options are given)

CLA	INS	P1	P2	Le	DATA	Meaning
'00'	'DC'	Y	'04'	n	Z	Overwrite the whole n-byte long record Z with the number Y (Y≠0) which is in a preselected linear fixed/variable or cyclic EF.
...	...	Y	XXX XX10 0	n	Z	Overwrite the whole n-byte long record Z with the number Y (Y≠0) which is in a linear fixed/variable or cyclic EF implicitly selected by this instruction. The EF's short FID is XXXXX.

Figure 15.18 Coding for the instruction UPDATE RECORD after ISO/IEC 7816-4 (only the main options are given)

15.4.11 Selected chips for memory cards

The following abbreviations are used:

- Vcc: Voltage range
- Current: Chip's current consumption at the specified clock frequency (first figure in operation, second figure in energy-saving status with clock, third value in energy-saving status without clock)
- W/E cycle: Numbered of guaranteed write/erase cycles per EEPROM section
- W/E duration: Duration of erasing or writing one EEPROM section
- HW: Additional hardware

Manufacturer and type	Memory		Further data	
Philips	ROM:	—	Vcc:	5 V
PCB 2032	PROM:	—	Current:	3 mA
	EEPROM:	256 bytes	W/E cycle:	10 000
			W/E duration:	5 ms
			HW:	---
Philips	ROM:	16 bits	Vcc:	5 V
PCB 2036	PROM:	177 bits	Current:	2.5 mA
	EEPROM:	44 bits	W/E cycle:	10 000
			W/E duration:	5 ms
			HW:	221 bit counter, unidirectional authentication
Philips	ROM:	—	Vcc:	5 V
PCB 2042	PROM:	—	Current:	3 mA
	EEPROM:	256 bytes	W/E cycle:	10 000
			W/E duration:	5 ms
			HW:	Write-protect by password
Philips	ROM:	16 bits	Vcc:	5 V
PCB 7960	PROM:	48 bits	Current:	3 mA
	EEPROM:	40 bits	W/E cycle:	10 000
			W/E duration:	5 ms
			HW:	104 bit counter
SGS-Thomson	ROM:	—	Vcc:	3 to 5.5 V
ST14C02C	PROM:	—	Current:	2 mA
	EEPROM:	2 kbits	W/E cycle:	1 000 000
			W/E duration:	10 ms
			HW:	I^2C bus interface
SGS-Thomson	ROM:	---	Vcc:	3 to 5.5 V
ST14C04C	PROM:	---	Current:	2 mA
	EEPROM:	4 kbits	W/E cycle:	1 000 000
			W/E duration:	10 ms
			HW:	I^2C bus interface

SGS-Thomson	ROM:	---	Vcc:	2.7 to 5.5 V
ST15E32F	PROM:	---	Current:	2 mA
	EEPROM:	32 kbits	W/E cycle:	100 000
			W/E duration:	10 ms
			HW:	I^2C bus interface
SGS-Thomson	ROM:	---	Vcc:	5 V
ST1305	PROM:	---	Current:	2 mA
	EEPROM:	192 bits	W/E cycle:	1 000 000
			W/E duration:	5 ms
			HW:	262 144 unit counter
SGS-Thomson	ROM:	16 bits	Vcc:	5 V
ST1333	PROM:	---	Current:	2 mA
	EEPROM:	272 bits	W/E cycle:	1 000 000
			W/E duration:	5 ms
			HW:	32 767 unit counter, unidirectional authentication
Siemens	ROM:	16 bits	Vcc:	5 V
SLE 4404	PROM:	48 bits	Current:	1.5 mA
	EEPROM:	352 bits	W/E cycle:	10 000
			W/E duration:	5 ms
			HW:	—
Siemens	ROM:	16 bits	Vcc:	5 V
SLE 4406	PROM:	48 bits	Current:	1.5 mA
	EEPROM:	40 bits	W/E cycle:	10 000
			W/E duration:	10 ms
			HW:	20 000 unit counter
Siemens	ROM:	96 bits	Vcc:	5 V
SLE 4412	PROM:	160 bits	Current:	10 mA
	EEPROM:	—	W/E cycle:	10 000
			W/E duration:	3 ms
			HW:	160 unit counter
Siemens	ROM:	—	Vcc:	5 V
SLE 4418	PROM:	—	Current:	3 mA

	EEPROM:	1 kbyte	W/E cycle:	10 000
			W/E duration:	5 ms
			HW:	—
Siemens	ROM:	—	Vcc:	5 V
SLE 4428	PROM:	—	Current:	3 mA
	EEPROM:	1 kbyte	W/E cycle:	10 000
			W/E duration:	5 ms
			HW:	Additional write-protect via PIN logic
Siemens	ROM:	—	Vcc:	5 V
SLE 4432	PROM:	—	Current:	3 mA
	EEPROM:	256 bytes	W/E cycle:	10 000
			W/E duration:	2.5 ms
			HW:	Additional write-protect via PIN logic
Siemens	ROM:	—	Vcc:	5 V
SLE 44R35	PROM:	—	Current:	3 mA
	EEPROM:	1 kbyte	W/E cycle:	10 000
			W/E duration:	2 ms
			HW:	Additional write-protect via PIN logic; unidirectional authentication for contactless memory cards
Siemens	ROM:	16 bits	Vcc:	5 V
SLE 4436	PROM:	177 bits	Current:	2.5 mA
	EEPROM:	44 bits	W/E cycle:	10 000
			W/E duration:	5 ms
			HW:	20 000 unit counter, unidirectional authentication
Siemens	ROM:	---	Vcc:	5 V
SLE 4442	PROM:	---	Current:	3 mA
	EEPROM:	256 bytes	W/E cycle:	10 000
			W/E duration:	2.5 ms
			HW:	—

Figure 15.19 Summary of selected chips for memory cards

15.4.12 Selected Smart Card microprocessors

The following abbreviations are used:

- Vcc: Voltage range
- Clock: Clock frequency range
- Current: Chip's current consumption at the specified clock frequency (first figure in operation, second figure energy-saving status with clock, third value energy-saving status without clock)
- Size: Dimensions of a die
- Min. width: Minimum width of structure on the chip
- EEPROM Pg: Size of an EEPROM section (for write access)
- W/E cycle: Numbered of guaranteed write/erase cycles per EEPROM section
- W/E duration: Duration of erasing or writing one EEPROM section
- HW: Additional hardware
- : No data available

Manufacturer and type	CPU	Memory		Further data	
Hitachi	H8	ROM:	16 kbytes	Vcc:	4.5–5.5 V and 2.7–3.3V
H8/3102		EEPROM:	8 kbytes	Clock:	1–10 MHz or 1 –5 MHz
		RAM:	512 kbytes	Current:	20 mA/10 MHz, 100 μA/Sleep
				Size:	18.0 mm^2 (3.5 ×5.2 mm)
				Min.width	0.8 μm
				EEPROM Pg.:	1–32 bytes
				W/E cycle:	10 000
				W/E duration:	max. 15 ms
				HW:	RISC-like CPU 2 I/O lines

Hitachi	H8	ROM:	12 kbytes	Vcc:	4.5–5.5 V and 2.7–3.3 V
H8/3111		EEPROM:	8 kbytes	Clock:	1–10 MHz or 1–5 MHz
		RAM:	800 bytes	Current:	20 mA/10 MHz, 100 μA/Sleep
				Size:	18.0 mm^2 (3.5 ×5.2 mm)
				Min.width	0.8 μm
				EEPROM Pg:	1–32 bytes
				W/E cycle:	10 000
				W/E duration:	max. 15 ms
				HW:	RISC-like CPU NPU
Motorola	6805	ROM:	6 kbytes	Vcc:	5.0 V±10%
MC 68HC05		EEPROM:	3 kbytes	Clock:	1–5 MHz
SC21		RAM:	128 bytes	Current:	1.2 mA/5 MHz, 41 μA
				Size:	14.8 mm^2 (2.9×5,1 mm)
				Min.width:	1.2 μm
				EEPROM Pg:	4 bytes
				W/E cycle:	10 000
				W/E duration:	10 ms
				HW:	5 I/O lines
Motorola	6805	ROM:	3 kbytes	Vcc:	5.0 V±10%
MC 68HC05		EEPROM:	1 kbyte	Clock:	1–5 MHz
SC24		RAM:	128 bytes	Current:	2.5 mA/5 MHz, 200 μA
				Size:	10.4 mm^2 (2.8 ×3.7 mm)
				Min.width:	1.2 μm
				EEPROM Pg:	4 bytes
				W/E cycle:	10 000
				W/E duration:	10 ms
				HW:	5 I/O lines
Motorola	6805	ROM:	16 kbytes	Vcc:	3–5.5 V
MC 68HC05		EEPROM:	3 kbytes	Clock:	1–5 MHz
SC27		RAM:	240 bytes	Current:	5 mA/5 MHz, 500 μA, 50

					μA
				Size:	21 mm^2 (4.2×5.0 mm)
				Min.width:	1.2 μm
				EEPROM Pg:	4 bytes
				W/E cycle:	10 000
				W/E duration:	10 ms
				HW:	5 I/O lines
Motorola	6805	ROM:	12.5 kbytes	Vcc:	3–5.5 V
MC 68HC05		EEPROM:	8 kbytes	Clock:	1–5 MHz
SC28		RAM:	240 bytes	Current:	5 mA/5 MHz, 50 μA
				Size:	26 mm^2 (4.9×5.3 mm)
				Min.width:	1.2 μm
				EEPROM Pg:	4 bytes
				W/E cycle:	10 000
				W/E duration:	10 ms
				HW:	5 I/O lines
Motorola	6805	ROM:	12.8 kbytes	Vcc:	5.0 V±10%
MC 68HC05		EEPROM:	4 kbytes	Clock:	1–5 MHz
SC29		RAM:	512 bytes	Current:	
				Size:	26 mm^2
				Min.width:	
				EEPROM Pg:	4 bytes
				W/E cycle:	100 000
				W/E duration:	10 ms
				HW:	NPU, PLL (only for NPU), 5 I/O lines
Philips	8051	ROM:	6 kbytes	Vcc:	5 V
P 83C852		EEPROM:	2 kbytes	Clock:	1–8 MHz
		RAM:	256 bytes	Current:	10 mA/8 MHz,
				Size:	22.5 mm^2
				Min.width:	1.2 μm
				EEPROM Pg:	1/8 byte
				W/E cycle:	100 000

				W/E duration:	5 ms
				HW:	2 I/O lines
					NPU
Philips	8051	ROM:	20 kbytes	Vcc:	5 V
P 83C855		EEPROM:	2 kbytes	Clock:	1–8 MHz
		RAM:	512 bytes	Current:	10 mA/8 MHz,
				Size:	
				Min.width:	1.2 μm
				EEPROM Pg:	1/8 byte
				W/E cycle:	100 000
				W/E duration:	5 ms
				HW:	2 I/O lines
					NPU
Philips	8051	ROM:	16 kbytes	Vcc:	5 V
P 83C858		EEPROM:	8 kbytes	Clock:	1–10 MHz
		RAM:	512 bytes	Current:	10 mA/8 MHz,
				Size:	
				Min.width:	0.8 μm
				EEPROM Pg:	1/8 byte
				W/E cycle:	100 000
				W/E duration:	2.5 ms
				HW:	2 I/O lines, NPU
Philips	8051	ROM:	16 kbytes	Vcc:	5 V
P 83C864		EEPROM:	4 kbytes	Clock:	1–10 MHz
		RAM:	256 bytes	Current:	10 mA/8 MHz,
				Size:	
				Min.width:	0.8 μm
				EEPROM Pg:	1/8 byte
				W/E cycle:	100 000
				W/E duration:	2.5 ms
				HW:	2 I/O lines

SGS-Thomson ST 16 601	6805	ROM:	6 kbytes	Vcc:	2.7–5.5 V
		EEPROM:	1 kbyte	Clock:	1–5 MHz
		RAM:	128 bytes	Current:	10mA/ 5MHz,
				Size:	
				Min.width:	1 μm
				EEPROM Pg:	1–16 bytes
				W/E cycle:	100 000
				W/E duration:	2.5 ms
				HW:	2 I/O lines
					simple MMU
SGS-Thomson ST 16SF42	6805	ROM:	16 kbytes	Vcc:	2.7–5.5 V
		EEPROM:	2 kbytes	Clock:	1–5 MHz
		RAM:	384 bytes	Current:	10mA/ 5MHz,
				Size:	
				Min.width:	1 μm
				EEPROM Pg:	1–32 bytes
				W/E cycle:	100 000
				W/E duration:	2.5 ms
				HW:	2 I/O lines
					simple MMU
SGS-Thomson ST 16SF44	6805	ROM:	16 kbytes	Vcc:	2.7–5.5 V
		EEPROM:	4 kbytes	Clock:	1–5 MHz
		RAM:	384 bytes	Current:	10mA/ 5 MHz,
				Size:	
				Min.width:	1 μm
				EEPROM Pg:	1–32 bytes
				W/E cycle:	100 000
				W/E duration:	2.5 ms
				HW:	2 I/O lines

					simple MMU
SGS-Thomson ST 16SF48	6805	ROM:	16 kbytes	Vcc:	2.7–5.5 V
		EEPROM:	8.2 kbytes	Clock:	1–5 MHz
		RAM:	384 bytes	Current:	10mA/ 5MHz,
				Size:	
				Min.width:	1 μm
				EEPROM Pg:	1–32 bytes
				W/E cycle:	100 000
				W/E duration:	2.5 ms
				HW:	2 I/O lines
					simple MMU
SGS-Thomson ST 16CF54	6805	ROM:	16 kbytes	Vcc:	5V±10%
		EEPROM:	4 kbytes	Clock:	5 MHz
		RAM:	480 bytes	Current:	10mA/ 5MHz,
				Size:	
				Min.width:	1 μm
				EEPROM Pg:	1–32 bytes
				W/E cycle:	100 000
				W/E duration:	2.5 ms
				HW:	NPU, 2 I/O lines, simple MMU, PLL for NPU
Siemens SLE 44C10S	8051	ROM:	7 kbytes	Vcc:	2.7–5.5 V
		EEPROM:	1 kbyte	Clock:	1–7.5 MHz
		RAM:	256 bytes	Current:	10 mA/5 MHz, 100 μA
				Size:	6.1 mm^2 (2.62×2,31 mm)
				Min.width:	0.8 μm
				EEPROM Pg:	1–16 bytes
				W/E cycle:	100 000
				W/E duration:	3.5 ms
				HW:	—

Siemens SLE 44C42S	8051	ROM:	15 kbytes	Vcc:	2.7–5.5 V
		EEPROM:	4 kbytes	Clock:	1–7.5 MHz
		RAM:	256 bytes	Current:	10 mA/5 MHz, 100 μA
				Size:	8.1 mm^2 (2.66×3.06 mm)
				Min.width:	0.8 μm
				EEPROM Pg:	1–16 bytes
				W/E cycle:	100 000
				W/E duration:	3.5 ms
				HW:	—
Siemens SLE 44C80S	8051	ROM:	15 kbytes	Vcc:	2.7–5.5 V
		EEPROM:	8 kbytes	Clock:	1–7.5 MHz
		RAM:	256 bytes	Current:	10 mA/5 MHz, 100 μA
				Size:	10 mm^2 (2.66×3.76 mm)
				Min.width:	0.8 μm
				EEPROM Pg:	1–32 bytes
				W/E cycle:	100 000
				W/E duration:	3.5 ms
				HW:	—
Siemens SLE 44CR80S	8051	ROM:	15 kbytes	Vcc:	2.7–5.5 V
		EEPROM:	8 kbytes	Clock:	1–7.5 MHz
		RAM:	256 bytes	Current:	10 mA/5 MHz, 100 μA
				Size:	10 mm^2 (2.66×3.76 mm)
				Min.width:	0,8 μm
				EEPROM Pg:	1–32 bytes
				W/E cycle:	100 000
				W/E duration:	3.5 ms
				HW:	NPU

Figure 15.20 Summary of selected Smart Card microprocessors

Index

A - PET 37
ABS 36
Abstract syntax notation one (ASN.1) 89
Access authorization 122
Access control 381
Addressing 117
Acrylonitrile - Butadiene - Styrene 36
Addition of new application 220
AFNOR 310
AID coding and structure 115
Algorithm execution speeds 372
Answer to reset 164
Anti-collision Algorithm 33
APDU 178
Application identifier 116
Application specific command (ASC) 129
Application's protective mechanisms 266
Application protocol data unit 178
Application-specific Instructions 233
Arimura, Kunitaka 3
Arrangement of card elements 24
Ask - Random 212
ASN 89
Asymmetric Crypto-algorithm 67
Asynchronous Transmission protocols 163
Atomic Routines 127
ATR164
ATR-file 174
Authentic Procedure 185
AUTHENTICATION 246
Authenticity 246
Authentication instruction 210
Automata theory 93
Availability 49
Availability of microprocessors 49
Average EEPROM 370
Background system 324
Base Station 363
Base station System 363
Basic encoding rules 90
BER 90
BGT 159
Birthday paradox 84
Blackbox test 286
Block guard time 159
Block waiting time 158
Block - oriented 67
Block waiting time 159
Body of the card 35
Bonding 298
Bond-Out-Chip 295
Booting Sequence 45
Brute force attack 69
Bug-fixing 128
Bump 38
BWT 159
BWT conversion table 411
Byte Repetition 148

CAD 307
CALCULATE EDC 232
Calculation times in DSA 80
Card-based payment transactions 318
Card format 18
Card material 36
Cards with contacts 21
Card without contacts 23
CBC - Modus 73
CCR 307
CCS 82
Chaining 161
Challenge-response procedure 246

Change Attributes 220
Change CHV 209
Character waiting time 158
Check character 173
Chi2 - test 88
Chinex remainder theorem 77
Chip accepting device 307
Chip card reader 307
Chip containing credit cards 347
Chip holder verification 366
Cipher block chaining 73
Chosen cipher-text attack 69
CHV 366
Cipher-text 67
Cipher-text only attack 69
Circuitry 41
Clearing 324
Clock supply 44
Close Coupling Card 26
Closed system architectures 320
Code Inspection 286
Code programmed in circuit 128
Code walk through 286
Coding and testing 285
Combined procedure 186
Compare EEPROM 233
COMPARE KEY 229
Completion instructions 230
COMPLETION END 229
Completing the operating system 228
Completion sequence 231
Concept 284; 375
Confusion 70
Contact based cards in the telephone system 15
Contact relation 21
Coupon collector test 88
Convention 135
Cost of error recovery 285
Credit cards 318
Credit card instructions 227
CRC 64;156
CRC-Checksum 64
CREATE FILE 219
Crypto-algorithm used in Smart Card 75
Cryptographic checksum 82; 185
Cryptography 66
Cryptology 66
CWT 158
CWT convention table 410
Cyclic 120
Cyclic redundancy check 64; 157

DATA 373
DATA Encryption algorithm 70
DATA encryption Standard 70
DATA exchange 374
DATA Structuring 89
Data transmission 45; 133
DCS1800 372
DEA 70
DEA - Random number generator 85
Deadlock 93
Debit procedure 327
Decision Coverage 288
DECREASE 206
DECRYPT 216
Decryption times for DEA 71
Decryption times for IDEA 72
Dedicated file 112
Definition of the card format 18
Definition of standard 7
DELETE EEPROM 233
DELETE File 223
DES 70
Description and function of contracts 42
Dethoff, JŸrgen 3
DF 112
Dice 298
Die 39
Die-bonding 39
Diffusion 70
Digital Signature 78; 255
Direct convention 134
Diskette Terminal 307
Discrete logarithm problem 78
Diversity 184
Downloading an EEPROM CELL 54
DRAM 56
DSS Algorithm 79
Dynamic asymmetric authentication 253
Dynamic keys 258
Dynamic signature 245
Dynamic test 289

ECB - Modus 72
ECC 62
EDC 62
EEPROM 52
EFs 112
Ejection 311
Electronic cheque 318
Electronic code book 73
Electronic funds 322
Electronic payment transactions 318
Electronic properties 40
Electronic purses 319
Electronic tool systems 358
Elementary file 112
Elementary time unit 136
Embossed cards 9
ENCRYPT 216

Encryption 66
Encryption times for DEA 71
Encryption times for IDEA 72
ENVELOPE 235
Epilogue field 157
EPROM 52
Error correction codes 61
Error detection codes 61; 291
Error processing 162
Error recovery 124
Etu 136
Extension control 126; 231
EXTERNAL AUTHENTICATE 213

FFF
FID 115
File identifiers 115
File management 217
File operations 205
File Selection 194
Files with prEN1546 328; 334
Fixed denomination 318
Flash - EEPROM 55
Float 319
Floating Gate 53
Flow of payments 329
Foil-printing 299
Format character 165
Fowler-Nordheim effect 53
FRAM 56
Freedom from noise 69
Functional microprocessor 46

GGG
Generate application cryptogram 227
Generating polynomial 64
Get challenge 212
Get chip number 211
Get processing options 227
Get response 234
Global positioning system 360
GPS 360
Graph theory 93
Greybox test 289
Grštrupp, Helmut 3
Ground state 54
GSM network 362

HHH
Hash - functions 83
History of Smart Card 2
Hot electron injection 55

III
I2C - Bus 145
ID - 00 21
ID - 000 21
ID - 1 21
IDEA 72
Identification numbering systems 373
Identification instruction 208
IFCD 156
IFD 307
IFSC 157
Important Smart Card return codes 418
IMSI 364
In vehicle unit 359
INCREASE 206
Independence of Events 87
Inductive coupling 25
Information field size of the card 157
Information field size for the interface device 156
Information field 156
Initial character 165
Initialization 302
Instruction APDU 178; 179
Instructions for Electronic purses 223
Instruction processing 102
Instruction set 191
Interface characters 167
Internal Authenticate 212
Invalidate 221
IPES 72
ISDN 324
ISO - layer model 133
ISO - Position 310
IVU 359

KKK
Kerckhoff, Auguste 67
Kerckhoff principle 67
Key management 259
Key version 260
Known plaintext attack 70

LLL
Lamination 38
Lasergravure 33
Lc 180
Le 180
LEN field 154
Life cycle of a Smart Card 296
Life expectancy 370
Linear fixed 118
Linear variable 118
Loading an EEPROM cell 54
Lock 224
Logical channels 188

MMM
Magnetic stripe cards 10

The magnitude of cipherspace 69
MANAGE CHANNEL 235
Manufacturing costs 47
Master file 112
Meet-in-the-middle attack 75
Memory management unit 109; 129
Memory cards 4; 12
Memory organization 107
Memory types 50
Message authentication code 82
Message structure 177
MF 112
Micro automata 94
Microprocessor card 5
Microprocessor hardware testing 279
Microprocessor security 48
Microprocessor software testing 280
Mifare - system 33
Micron 33
Mini card 18
MMU 109; 130
Mobile equipment 363; 364
Mobile station system 363; 364
Module 109
Module testing 299
Mondex system 342
MOSFET 53
Multiple laser image 34
Multiple storage 65
Mutual authenticate 214
Mutual authentication 253

NNN
NAD 154
Node address 154

OOO
OBU 359
Odd parity 62
On board unit 359
Open and closed system architecture 320
Operating system 102
Optical memory card 15
OSI - Communication Model 134
OSI-Layer model 133

PPP
Packing 304
Padding 81
Parity bit 62
Passive protective mechanisms 262
Pattern test 88
PC 37
PCB field 155
PCMCIA 308
PCMCIA Smart Card terminal 308
Personal communication network (PCN) 370
Personification 303
PES 72
PET 37
PETP 37
Physical Properties 17
Physical transmission layer 135
PIX 116
PLL - phase locked loop 58
Plaintext 67
Plug-in cards 18
Polycarbonate 37
Polyethylenterephtalat 36
Polyvinyl chloride 36
Portable terminals 308
Post paid 361
prEN1546 328
Prepaid 361
Prepaid memory cards 326
Primary flat 296
Production methods 35
Production tests 274
Programming voltage contact 41
PROM 52
Proprietary application identifier extension 117
Protocol control byte 156
Protocol type selection 173
Processor type 49
PTS 174
PTS0 175
PTS1 175
PTS2 175
PTS - request 175
PTTS 175
PUK 209
Purse to purse transaction 319
PVC 36

QQQ
Qualification tests 273
Quality assurance and testing 273

RR
RAM 56
Random numbers 84
READ BINARY 197
Read instructions 196
Receiving single bytes 147; 151
Reception sequence counter 157
Recycling 305
REGISTER 219
REHABILITATE 222
Relationship between the different card formats 21
Remote Coupling Cards 32
Removal options 312

Response APDU structure 180
Return code manager 102
RID 115
ROM 51
RSA 75
RSA - Algorithms 75
RUN GSM ALGORITHM 234

SSS
SAM 325
Sample Coding of Smart Card instructions 422
SDL - symbolism 96
Search instructions 204
Secure application module 325
Secure messaging 182
Security features 33
Security methods 237
Security modules 315
Security test 281
Security through confidentiality 67
SEEK 205
SELECT FILE 195
Security patterns 34
Size of contact 22
Selected chips for memory cards 426
Serial test 88
Short - circuit resistance 313
Short messages 373
Shutdown sequence 45
Shrink-procedure 47
SIGN DATA 217
Signature 82
SIM - Subscriber Identity Module 366
Single Lane 363
Sliding contacts 311
Smart Card 12
Smart Card communication at T = 1 153
Smart Card microprocessors 46
Smart Card profiler 104
Smart Card return codes 413
Smart Card security 261
Smart Card terminals 307
Software protection mechanism 264
Specific character 172
Specification 285
SRAM 56
Standardization 2
State automata 93
Statement coverage 288
Static asymmetric authentication 253
Static authentication 249
Static Test procedures 286
Stationary terminals 307
Stresses 36
Stripe's storage capacity 12
Structural formula for materials 36
Subscriber identity 363
Subscriber identification Module 363
Supplementary hardware 56
Supply voltage 42
Supply current 42
Switching system 369
Symmetric crypto-algorithms 70
Synchronous data transmission 141
System and return codes 181
System integration 285

TTT
T = 1 153
T = 1 protocol 155
TA1 168
TA2 172
TAB technique 38
Table of most important Smart Card instructions 413
Table of Important Smart Card return codes 418
Table of class bytes used 413
TAi 167; 168
TB1 167; 168
TB2 168; 170
Tbi 168
TC1 167; 168; 170
TC2 171
TCi 171
TCSEC 282
Temporary key 260
Temporary mobile subscriber identity 365
Terminal 307
Terminal connection 320
TEST EEPROM 232
Test key 288
Test methods for software 283
Test procedures 286
TEST RAM 231
Testing of systems and applications 289
Testing procedure 290
Testing strategies 286
Testing the card's body 274
TLV - length 89
TLV - structure 89
TLV - tag 89
TLV - VALUE 89
TMSI 365
Top - down - system 383
TPDU 140
Transmitability 320
Transmission protocol T=0 140; 147
Transmission protocol T=1 140; 153
Transmission protocol T=14 140
Transmission protocols 139
Transmission protocol data unit 140

Transmission sequence counter 156; 189
Transmission spread of algorithm 379
Transmission speed 411
Transparent 118
Triple-DES 74
Tunnel effect 53
Tunnel oxide layer 53
Two-chip solution 59
Types of operations for block encryption Algorithms 73
Typical attack 267
Typical Smart Card applications 4

UUU
UART Module 58
UNBLOCK CHV 209
Unidirectional authentication 247
UPDATE BINARY 197
User interface 375

VVV
VERIFY PIN 208
VERIFY SIGNATURE 217

WWW
Waiting times 157
Waiting time extension 160
Warm reset 45
The waterfall model 284
Whitebox test 287
Work waiting time 170
WRITE BINARY 197
WRITE DATA 229
Write instructions 196

XXX
X.25 324

XOR - checksum 62